ENVIRONMENTAL SCIENCE, ENGINEERING AND TECHNOLOGY

FOREST PLANTATION DEVELOPMENT AND MANAGEMENT IN GHANA

Environmental Science, Engineering and Technology

Additional books in this series can be found on Nova's website
under the Series tab.

Additional e-books in this series can be found on Nova's website
under the e-book tab.

ENVIRONMENTAL SCIENCE, ENGINEERING AND TECHNOLOGY

FOREST PLANTATION DEVELOPMENT AND MANAGEMENT IN GHANA

DAVID MATEIYENU NANANG
AND
THOMPSON K. NUNIFU

nova publishers
New York

Library of Congress Cataloging-in-Publication Data

ISBN: 978-1-63483-205-2

Published by Nova Science Publishers, Inc. † New York

CONTENTS

PREFACE

The forestry sector in Ghana is in a crisis, irrespective of whichever way one looks at it. Ghana's forest cover declined from 9.6 million hectares in 1961 to 4.9 million hectares in 2010. This is a reduction of about 49% in 50 years! If this is not a crisis, we do not know what else is. To address this alarming situation, Ghana has three main policy options. These options, which are best implemented simultaneously, include: arresting the current high rates of deforestation; expanding forest plantations; and developing and implementing a national sustainable forest management strategy. Of the three prongs, forest plantation development offers the best hopes of addressing the forest sector crisis.

A necessary condition for successful plantation forestry development is the availability of scientific information and tools to those who need it, be it private forest developers or the students who are studying to become foresters to advise clients to successfully manage plantations. Even more basic than that, there seems to be little understanding of the key issues affecting plantation forestry such as the importance of forest plantations, the barriers and opportunities for their development, the impacts of international climate change policy on plantation forestry, and the role of plantations in the national development context.

Forest Plantation Development and Management, is a sequel to the earlier book by the first author on *Plantation forestry in Ghana: theory and applications.* The present book has been expanded from the previous one to include five new chapters, re-focused the discussion and re-written in many places to the extent that it warranted a new title.

The book is particularly appropriate for undergraduate courses in plantation forestry, mensuration, inventory, forest economics and rural development, though graduate students and other forestry practitioners and general interest readers would find the material as a very useful reference.

We recognise that some of the content is beyond the undergraduate level, but this was intended to provide additional material for graduate students looking for guidance on research methods for either their course work or for their theses.

Throughout the book, several case studies are presented, using growth and yield information on teak and neem plantations in the guinea savannah and semi-deciduous vegetation zones of Ghana. We used these two species because of the availability of detailed data on them through previous studies. While these case studies are meant to be illustrative, they are useful in demonstrating how the principles described in the theoretical parts of the book are applied.

Our appreciation goes to our families: Bernice, Baaloon, Yennumi, Yarpaak and Tiiloon Nanang and the Nunifu family, especially Suguru, Yalinan, and Damian. To all our families and friends and everyone who holds dear the conviction to promote sustainable management of natural resources in Ghana and around the world, we dedicate this book.

It should be noted that while Dr. Nanang and Dr. Nunifu are employees of the Governments of Canada and Alberta respectively, they authored this book in their personal capacities, and not as part of their official government duties.

BOOK SUMMARY

This book is divided into five main parts: plantation forestry in context; forest growth dynamics and assessments; plantation silviculture and management; forest plantation economics and international climate change policy; and conclusions. The introductory chapters provide the context for plantation development by highlighting the importance of Ghana's forest resources and their current state, why plantations are a major part of the solution to present and future wood scarcity and the barriers, and opportunities and incentives for plantation development in Ghana. Part II of the book is dedicated to describing forest growth dynamics, forest measurements and forest resource inventory methods. Part III begins with descriptions of the critical silvics of six tree species that have been grown and /or promoted for forest plantation development in Ghana, followed by a chapter on plantation silviculture, including descriptions of seed collection, nursery practices and planting. The final chapter in Part III focuses on managing forest plantations for rural development and environmental protection. An important consideration in plantation establishment is the economics of forest plantations vis-à-vis other land uses. Part IV concentrates on the economics of plantations, and describes the tools and data requirements for carrying out economic analyses of plantation forestry investments, the alternative models for determining the optimal rotation age of plantations, and a final chapter on the economics of managing plantations within the context of international climate change policies. The concluding Part V contains two chapters. The first one describes the approaches to achieving sustainable forest management in Ghana and the final chapter discusses recommendations on the necessary conditions for accelerating plantation development in Ghana.

PART I. PLANTATION FORESTRY IN CONTEXT

Chapter 1

OVERVIEW OF GHANA'S FOREST SECTOR

This Chapter provides the context for plantation forestry development within Ghana's forestry sector. It gives the background information on the importance of the forestry sector, a description of the vegetation zones and ecology of the forests, forest sector governance, rights to and benefits from forest resources, forest products trade, the challenges facing the forestry sector, and a forecast of the future demand and supply of wood products in Ghana. The final part of the chapter gives a brief overview of plantation forestry in Ghana, describing the history of plantation development, the governance and legislative framework underpinning plantation forestry, and the current extent of plantation forests in Ghana.

1.1. IMPORTANCE OF GHANA'S FORESTRY SECTOR

The forestry sector continues to play a significant role in Ghana's social and economic development agenda. Historically, this sector has accounted for about 5 – 6% of Ghana's gross domestic product (GDP) and approximately 11% of total commodity export earnings (Owusu 1998). In 2006, forest products exports contributed about 9% to total export earnings and 3% of GDP (Ghana Forestry Commission, 2007; World Bank 2007). In 2013, Ghana earned US$164 million from wood products exports which constituted about 1.2% of total export earnings for that year. The contribution of forest products to export earnings has been decreasing since 2008 as a result of consistent annual declines in export volumes during this period.

The forest sector also plays a significant role as an employer in both rural and urban communities across the country. The formal wood sector employs about 120,000 people across the country, and approximately two million people depend on the formal and informal forestry sectors.

Forests and woodlands serve as the primary source of fuelwood energy for at least 75% of the total population of Ghana, and about 90% of the rural population (Benhin and Barbier 2001). Also, forests and woodlands provide food, medicines, spiritual, cultural and environmental services to over two million Ghanaians. The forests also play environmental roles such as soil conservation, watershed functions, carbon sequestration and minimising extreme weather damage to human and animal lives and property.

Furthermore, Ghana's forests and woodlands are home to several animal and plant species and serve to maintain biodiversity. Non-timber forest products such as bush meat, wild fruits and tubers, honey, oils and construction materials are derived from Ghana's forests and woodlands. The bush meat sector is estimated to include about 300,000 hunters at the local community level who produce about 385,000 tonnes of bush meat annually, mainly from forests, valued at about US$350 million (Adu-Nsiah 2009), making the trade an important contributor to household and national incomes. Of the total quantity of bush meat harvested, Ghanaians are estimated to consume 225,287 tonnes, worth US$205 million annually (Adu-Nsiah 2009).

Ghana's forests and woodlands also support several plant and animal species with enormous recreational and tourism potential. Some of these areas have been developed into national parks and other tourist areas, that generate significant revenues from user fees, accommodations, meals, and souvenir shops located in and around them. Also, they generate employment in surrounding communities and are major contributors to the national economy. It is estimated that in 2012, tourism contributed about 7% to Ghana's GDP and created over 300,000 direct and indirect jobs (Ghana News Agency 2012).

1.2. Natural Vegetation and Ecology

Ghana occupies an area of about 23.9 million ha and lies north of the equator (between 4°45' and 11°11' North latitude and between 1°14' East and 3°07' West longitude) and wholly within the tropics. The high forest zone (HFZ) is found in the south-western third of the country (Figure 1). This zone covers an area of 8.1 million ha, with four broad ecological types - wet evergreen, moist evergreen, moist semi-deciduous, and the dry semi-deciduous ecological types. These zones have been identified to be floristically synonymous with the *Cynometra-Lophira-Tarrientia*, *Lophira-Triplochiton, Celtis-Triplochiton*, and the *Antiaris-Chlorophora* associations respectively, recognised by Taylor (1952). According to Taylor (1952), there is no distinct line of demarcation between these associations as one imperceptibly merges into another. The main timber-producing areas are the deciduous and evergreen forests in the southwest.

Within The HFZ, the moist evergreen forest contains about 27% of the commercial/economic species, whilst the moist semi-deciduous forest has up to 17% of such species; the wet evergreen forest is relatively poor in economic species (only 9%) (FAO, 2002a). The HFZ has a two-peak annual rainfall from April to July and September to November respectively. The rainfall varies between 1,200 - 2,200 mm, with a comparatively short dry season during January and February and a high relative humidity that is seldom below 85% (FAO, 2002a). The soils in the HFZ are highly leached and acidic (pH 4.0-5.5) due to the high rainfall. They have a low cation exchange capacity, available phosphorus, nitrogen and organic matter (FAO 2002a). In the wettest zones, the soils are very infertile, strongly acidic and often have high aluminium content. FAO (2002a) notes that the relatively short dry seasons coupled with the high humidity of the HFZ reduce the risk of fire in forest plantations.

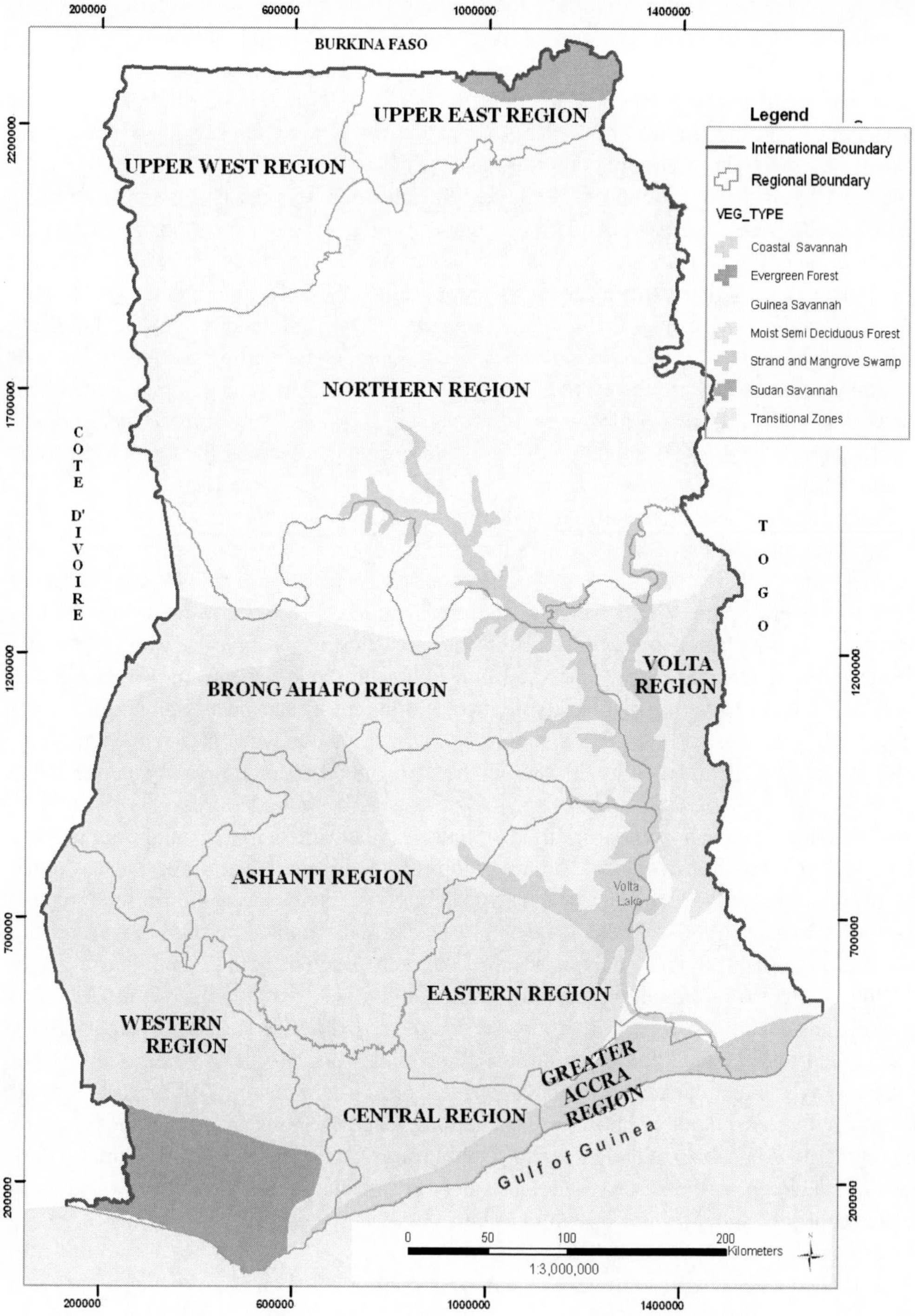

Figure 1. Natural vegetation zones of Ghana.

The savannah zone is classified into the southern (coastal) and the northern savannahs, based on their location. The largest part of the savannah zone is found mainly in the northern part of Ghana and occupies an area of about 14.7 million ha and a forest-savannah transition zone (in the middle belt) of about 1.1 million ha (ITTO, 2005a).The northern savannah is further divided into the Guinea and Sudan savannah zones. The Sudan savannah is restricted to a small area in the north-eastern corner of the Upper East Region.

The characteristic vegetation of the Guinea Savannah zone consists of short deciduous, widely spaced, fire-resistant trees. These do not form a closed canopy and overtop an abundant ground flora of grasses and shrubs of varying heights (Taylor 1952). The most frequent and characteristic tree species are *Isoberlina doka*, *Monites kerstingii*, *Burkea africana*, *Danielia oliveri* and *Terminalia avecinoides*. Two indigenous species, *Vitellaria paradoxa* (shea tree) and *Parkia biglobosa* (dawadawa) are conserved by farmers because of their economic value and are therefore common on farmlands. The ground vegetation, which includes *Panicum maxima*, *Andropogon gayanus* var. *gayanus* and *Cassia mimosoides*, desiccates during the dry season and predisposes the savannah to annual fires which leave the soil surface bare.

The savannah zone is characterised by distinct wet (rainy) and dry seasons of about equal duration. Two air masses of very contrasting characteristics determine the climate in this zone. These are the harmattan winds generally called the North East Trade Winds that usher in the dry season and the South Atlantic Maritime Air Mass referred to as the south west monsoon winds which transport moisture into the area during the rainy season. There is a moderate mean annual rainfall of 960-1200 mm falling in one season from March/April to October and showing a very irregular distribution within a rainy season and great differences from year to year (Fisher 1984). Maximum rainfall during the year is achieved in July-August. Mean annual temperature is 28.3°C that does not vary significantly between the seasons.

Soils of the savannah zone are varied because of the varied nature of the underlying geology. In general, however, two broad groups of soils are recognised: the savannah ochrosols and the groundwater laterites. The savannah ochrosols are found on the Voltaian sandstones (Boateng 1966). They consist of well-drained, friable, porous loams and are mostly red or reddish-brown in colour. Most of the area covered by these soils has a gently undulating topography. Soils in the depressions are quite thick, but upland soils usually have a zone of ironstone concretions from 10 cm to one metre below the surface (Boateng 1966). The soils tend to be eroded and form surface crusts under the impact of strong rainfall, but they have only a small capacity to keep water. According to Boateng (1966), despite their deficiency in nutrients, notably phosphorous and nitrogen, these soils are among the best soils in the northern savannah zone and are extensively farmed. A typical soil profile shows a dark greyish humus loam on the surface and subsequent layers show from grey to brownish loam with quartz gravel through light brown clay into moderately compact clays at about 70 cm below around level (Lawson et al. 1968).

The groundwater laterites are very extensive and are formed on the Voltaian shales and granites. They consist of a pale-coloured, sandy or silty loam with a depth of up to 65 cm underlain by an iron pan or a mottled clayey layer so rich in iron that it hardens to form an iron pan on exposure (Boateng 1966). Drainage on these soils is poor as a result they tend to get waterlogged during the rainy season and to dry out during the long dry season. These soils, especially those developed on the Voltaian shales are considered to be among the poorest soils in Ghana, and little cultivation takes place on them (Boateng 1966).

1.3. Forest Sector Governance

1.3.1. What Is Forest Sector Governance?

In general, governance is a system of rules and institutions that provide the basis for societies to make decisions and implement them. Forest Governance describes the process by which decisions about forest resources are taken and implemented. It also specifies accountability and clarifies who makes these decisions. Most problems of forest sustainability are often blamed on poor or bad forest governance. For example, illegal logging is often considered as a sign of poor forest governance. While it is not true that all forest problems are due to governance failures, it is also clear that good forest governance is a necessary condition for sustainable forest management.

Good forest governance has several elements. These include transparency; inclusiveness; accountability; equitable rule of law; responsiveness; effective and efficient; and consensus-oriented.

- *Transparency*: people should have access to clear and complete information regarding forest resources. Transparency will reduce the potential for bribery and corruption by more powerful corporate players. Communities and stakeholders should be able to monitor performance of forest sector governance based on the publicly available information.
- *Participatory/inclusiveness*: every person who has a stake in forest resources, including the marginalised, poor, indigenous peoples, women and the powerful in society should have a say in how decisions are made and implemented regarding forest resources. Participation by all stakeholders ensures that benefits are equitably distributed, and this contributes to good forest governance. Participation is further enhanced by developing regulations that are easy to understand, implement and enforce.
- *Accountability*: Accountability refers to holding decision makers responsible for their actions. If decision makers are not accountable, then the chances of arbitrary decisions are increased. Accountability reduces abuse of power by those in authority.
- *Equity rule of law* implies that the law applies to everyone in a fair way to help reduce conflict. Incentives and penalties have to be applied as needed equally to all citizens and corporations.
- *Responsiveness:* forest governance must be responsive to the needs of the population, changes in climate and changing socio-economic and political circumstances of the country.
- *Effective and efficient:* it is important that forest governance be conducted in efficient and effective ways, which means that bureaucratic layers should be minimised. If large amounts of financial resources are dedicated to governance, then there will be

less available to engage effectively stakeholders. Also, efficient methods such as modern communication tools should be applied in ensuring participation.

- *Consensus oriented:* as much as possible, the goal of forest governance is to reach consensus on decisions regarding forest resources. In order not to marginalise some of the stakeholders, it is critical to use processes that result in consensus on how the resources are used for the benefit of all. The chances of achieving consensus are improved if the forestry sector proactively engages other land-based ministries or private-sector actors.

1.3.2. Forest Sector Institutions

Institutions are the formal and informal rules that govern human behaviour. However, we can also think of institutions as organisations, such as government bureaucracies or non-governmental organisations (NGOs). Institutions are the major vehicles through which forest governance is delivered. Within the forestry sector in Ghana, the main institutions that influence forest management and governance include: the policy and legislative framework; the Forestry Commission; the Ministry of Lands and Natural Resources; the Ghana Timber Association; NGOs; and other government-wide institutions such as other government ministries and the judiciary.

Historically, forest management was the responsibility of the Forestry Department, which was created in 1909 with the aim of encouraging the reservation of 20-25% of the country's land area (Smith 1999). Since 1999, management of forests and other natural resources in Ghana became the responsibility of the Ministry of Lands and Natural Resources[1]. Forest management is now led by the Forest Commission (FC), which was created under the *Forestry Commission Act*, 1999 (Act 571). The FC is responsible for regulating the utilisation of forest and wildlife resources, the conservation and management of those resources and the coordination of policies related to them. The FC is also tasked to manage the nation's forest reserves and protected areas and assist the private sector and the other bodies with the implementation of forest and wildlife policies. According to the Act, the FC shall undertake the development of forest plantations for the restoration of degraded forests areas, the expansion of the country's forest cover and increase the production of industrial timber.

The FC consists of 12 members, including a Chairperson, the Chief Executive Officer of the FC, and representation from the forest and wildlife industries, national house of chiefs, professional foresters' association, NGOs, etc. The FC embodies the various public bodies and agencies that were individually implementing the functions of protection, management and regulation of forest and wildlife resources. These agencies currently form the divisions of the Commission: Forest Services Division, Wildlife Division, Timber Industry Development Division, Wood Industries Training Centre, and Resource Management Support Centre. The FC has responsibilities for law enforcement, monitoring, policy development, forest management and revenue collection, all at the same time.

[1] The name of the ministry responsible for forestry has often changed with changes in governments. As at 2014, it was called the Ministry of Lands and Natural Resources. Between the years 2000 - 2008, it was called the Ministry of Lands and Forestry.

These functions of the FC in forest sector governance are potentially conflicting in nature and may result in inefficiencies in the implementation of forest sector plans. Planning and managing forests pose a particular challenge because of the multiple benefits produced by forests and also because of the fact that forestry practices have impacts on other land uses. Forest governance in Ghana is centralised within the Forestry Commission, and this centralisation is a consequence of the fact that forests are publicly owned. When the planning agency is the same as the implementation and regulatory agency, then there are no external controls and goals that are set out in the plans can be changed internally at the implementation stage. A case in point was when the FC reduced the annual planting target under the National Forestry Plantation Development Programme from 20,000 ha to 16,250 ha in 2004 and further to 10,000 in 2005. This change was possible because the FC had the mandate to plan how much to plant, and the mandate to carry out the planting as well. If a separate agency handled implementation, it would find efficient ways to deliver on them and become accountable to the planning agency. With the same agency responsible for both planning and management, there is no accountability, and this results in inefficiency. As a result, there is a need to separate the functions of planning from those of implementing forest management plans to eliminate such inefficiencies.

The separation of functions may neither be easy to achieve nor cheap to implement. Separation may come at an additional cost in terms of physical infrastructure and personnel. As a result, any opportunities to exploit economies of scale in designing planning and management institutional structures should be utilised. In cases where this separation is not possible, the organisation needs to ensure that the management (implementation) branch has an arms-length relationship with the planning branch, and that the management branch is given clear goals and direction to operate with.

The FC has ten regional offices (one for each of the ten regions of Ghana), each headed by a Regional Manager. In each region, there are District offices, led by the District Manager, supported by technical officers, labourers and other administrative staff. Ghana currently has 216 Administrative Districts and only 51 forest districts. For example, the three northern regions have a total of ten forest districts, covering an area of about 14.7 million ha (two-thirds of the land area of Ghana). It is practically impossible for these 10 forest districts to effectively carry out plantation forestry development for such a large area. In order to increase the effectiveness of plantation development in Ghana, there needs to be an increase in the number of District Forestry offices A second best solution is to have a senior technical officer and labourers stationed in the districts without forestry offices to manage nurseries to provide seedlings and technical support to clients within the District. Increasing the presence of forestry officers across the country is absolutely necessary if plantation forestry is to be effectively delivered to all parts of the country.

Table 1. Distribution of District Forestry offices in Ghana

Ashanti	**Brong Ahafo**	**Central**	**Eastern**	**Greater Accra**	**Northern**	**Upper East**	**Upper West**	**Volta**	**Western**	**Total**
7	6	4	6	3	5	3	2	5	7	**51**

Forestry research is led by the Forestry Research Institute of Ghana (FORIG) and other forestry faculties at the universities. The Faculty of Renewable Natural Resources of the Kwame Nkrumah University of Science and Technology (KNUST), The School of Natural Resources of the University of Energy and Natural Resources (UENR), Sunyani and Faculty of Renewable Natural Resources at the University for Development Studies (UDS) in Tamale offer training in all aspects of forestry at the undergraduate and graduate levels. The School of Natural Resources at the UENR owns 250ha of a teak plantation and a field station at Brosanko in the Brong Ahafo Region.

Community participation in forest management is facilitated by the Collaborative Forest Management Unit of the FC. In addition, there are NGOs such as Conservation International, World Vision, Forest Watch Ghana and the Adventist Development and Relief Agency (ADRA) that are active in forestry activities. The Ghana Timber Association represents timber loggers and millers in Ghana and constitutes an active lobby group for timber industry interests. The Timber and Wood Workers Union of the Trade Union Congress of Ghana is also an important stakeholder in forestry issues in Ghana (ITTO 2005a).

1.3.3. Policy and Legislative Framework

The forestry sector is governed by the 1992 Constitution of the Republic of Ghana (and subsequent amendments) and a suite of laws, regulations and policies. The main ones include: *Administration of Stool Lands Act*, 1962 (Act 123), *Forest Protection Decree*, 1974 (NRCD 243), *Trees and Timber Management (Amendment) Act*, 1994 (Act 493); *Local Government Act*, 1993 (Act 462), *Forestry Commission Act*, 1999, (Act 571), *Timber Resources Management Act*, 1997 (Act 547), *Timber Resources Management Regulations*, 1998, (LI 1649 and LI 1721 of 2003), *Forest Plantation Development Fund Act*, 2000 (Act 583) and subsequent amendments in 2002 (Act 623), *Forest Protection (Amendment) Act*, 2002 (Act 624), the *Timber Resources Management Act*, 2002 (Act 617) and the Forest and Wildlife Policy (2012).

This long list of legislations and policies may suggest a robust system for managing forest resources. But it also suggests something else – the over-centralisation of forest sector governance within the central government. These policies and laws are largely ineffective in ensuring sustainable forest management due to lack of enforcement. There are significant bottlenecks in implementing many aspects of the policy initiatives. For example, a closer look at these policies and laws shows that they tend to be piecemeal in their approaches, and hence lead to confusion. For consistency and ease of application, it is preferable to clarify and consolidate these laws and regulations. Implementation at the various levels remains a nightmare in most cases, and penalties for breaches of forestry-related laws are ridiculously low.

1.3.4. Policy and Legislative Evolution

The British ruled Ghana as a colony for part of the 19th and much of the 20th century, naming the then Gold Coast as a British Crown colony in 1874. Timber exploitation mainly of African mahogany (*Khaya* and *Entandrophragma* spp.) started in 1891 when about 3,000 m^3 of mahogany was exported (Taylor 1960). Since then the trade in timber grew and expanded into a major economic activity in Ghana.

With export came an expansion of timber extraction and trade and the need to regulate harvesting activities. Table 2 summarises the forest policies in Ghana during the colonial era. The first two policies in 1894 and 1897 were intended to gain control over lands. Forest management practice started with the granting of concessions to companies for timber exploitation in 1900 but it was not until 1927 that the legal power to enforce reservation was secured through the *Forests Ordinances* (cap 157) (FAO 1997).

The colonial forest policies were primarily geared towards the sustained supply of timber for the wood industry, which led to the over-exploitation of forest resources, especially the desirable timber species. In fact, even the silvicultural systems that were implemented at that time were also intended and focused mainly on regenerating only the desirable tree species. It is no doubt then that subsequent post-colonial policies and efforts were directed towards the creation of permanent forest reserves to restore or at least protect the remaining forests. By the end of 1978, the government had placed about 3.3 million ha of forests under permanent forest estate.

Table 2. Colonial Forest Policies in Ghana from 1894 – 1951

Policy/Act	Category	Purpose
The Crown Land's Bill 1894	Land tenure policy	Colonial control over lands and forestry
The Land's Bill 1897	Land tenure and Pseudo-conservation policy	Colonial control over lands and forestry
The Concessions Ordinance 1900	Pseudo-conservation policy	Colonial administration of contracts between the government and British timber merchants
The Forest Ordinance 1901	Extractive policy and Pseudo-conservation policy	Appointment of forestry officers, constitution of reserve acquisition of lands and appointment of forestry commissioner
Timber Protection Ordinance 1907	Extractive policy and Pseudo-conservation policy	Prevent cutting of immature timber for the purpose of guaranteeing future supplies of timber
The Undersized Timber Trees Regulation 1910	Pseudo-conservation policy	Enforcement of the Timber Protection Ordinance
The Forestry Ordinance 1911	Extractive policy and Pseudo-conservation policy	Creation, control and management of forest reserves
The Forest Ordinance 1927	Extractive policy and Pseudo-conservation policy	Involvement of traditional rulers and enforcement of forestry laws and regulations
Concessions Ordinance, 1939	Extractive and conservation	Timber harvesting rights and revenues; conditions and limitations on forest operations
Forest Policy, 1948	Extractive and conservation	Creation of permanent forest estate; Protection of forests; Protection of forest catchment areas; Environmental protection for ecological balance
Forests Ordinance, 1951	Protection	Protection of Forests Protection of Forest reserves

Source: Adapted and modified from Asante (2005)

The first post-colonial forest legislations were the *Forest Commission Act* of 1960; *Forest Improvement Fund Act* of 1960; and the *Concessions Act* of 1962. Over several decades, successive governments have pursued many legislative and policy initiatives to ensure management and protection of forest resources and optimise the benefits that Ghanaians derive from their forest resource endowments. These changes have not significantly reduced the rate of deforestation nor improved forest management practices.

One of the most comprehensive restructuring in the industry's history was the phased ban on exports of unprocessed timber logs, which began in 1979. In fact, the genesis of restricting exports of unprocessed logs can be traced to provisions in the *Trees and Timber Act*, 1974 (NRCD 273) which imposed levies on selected species exported in unprocessed form. Ghanaian governments have sought to control external demand for Ghana's timber using log export bans in an effort to encourage value-added economic activities. In 1979, 14 timber species including the traditional redwood species were banned from export in log form (Richards 1995). The list of banned species was increased to 18 in 1989. In a follow-up to this policy, the government extended the number of restricted timber species, imposed higher duties on other species, and completely phased out log exports by 1995 (Richards 1995). The government hoped that with these measures, increased sales of wood products would replace earnings from logs.

In 1994, a new Forest and Wildlife Policy was introduced to replace the previous forest policy that was adopted in 1948. The objectives of the new policy, among other things, were to manage and enhance Ghana's permanent estate of forest and wildlife resources for preservation of vital soil and water resources; promote the development of viable and efficient forest-based industries, particularly in secondary and tertiary processing; and promote public awareness and involvement of rural people in forestry and wildlife conservation so as to maintain life-sustaining systems. The promulgation of the 1994 policy resulted in the development of the Forest Sector Development Master Plan (1996-2020) and a medium-term plan. This Master Plan was prepared by the Ministry of Lands and Forestry to achieve sustainable utilisation and development of forest and wildlife resources, modernisation of the timber industry, and the conservation of the environment thereby ensuring the realisation of the objectives of the 1994 Forest and Wildlife Policy (Ghana Forestry Commission 1998). Under the medium-term plan, the Ministry implemented the Comprehensive Sector Investment ten-year (1999 to 2009) programme called the Natural Resources Management Programme (NRMP). The NRMP consisted of five components: high forest resource management, savannah resource management, wildlife resource management, biodiversity conservation, and environmental management.

The 1994 policy was developed through broad consultations with stakeholders and took five years to complete (1989-1994). This policy was arguably the most comprehensive forest policy Ghana has had up till 1994. A major shortcoming of the policy was that while it clearly articulated its objectives, neither strategy nor specific government agencies were identified to achieve these objectives. Secondly, the implementation of the policy was challenged by the complex land tenure system in Ghana, weak institutional and governance structure of the FC, and the lack of effective engagement with relevant shareholders. A final weakness of the 1994 forest policy was the fact that it did not provide any incentives to conserve trees on off-reserve lands or farmlands by farmers. As a result, since the adoption of the policy, the rate of deforestation had actually increased, plantation forestry development was not widely adopted, the wood harvesting and processing sectors remained inefficient, and community participation in forest management was minimal.

Despite its shortcomings, the 1994 Forest and Wildlife Policy resulted in replacing timber concessions (leases) with timber utilisation contracts that are awarded based on competitive bidding. This policy also led to a more realistic Annual Allowable Cut (AAC) of one million m^3 (on and off reserve) which was implemented in 1996 and provided the foundation for the initiation of a forest management certification system project in 1997.

An important policy related to levies was introduced in 1996. An air-dry levy on nine selected species ranging from 10 – 30% was introduced. These nine species accounted for 80% of Ghana's exported forest products (Donkor 2003). The levy was intended as a disincentive for exporting air-dried sawn wood. Moisture accounts for at least 75% of wood manufacturing problems; therefore, reduction of wood-related problems correlates with a reduction in moisture content (Wengert 2001), and consequently results in increased value of the wood. The government also embarked on the promotion of the use of lesser-used species (Donkor 2003). These policy initiatives resulted in increased value-added processing, a reduction in the export of air-dried sawn wood and an increase in the export of plywood and veneer (Donkor et al. 2006; Nanang 2010).

Whilst export trade in timber and wood products has been encouraged over the years, little attention has been paid to the supply of wood products to the local market, and this has encouraged illegal logging by chainsaw operators. As a result, in 2005, the government directed sawmills to sell at least 20% of their timber production on the domestic market. The Timber Industry Development Division (TIDD) of the FC was asked to ensure that permits for timber exports are only approved after sawmillers show evidence that they have supplied 20% of their production to the domestic market. However, this 20% is inadequate to meet the demand for wood products in the local market. The need to fill the gap between demand and supply of sawn wood in the local market is one main cause of illegal logging activities (Gayfer et al. 2002; Forest Research Programme 2006). A few studies have examined the link between illegal logging and export of forest products from, and deforestation in, Ghana (Sarfo-Mensah, 2005; Owusu 1998). These studies conclude that illegal logging supplies the local timber markets and is not a major source of wood for the export market.

In recognition of the important role of forest plantations in providing wood products to the population, a Forest Plantation Development Fund (FPDF) was established through the *Forest Plantation Development Fund Act*, 2000 (Act 583) to provide financial assistance for the development of private commercial forest plantations and for research and technical advice to persons involved in commercial plantation forestry on specified conditions. In May 2002, this Act was amended through the *Forest Plantation Development Fund (Amendment) Act* (Act 623) to enable plantation growers, both in the *public* and private sectors, to participate in forest plantation investments. Money to support the Fund is derived from the proceeds of the timber export levy imposed under the *Trees and Timber Decree 1974* (NRCD 273) as amended by the *Trees and Timber Management (Amendment) Act*, 1994 (Act 493); grants and loans for encouraging investment in plantation forestry; grants provided by international environmental and other institutions to support forest plantation development projects for social and environmental benefits; and moneys provided by Parliament for private forest plantation purposes.

The latest major change in the policy evolution in Ghana was the introduction of the new Forest and Wildlife Policy in 2012 to replace the 1994 Forest and Wildlife Policy. The new policy *aims at the conservation and sustainable development of forest and wildlife resources for the maintenance of environmental stability and continuous flow of optimum benefits from the socio-cultural and economic goods and services that the forest environment provides to the present and future generations while fulfilling Ghana's commitment under international agreements and conventions.* The policy framework espouses five specific policy objectives as follows:

1. To manage and enhance the ecological integrity of Ghana's forests, savannah, wetlands and other ecosystems for the preservation of vital soil and water resources, conservation of biological diversity, and enhancing carbon stocks for sustainable production of domestic and commercial products;
2. To promote the rehabilitation and restoration of degraded landscapes through forest plantation development, enrichment planting, and community forestry informed by appropriate land-use practices to enhance environmental quality and sustain the supply of raw materials for domestic and industrial consumption and environmental protection;
3. To promote the development of viable forest and wildlife based industries and livelihoods, particularly in the value-added processing of forest and wildlife resources that satisfy domestic and international demand for competitively-priced quality products;
4. To promote and develop mechanisms for transparent governance, equity sharing and citizens' participation in forest and wildlife resource management; and
5. To promote training, research and technology development that supports sustainable forest management while promoting information uptake both by forestry institutions and the general public.

The policy document then goes on to identify up to a total of 72 policy strategies to help achieve the above five policy objectives. The provision of some detail as to how the policy objectives will be achieved addresses a major criticism of the 1994 policy. The document justifies extensively why the 1994 policy had to be revised to take into account recent environmental, global and international policy challenges. It also tries to address the imbalance between the supply of wood for export and the domestic demand that is currently being filled through illegal logging.

Another good aspect of the policy is that it places forestry and wildlife issues in the context of the national development agenda. By so doing, forestry and wildlife development will be carried out in a way that recognises the inter-linkages with, and benefits from the synergies of, other sectors of the economy.

Despite these positives, the new policy provides very little changes in the overall direction, aims and objectives compared to the 1994 policy. Was the ineffectiveness of the 1994 policy due to a flawed policy or the lack of implementation? There were certainly some deficiencies in the 1994 policy as identified in the preceding paragraphs, but a closer look at the previous policy and comparisons with the current forest-sector situation reveals that most of the policy objectives were never implemented and hence their effectiveness or lack thereof could not be objectively ascertained.

The new policy has 14 legislative proposals and another 13 proposals to produce plans, guidelines, procedures or standard documents as part of the implementation strategies. A policy document that relies so heavily on legislation for successful implementation runs the risk of not getting much done. Even more worrying is the fact that most of what is intended to be achieved by these proposed legislations can be achieved by enforcing existing laws. If the existing legislations are not enforced, what makes the government think more legislation will solve the problems of the forestry sector? For example, the new policy provides strategic direction for eliminating illegal logging and chainsaw operations. It plans to achieve this using two policy strategies:

a) Develop a comprehensive national strategic plan to address all issues relating to illegal logging and chainsaw activities including trade in illegal timber and wood products; and
b) Develop the necessary legislation to support the implementation of the national strategic plan on illegal logging.

While there may still be a need to strengthen legislation, it is incorrect to blame the inability to control illegal logging on the lack of legislation. There is currently legislation against illegal logging. First, it is called "*illegal logging*" because the activity is being done against an existing legislation; otherwise it would not be illegal. Under the Timber Resources Management Act – Act 547 (1997), Article (1) states that "*no person shall harvest timber from any land to which section 4 of this Act applies unless he holds timber rights in the form of a timber utilisation contract entered into under this Act in respect of the area of land concerned*" (Section 4 defines lands on which a timber utilisation contract can be allocated).

The problem in the forest sector is well known and simple. Plans and strategies are not effectively implemented, and our forest-sector laws are not being enforced and we somehow believe that by developing even more policies and enacting even more laws, we will work ourselves out of the problem.

The new policy claims that there is a paradigm shift from the past policies, as the new policy places emphasis on non-consumptive values of the forest and creating a balance between timber production and marketing to satisfy particularly domestic demands (Ghana Forest and Wildlife Policy 2012). In reality, these same ideas were expressed in the 1994 Forest and Wildlife Policy and hence hardly a paradigm shift. For example, Section 5.4.9 of the 1994 policy states that "*promotion and development of a well-structured local market as an essential component of the timber industry in order to satisfy domestic needs and to maximise utilisation of harvested timber.*" For non-consumptive values, Section 3.2.1 of the 1994 policy states "*promote public awareness and involvement of rural people in forestry and wildlife conservation so as to maintain life-sustaining systems, preserve scenic areas enhances the potential for recreation, tourism and income-generating opportunities.*"

There is also an underlying premise that government cannot make forest policy or make major decisions affecting the forest sector outside of the policy document that only changes every two decades or so. In reality, governments make policies every day, and so the government of Ghana should not feel compelled, neither is it necessary to wait until the policy document is reviewed, before tackling new challenges that emerge in the forestry sector.

In addition to the Forest and Wildlife Policy, the Government of Ghana also introduced a new Ghana Forest Plantation Strategy 2015-2040 in October 2013. This strategy contains an implementation plan and targeted planting with an estimated budget for the 25-year period of US$3.5 billion. Details of this strategy are discussed in Section 1.8 of this Chapter.

Ghana has been working towards a Voluntary Partnership Agreement with the European Union since 2003. This is a policy decision that has implications for forest governance in Ghana to curb illegal logging through Forest Law Enforcement, Governance and Trade (FLEGT). The cornerstone of this policy is the FLEGT Voluntary Partnership Agreement (VPA). In November 2009, Ghana was the first country in the world to formally sign the FLEGT-VPA, which is a bilateral agreement between the European Union (EU) and wood exporting countries, and aims to improve forest governance and ensure that the wood imported into the EU has complied with the legal requirements of the partner country. Under this agreement, Ghana will develop systems to verify the legality of their timber exports to the EU (European Union Commission/VPA 2009).

Ghana expects that the VPA will help further its governance reforms of the forestry sector, contribute to sustainable forest management, provide conditions that encourage investment in forest restoration and thus improve the resource base, realise the full economic value of forests and ensure that the forest sector contributes to poverty alleviation (European Union Commission/VPA 2009). Ghana decided to enter into the VPA to demonstrate its commitment to good forest governance, and as a means to maintain access to valued markets and open up new ones. With the VPA, Ghana will also promote investment in the sector to ensure the future viability of its industry (European Union Commission/VPA 2009).

While Ghana has been showered with praises for being the first African country to sign the VPA, implementing a licensing system to fully operationalise the Agreement has been more difficult. The latest information from the FC is that the timber validity licensing system has been developed and is undergoing testing. The latest policy initiative was the revision of the stumpage fees for harvesting naturally grown timber that became effective March 1, 2014. The new fees are approximately a 100% increase over the old fees. This increment is in conformity with the Timber Resources Management Regulations, 1998 (LI 1649) and subsequent amendments reflected in the Timber Resources Management (Amendment) Regulations, 2003 (LI 1721).

Though the action of the FC to revise the stumpage fees is commendable, this was long overdue. Forest Watch Ghana estimated that from 2003 to 2013, the failure to revise stumpage fees since 2003 led to a loss of $16 million in revenue to the Government of Ghana (Forest Watch Ghana 2014). The Minister responsible for forestry is required to determine stumpage fees partly to the free-on-board price of air-dried lumber. The study observed that 80 percent of the loss did not originate from adjusting the free-on-board price but rather due to inflation. The failure to adjust stumpage fees regularly is being seen as a weakness of the FC against the much stronger forest industry lobby.

1.4. Rights to and Benefits from Forest Resources

The 1992 Constitution provides under section 267 that all forest lands shall be vested in the appropriate stool/skin on behalf of and in trust for the subjects of the stool/skin in accordance with customary law and usage. Until 1994, rights to exploiting forest resources were transferred to timber firms in the form of conventional forest concessions. This system followed basic forest-management requirements such as harvest planning, standards for road-building and tree marking, pre-harvest operations, environmental conservation and enrichment planting (ITTO 2005a).

The Forest and Wildlife Policy of 1994 abolished the concession system and replaced it with a new timber utilisation contract (TUC) system (ITTO 2005a). The right to harvest trees is granted by the Forestry Commission in the form of permits that detail the area, volume and species to be harvested. There are three types of harvesting permits: Timber Utilisation Contracts (TUCs); Timber Utilisation Permits (TUPs); and Salvage Permits (SPs). TUCs are issued by the government for the commercial exploitation of timber. Contract holders must conclude a Social Responsibility Agreement with land-owning communities. TUPs are harvesting rights given to communities for the exploitation of timber for non-commercial and development purposes.

Ghana's annual allowable cut (AAC) is based on a polycyclic selection felling system using a cutting cycle of 40 years in natural forests. In 2005, concerns about the sustainability of the forest resource motivated the Ministry of Lands and Forestry to direct that Ghana's AAC be reduced from 1.2 million to 1.0 million m^3 (for both on and off reserve). The Forest Services Division (FSD) now has the mandate to set the AAC each year. The Forestry Master Plan (1996 – 2020) directs that only 0.50 million m^3 and 0.30 to 0.50 million m^3 of the resource can be harvested annually from the reserves and off-reserves, respectively. Unofficial estimates, however, show that more than twice the AAC is harvested every year.

Due to the inefficiencies of the concession system, in 2002, a competitive bidding system was introduced to allocate timber-resource-use rights. This move was intended to promote efficiency, transparency and accountability in timber resource management and revenue collection. This reform was based on the *Timber Resource Management Act* (Amendment) 2002 (Act 617) and its subsidiary legislation the *Timber Resource Management (Amendment) Regulations,* 2003 (LI 1721). Under the new system, allocation of timber utilisation contracts (TUCs) is based on public bidding for rights to harvest timber in each area on the basis of an annual Timber Rights Fee (Ghana Forestry Commission 2008). Act 617 also excluded from its application, land with private forest plantation, provided for the maximum duration of timber rights, and for incentives and benefits applicable to investors in forestry and wildlife sectors.

According to the 1992 Constitution of the Republic of Ghana, revenues accruing to forest resources in forest reserves would be distributed as 40% to the stools and 60% to the State (through the Forestry Commission). For off-reserve lands, the revenues are distributed as 60% to stools and 40% to the State. With 60% of the revenue from on-reserve forest resources going to the government, communities and landowners had always considered the distribution to be inequitable and this had implications for the government's ability to protect forest reserves from encroachment and illegal logging. This concern was mitigated by the introduction of social-responsibility agreements between TUC holders and the communities where they operate to provide them with negotiated social amenities. The cost of such amenities should not exceed 5% of the annual royalty accruing from the operations under the TUC according to the *Timber Resources Management Regulations*, 1998 (LI 1649, section 13(1) b). Secondly, in 2006, a recommendation was made to change the 60/40 stumpage sharing ratio between the FC and other stakeholders to a 50/50 ratio. This recommendation was adopted and implemented by the then Minister for Lands and Forestry. On December 20, 2006, FC and the Office of the Administrator of Stool Lands (OASL) agreed to work out a framework for the implementation of the 50/50 stumpage sharing formula. This new formula was implemented with effect from January 1, 2007 (Ghana Forestry Commission, 2009b).

The existing scope of stumpage disbursement therefore stipulates that after the 10% administrative fee for the OASL has been deducted, the remaining stumpage payable shall be shared by a 50/50 ratio between the FC and the other stakeholders. FC's portion of the stumpage is to be applied to cover cost and expenses of staff remuneration, administration, operations and investment. The formula applies to both on-reserve and off-reserve situations. The 50% share for the other stakeholders is deemed 100% and then distributed based on the proportions spelt out as follows (Ghana Forestry Commission 2009b):

- twenty-five percent (25%) to the stool through the traditional authority for the maintenance of the stool in keeping with its status;
- twenty-five percent (25%) to the traditional authority; and
- Fifty percent (50%) to the District Assembly, within the area of authority of which the stool lands are situated.

1.5. Forest Products Trade

Over many decades, Ghana has been exporting a variety of wood products to about 80 countries worldwide, spanning most continents, and countries from Angola to Yugoslavia. These included logs, air-dried and kiln-dried sawn wood, rotary-cut and sliced veneer, plywood, machined timber (including mouldings and profiled boards), flooring and furniture parts. About 90% of all exports are made up of sawn wood, veneer and plywood. Ghana's forest products sector continues to be characterised by very low lumber recovery factors that range from 20 to 40% of the log input (FAO 2005). The total annual log requirement of sawmills in the country is about 1.3 million m^3 for lumber alone (Odoom and Vlosky 2007).

Both past and recent trends of forest products exports show continuous shifts in products mix over time and additions of new destinations for Ghana's wood products. These trends may be the result of the policy and regulatory changes in Ghana or due to international competition among the producers of tropical forest products.

Figures 2 and 3 show the trends in the quantities of wood production and exports for logs, sawn wood, plywood, and veneer from 1961 to 2013. Log production and exports reached their lowest levels in 1983, during a period of serious economic decline in Ghana. The implementation of the Economic Recovery Programme (ERP) in 1983 resulted in a timber rehabilitation credit extended by the World Bank to revitalise the forestry sector. The credit provided much-needed capital equipment to increase value-added wood processing that resulted in a major boost that increased production of all forest products, especially sawn wood (Owusu 2001).

Exports of plywood and veneer have fluctuated since 1961 but showed a consistent upward trend after the log export ban on 14 timber species in 1979. The quantity of veneer exported over the study period has averaged 61% of production, representing the highest percentage of any forest product exported. Most of the veneer produced in Ghana is used for either decorative furnishing or for making plywood in importing countries.

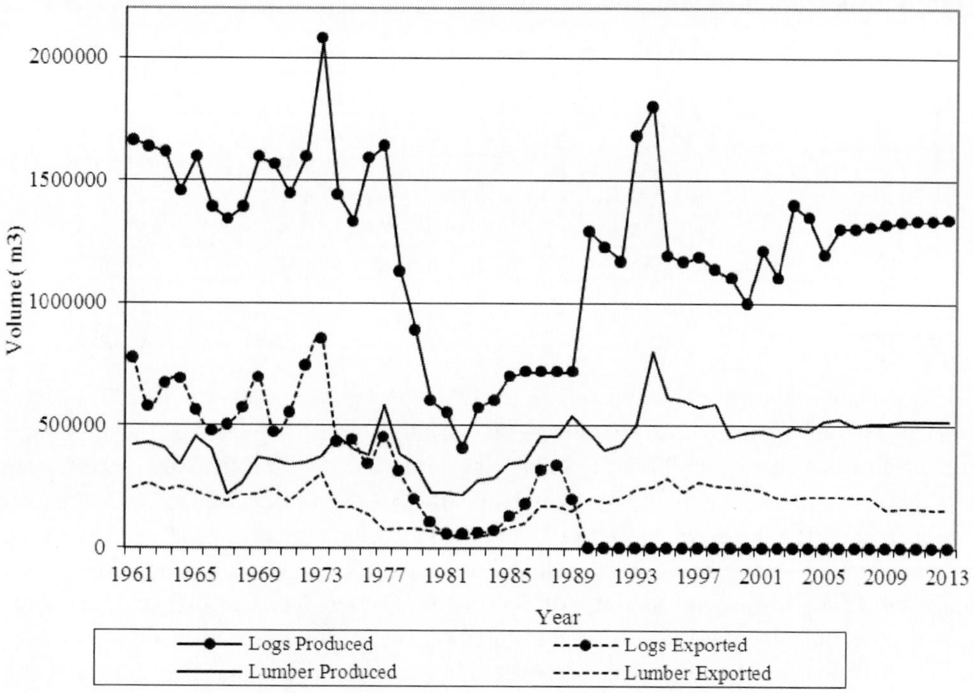

Figure 2. Quantities of logs and lumber produced and exported from 1961 – 2013.

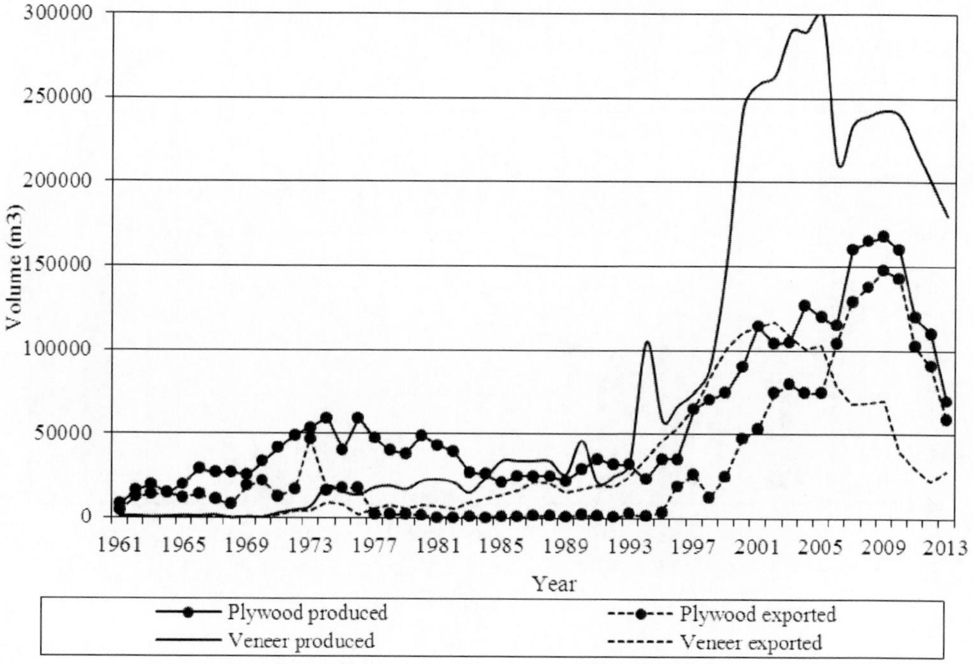

Figure 3. Quantities of plywood and veneer produced and exported from 1961 – 2013.

Table 3. Export volumes and value of wood products from Ghana from 2008 – 2013

Year	Export Volume (m^3)	Annual change (%)	Export value (Euro)	Annual change (%)
2008	545,965	-	186,611,984	-
2009	426,222	-21.93	128,226,984	-31.29
2010	403,254	-5.39	137,842,837	7.50
2011	319,843	-20.68	107,431,995	-22.07
2012	251,346	-21.45	99,836,394	-7.07
2013	271,772	8.17	119,327,780	19.52
2008-2013		**-50**		**-36**

Table 3 shows that export volume consistently decreased from 2008 to 2012. The slight increase in volume from 2012 to 2013 was due mainly to a 100% increase in the export volume of air-dried lumber in 2013. It is alarming to note that in the five-year period from 2008 to 2012, exports were reduced by more than half (54%) and from 2008 to 2013, the total reduction in export volume was 50%. Reduction in the volume alone would not have been bad if the reduction were a result of value added-processing. However, a review of the export earnings shows a decrease in earnings of 36%. The smaller decrease in export earnings compared to volume exported may be due to inflation (general increase in wood prices over time). The fact that export value has been declining as well is a signal that Ghana is not reducing the volume exported to add value. The reduction in export volume may reflect the reduction in available wood supply to the mills as a result of the degradation of the forest resource. Another possible explanation is that some of the wood supply is going through chainsaw operators (legal and illegal), which is not exported and hence not captured by these statistics.

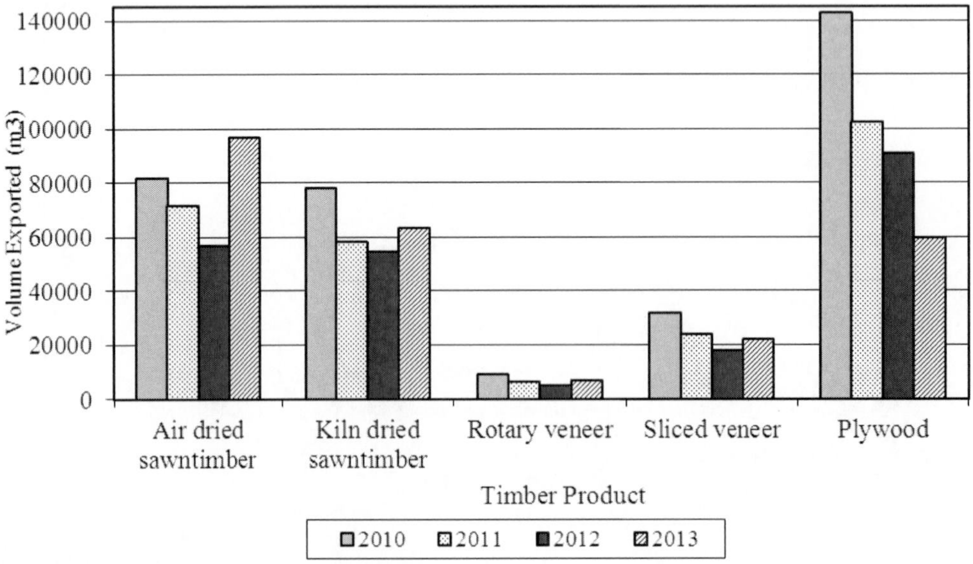

(Data source: Ghana Forestry Commission 2014).

Figure 4. Volume of wood product exports from Ghana from 2010-2013.

Table 4. Top 10 exported wood species from Ghana in 2013

Species	Export Volume (m3)	% contribution to export volume	Export value (Euro)	% contribution to export value
Rosewood	40,998	15.09	19,872,463	16.65
Teak	43,341	15.95	18,896,693	15.84
Ceiba	52,443	19.30	17,896,742	15.00
Wawa	43,274	15.92	15,666,377	13.13
Mahogany	15,824	5.82	9,333,109	7.82
Asanfina	5,577	2.05	4,399,888	3.69
Ofram	11,966	4.40	4,353,607	3.65
Sapele	5,546	2.04	3,704,035	3.10
Koto/Kyere	4,579	1.68	2,830,028	2.37
Chenchen	6,730	2.48	2,344,792	1.97
Totals	**230,278**	**84.73**	**99,297,734**	**83.21**

Source: Ghana Forestry Commission (2014). Total export volume and total export value for 2013 used for the calculations are taken from Table 3.

Table 4 displays the information on the top ten exported wood products from Ghana by value. It is clear that these 10 species constitute approximately 85% of volume exports and 83% of total value of wood exports. This means that the export sector relies heavily on a few tree species, especially the top four (teak, rosewood, ceiba and wawa) for its survival.

The most recent data from the Ghana Forestry Commission (2014) show that Ghana's export products consisted of primary, secondary and tertiary products. In 2013, the primary export products were poles and billets, which accounted for 1.98% of value and 3.34% of volume exports. Secondary Wood Products (mainly lumber, boules, veneers, and plywood) continue to dominate wood exports, constituting 91.59% in volume and 91.34% in value. The tertiary wood products (consisting of mainly processed lumber/mouldings, flooring, dowels and profile boards) on the other hand, accounted for 5.06% by volume and 6.58% by value of the total wood products exports during the year.

Forest products exports to Europe in 2013 accounted for €47.89 million (40.14%) from a volume of 84,210m^3 (30.98%). The key European markets included Germany, Italy, France, the U.K. and Belgium. Volume of exports to Asia and the Far East countries of China, India, Thailand, Russia, and Vietnam have been increasing over the years and have overtaken the European Union as the main destination for Ghana's timber. Exports to Asia accounted for 107,020 m^3 (39.38%) valued at €46.91million (39.31%) in 2013. Exports to Asian markets were mostly teak and gmelina poles/billets as well as a recent surge in rosewood lumber exports.

On the African continent, trade is still minimal, with Ghana exporting mainly plywood and air-dried lumber that amounted to €28.09 million (23.54%) from a volume of 81,470 m^3 (29.98%) in 2013. The major African destinations included South Africa, Morocco, Cape Verde and the ECOWAS countries. In the ECOWAS sub-region, exports accounted for 65,539 m^3 (80.44%) valued at €20.43 million (72.73%)) of the total African wood imports from Ghana. Ghana exports mainly plywood and lumber to ECOWAS countries by land transport.

Nanang (2010) examined the factors affecting the external demand for Ghana's forest products. The study revealed that external and domestic factors combine to influence the demand for Ghana's timber exports internationally. In the domestic forestry sector, initiatives that affect the timber supply (such as AAC restrictions), regulations that influence the domestic processing and export of forest products and the fiscal framework that determines the external value of the local currency (devaluation) all play important roles (Nanang 2010). On the international scene, the income of importing countries and world prices of the wood products are significant determinants of export volume (Nanang 2010).

Sawn wood and plywood face stiff competition in the international market, and this has revenue and tax policy implications for Ghana's forestry sector. The demand for Ghana's forest products is tightly linked to fluctuations in the economies of the importing countries (Nanang 2010). These fluctuations were observed from data from the Ghana Forestry Commission on forest products exports in 2009. Analysis of export figures for January to May 2009 showed decreases of 37.0% in revenue and 32.5% in volume of wood exports over the same period in 2008. The data showed considerable reductions in market demand for wood products especially in the major importing countries (Germany, Italy, Spain, U.K. and U.S.A) due to the credit crunch and the global economic downturn which generally affected the cash flow of most buyers of wood products (Ghana Forestry Commission 2009a).

The reliance of Ghana's forestry sector on external markets further implies that revenues from forest products would be closely tied to the international business cycle, as economic downturns in importing countries would decrease revenues and vice versa. The price elasticities for sawn wood and plywood showed that a significant degree of substitution for these products could occur with small price increases in timber prices. Consumer responses to changes in the prices of forest products and the potential for substitution by other wood or non-wood products represent two fundamental elements of long-term forest resource planning and development.

From a forest tax policy perspective, Nanang (2010) showed that demand for sawn wood and plywood is elastic, which implies that Ghana's exporting firms may not be able to shift a large proportion of any domestic tax increases on these products to importing country consumers, without losing market share or profits. Export tariffs on wood products would in general increase the prices of the final products in importing countries. For sawn wood and plywood, most of the tariff would be paid by the producers in Ghana, with a marginal increase in final price. If the producers were to shift the tax to consumers in importing countries in higher prices, the demand for Ghana's sawn wood and plywood would fall. Veneer has inelastic demand, and hence more of the tax burden can be passed on to consumers, with relatively little or no impact on exports.

Policies that encourage domestic processing and restrictions on both legal and illegal harvesting would work to ensure greater value-added benefits to Ghana. The focus on exported forest products at the expense of the domestic market is a major cause of illegal logging practices that are inconsistent with sustainable forest management principles. There is an urgent need to increase the supply of legally processed wood products to the domestic market.

1.6. CHALLENGES FACING THE FOREST SECTOR

Despite an encouraging and a strong indication of steps towards sustainable forest management through policy and legislative changes and international involvement, the forestry sector continues to face the following challenges.

- The most critical threat to Ghana's forestry sector is deforestation. Poor management of some forest reserves and off-reserve forests, increasing human population, uncontrolled wildfires, illegal logging, high dependence on wood energy, corruption, etc. all contribute to the high rates of forest cover loss in Ghana.
- Given that successive silvicultural systems since 1946 have been unsuccessful in achieving a desirable level of natural regeneration - including the current polycyclic felling system - there is a clear challenge for an innovative forest management system that would ensure sufficient regeneration of the natural forests. The most appropriate silvicultural system should be one that meets sustainable forest management goals.
- Thirdly, wood harvesting and processing continue to be inefficient, thereby leading to enormous waste in processing and destruction of residual forest stands. It is estimated that about 30 to 35% of the wood harvested are left in the forest as waste while the conversion rate of logs into lumber is about 40% (Dauda 2009). The combination of inefficient extraction methods and processing result in a final lumber volume that is only 25–40% of the total log volume extracted (Chachu 1989). These high levels of inefficiency reduce the international competitiveness of Ghana's forest industry and contribute to the excessive exploitation of the natural forest resource.
- The rising demand for wood products resulting from population growth will continue to pose challenges in terms of how to meet this demand from a diminishing forest resource base. In this case, improving the efficiency of forest resource use, diversification towards value-added wood products, increasing wood supply through plantation development and the use of lesser-known tree species will be essential.
- Ghana's timber products face high international competition from other tropical forest products, due to their elastic demand (Nanang 2010). The demand is also affected by the economic health of importing countries; hence fluctuations in the international business cycle partly drive export demand for Ghana's wood products.
- High poverty levels in Ghana in general and among forest fringe communities, in particular, constitute a major stumbling block to sustainable forest management by making it difficult to control illegal forest product exploitation activities. Forest management strategies that ensure effective community participation will mitigate this problem.
- Another major challenge to the forestry sector is the competition between forestry and other land uses. Ghana is mainly an agricultural country, and hence expansion of agricultural cultivation, mining and urbanisation into forests continue to contribute to deforestation and forest degradation.

- Finally, there are institutional constraints as well. Policy implementation failures are widespread in Ghana. Developing sound policies and legislations are good first steps in managing forest resources; however, without the capacity to implement and enforce those policies effectively, they amount to nothing. The Forestry Commission seems to be ineffective against the more powerful timber industry in enforcing forestry legislation and policies. It is questionable whether the FC can protect the public interest when it comes to forestry issues.

1.7. A Forest Sector in Crisis

Wood production and consumption in Ghana have been increasing steadily since 1961, the earliest date for which reliable data are available. This section projects the production and supply of roundwood for Ghana from 2010 to 2100. The analyses are based on the following information and assumptions. The historical roundwood production data is taken from FAO (2011), and this production is assumed to meet the current demand for roundwood in Ghana (though, in reality, the supply falls short of demand). The roundwood production includes all wood harvested from Ghana for fuelwood, charcoal, saw logs, veneer logs and wood for plywood and other industrial wood requirements. In 2010, roundwood production was 36.755 million m^3 (FAO 2011). Based on the historical data of roundwood production from 1961 to 2010, the average annual increase in roundwood production was about 3%, identical to the average population growth rate in Ghana over the same period. The growth in roundwood production was assumed to remain constant throughout the projection period.

We assume the supply of timber will come solely from the natural forest. Based on FAO (2011) estimations, the average growing stock of forests in Ghana was 49m^3/ha, with a mean annual increment (MAI) of 4m^3/ha/yr in forest reserves. Given that the MAI in off-reserve areas is likely less than that in the reserved areas, an average MAI of 3m^3/ha/yr is used and applied to the total forest area in Ghana. It is further assumed that all the growing stock can be profitably recovered at a reasonable cost (though this is not true, it ensures the supply projections are over-estimated).

The average deforestation rate in 2010 was about 1.9% of the forest area, and this was assumed to be constant throughout the projection period as well. Therefore, supply in each of the projection years was estimated as the forest area remaining in that year multiplied by the combined growing stock per ha and the MAI. Figure 5 presents the demand and supply curves. At the current rates of deforestation and natural forest growth, the demand will outstrip supply of roundwood by 2058 even with these optimistic projections of supply.

One of the key assumptions is that all timber stocks can be harvested, which, of course, is not the case. This inability to harvest all timber stocks, together with continuous forest degradation, means that Ghana would be unable to meet its demand from the natural forests much sooner than 2058. At the current deforestation rate, the total forest area will be reduced to about 0.90 million ha by 2100, from the 4.94 million ha in 2010. If we consider that the rate of deforestation is likely going to be higher than that estimated for 2010, then the timber supply outlook is even gloomier! Given that not all the wood in the forest can be recovered Ghana can run out of wood, even if we do not run out of forests.

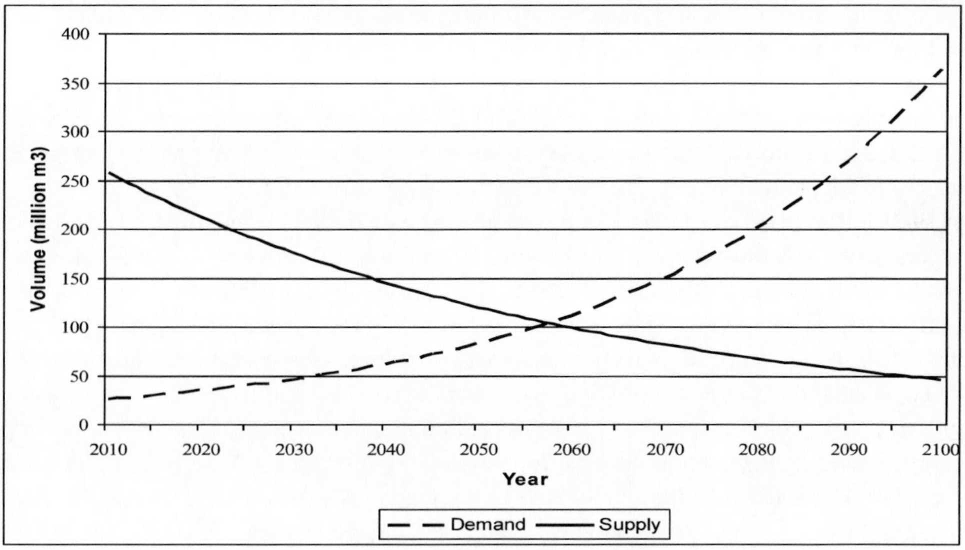

Figure 5. Projected demand and supply of roundwood in Ghana from 2010 – 2100.

It is important to highlight that the gap theory applied in this analysis assumes a continuous constant (linear) decrease in wood resources as human populations continue to grow and assumes that nothing is done to increase wood supplies through plantation development and natural regeneration. If the decrease in wood supplies is at a faster rate than assumed here, the shortages will occur sooner than projected. However, if wood supplies increase through plantation development or increased natural regeneration, it would mitigate the wood scarcity, and shortages would occur later than projected, all other things held constant. On the demand side, the analysis does not recognise the phenomenon of substitution whereby, as wood becomes scarce, people will use wood more efficiently or substitute with other materials. Substitution and efficiency have the potential to decrease the demand for wood supplies and hence would be critical to reducing the severity of the wood crises.

The above analyses suggest that Ghana's forestry sector is already facing a crisis. The forest stocks are being depleted at a faster rate than they are being renewed. The depletion of the forests will lead to serious ecological, economic, social and environmental consequences. New forest resources must be created to offset greater losses of tree stocks and minimise the impacts of the crisis.

1.8. Overview of Plantation Forestry in Ghana

1.8.1. Brief History of Plantation Development

Plantation forestry development in Ghana is the most promising option to reduce deforestation and ensure the availability of forest products to meet social, environmental and economic objectives. The need to develop forest plantations to meet the wood products requirements of the inhabitants of the Guinea savannah zone was recognised as early as 1956. A national plantation project was launched in 1970, but during a decade of economic decline there was little activity. Another plantation programme was launched in the Northern and Upper Regions (currently Upper East and Upper West Regions) in 1976, but also slowed down due to lack of funds in the early 1980s (FAO 1995). As of 1997, there were only 40,000 ha of forest plantations consisting of about 15,000 ha planted by the erstwhile Forestry Department and the rest by forest industry firms. There are also a large number of small holdings of forest plantations with teak as the main species (FAO 1997).

In 1989, a Rural Forestry Division was established within the Ghana Forestry Department to encourage the establishment of plantations to mitigate the effects of projected wood shortages identified by the World Bank (1988). This programme was implemented under the Rural Afforestation Programme (RAP) from 1989 to 1995. The mandate of the new Division was to establish and expand existing nurseries, to initiate and expand community and individual plantations, and to provide technical advice to farmers on establishment, management and protection of the trees. In addition, the new Division was to provide extension services and education on rural forestry and agroforestry. Of the planting that did take place, woodlots accounted for about 90% of the seedlings while the remaining 10% were aimed at agroforestry, boundary planting, windbreaks and home gardens (FAO 2002a).

The years between 1996 and 1999 were quiet on the plantation development front, with no major plantation forestry initiatives. The interest in plantation forestry development was renewed with the passing in 2000 of the *National Forest Plantation Development Fund Act*, (2000) followed by the launching in 2001 of the National Forest Plantation Development Programme (NFPDP) with a target planting of 20,000 ha/yr. Interestingly, while the NFPD Fund was empowered by legislation, the NFPDP was not. The Forestry Commission has enumerated some challenges that they face in the implementation of the NFPDP (Ghana Forestry Commission 2008b):

- The FC acknowledges that the late release of funds towards programme implementation continues to hamper the success of the NFPDP. The lateness results in late payment to communities for services rendered as well as the inability to acquire inputs in a timely fashion to implement the programme. Also, the limited funding results further in inadequate logistics (such as vehicles, accommodation and equipment) and capacity of existing staff to execute their duties effectively;
- A serious problem that needs to be addressed is how to ensure that farmers under the modified taungya system (MTS) continue to tend the farms. Despite the increase in benefits to farmers under the MTS, the success of the MTS is being hindered by the inability of taungya farmers to tend their farms. Some planted sites have apparently been abandoned by some taungya farmers one or two years after planting the tree seedlings;
- Many of the participating taungya groups lack adequate capacity to cope with the cost of tending large areas developed so far;
- Cattle herdsmen threaten the security of FC staff, set wildfires to the dry vegetation during the dry seasons, trampling of seedlings and/or browsing of the herds in young plantation resulting in damage to some of the growing trees has continued to affect the success of plantations in some areas of the country; and
- Wildfires have continued to be a major concern for the survival of planted seedlings, yet the FC has a limited fire management capacity to deal with annual wildfires that destroy forest plantations.

1.8.2. Plantation Development under the Savannah Accelerated Development Authority (SADA)

In 2010, the Government of Ghana passed an Act of Parliament (Act 805) to establish the Savannah Accelerated Development Authority (SADA) to provide a framework for the comprehensive and long-term development of the Northern Savannah Ecological Zone and to provide for related matters (SADA 2010). Though SADA was intended to be an all-encompassing development programme that includes agricultural development, industrialisation, economic and social infrastructure, all within a "Green and Forested North," the discussion here will be limited to the forestry component of SADA.

After passing the SADA Act, the Authority released the Strategy and Workplan for 2010 – 2030, with a very ambitious target of achieving a Forested North over a period of 20 years.

The vision was stated as *" to develop a healthy and diversified economy based on the concept of a "Forested North, "where food crops and vegetables are inter-cropped with economic trees that are resilient to weather changes, sustain a stable environment, and creating a permanent stake in land for poor people* (SADA 2010). According to the SADA Strategy, this vision of a "Forested and Green North" implies economic transformation is woven around afforestation using economic trees (perennials) that provide the opportunity for exporting and also the basis for secondary and tertiary processing along the value chain. In addition to the economic focus, the important roles that trees play in mitigating climate change, improving soil fertility within agroforestry systems and preventing droughts and flooding were also emphasised.

The overall objectives within SADA for planting trees are for environmental and economic reasons. The contract for the tree planting was awarded to Asongtaba Cottage Industries [ACI] Construction Limited through a sole-source process. ACI is a private company that had no experience in afforestation. Implementation of the afforestation component of the SADA strategy was fraught with allegations of financial and contractual malfeasance. In 2014, an audit report by the Auditor General's Department identified several corrupt practices. The Audit Report noted that even though management of SADA got approval from the Public Procurement Authority to single-source the afforestation project to plant 5 million trees in five regions (Northern, Upper East, Upper West, Volta and Brong Ahafo) valued at GH₵32.5 m to ACI Construction Limited, the project did not fall under any of the single-source procurement conditions to warrant the contract to be awarded under sole sourcing. The audit team noticed that the contractor did not execute the project satisfactorily because the work load on the contractor was too much for him to execute (Report of the Auditor General on SADA 2014).

The Audit Report recommended that SADA management should engage the services of the Forestry Commission to assess the trees planted and recover the excess payment made from the company. It also recommended that management should institute action to invoke sanctions under Section 92(1) of the Public Procurement Act (Act 663) against the former CEO, Alhaji Gilbert Iddi, who signed the contract (Report of the Auditor General on SADA 2014).

Beyond these financial issues, the afforestation project itself was not well planned and executed. Some shortcomings of the project included:

- *Contracting the project to a company with no expertise in forest management*: the company that was contracted to implement this project had no expertise in any aspect of forest management or plantation development. Experience in plantation development by the company is critical because tree growers need technical advice on how to take care of the trees after they are planted to ensure the success of the plantations. By choosing a company without this expertise, those who planted the trees would have no access to such expertise and hence jeopardised the success of the project.
- *Poor strategic planning for the afforestation project.* Before putting the first seedling in the ground, SADA should have developed an appropriate implementation strategy for the afforestation component. This plan would have identified what trees should be planted, where they should be planted, by whom, outline the benefit sharing plan, when trees should be planted, a fire management plan, and clearly delineate responsibility for technical monitoring and support to tree growers.
- *Sub-contracting seedling production to untrained individuals.* The company subcontracted the production of seedlings to individuals who had no training in forestry or expertise in raising seedlings. This led to several problems including poor quality seedlings and wrong timing of when seedlings were ready for out planting.
- *Out planting trees at the end of the rainy season.* Due to poor timing of raising seedlings, most of the planting was carried out at the end of the raining season. Overall, the project was poorly planned and executed. The timing of the planting resulted in poor survival of the seedlings, although there is controversy surrounding this observation (see discussion below on assessment of the project).
- *Lack of management plans to protect the planted seedlings*: There were insufficient plans in place to protect seedlings from damage by animals and fire, and lack of overall technical coordination and monitoring of the project. Planting trees is the easiest part, ensuring that they survive and are protected from destruction by human, biological and inorganic agents is more complicated and requires careful planning and effort.
- *Too much focus on job creation.* One of the aims of reforestation/afforestation projects is for rural development through job creation. However, under SADA, there was too much emphasis on job creation to the detriment of the planted trees. Hence, there was too little focus on ensuring that the planted trees were established and protected from damage. This view was evidenced by JoyFM (Radio Station) interview with the then Chief Executive Officer of SADA, Alhaji Gilbert Iddi. He is reported to have told JoyFM that he did not think the SADA afforestation project had failed. In his opinion, whether the trees survived or not, was not the most important issue. As far as he was concerned, the project had created employment, and the trees had also been planted. "These are the two focal issues that we will use to be able to determine the achievement or otherwise of the intervention," he said (JoyFM 2014).

Assessment of the SADA Afforestation Project

SADA contracted the Faculty of Renewable Natural Resources of the University for Development Studies (UDS) to evaluate the afforestation project awarded to ACI. The objectives were to determine the percentage survival of planted species on various plantations; ascertain the silvicultural and management practices implemented in the plantations; determine acreages planted on various plantations; and verify the performance of contractors engaged on the projects. The report indicates that the team surveyed a total of 145 plantations in all the four afforestation zones encompassing 45 political districts in five Administrative Regions of Ghana. The report further revealed that eight tree species were planted in the various operational zones namely; *Tectona grandis, Senna siamea, Albizia lebbeck, Khaya senegalensis, Mangifera indica, Anacardium occidentale, Eucalyptus spp. and Moringa oleifera.* The assessment concluded that the average percentage survival of all the planted species in the SADA plantations was very high (85%) with the highest (88%) recorded in the Eastern Zone and the least (76%) in the Southern Zone. The Western Zone had the highest acreage of 204 ha while the Central Zone had the least of 140 ha (Faculty of Renewable Natural Resources 2013).

This assessment was undertaken following public outcry about how the resources allocated to the tree planting project were used, and as a result, the independence of these conclusions have been questioned, with some suggesting that the evaluation was intended as a cover-up for the failure of the project. Outside of the politics of SADA, a careful review of the assessment report reveals some shortcomings in describing the methodology and the reporting of the assessment. First, the report does not say whether the 145 plantations assessed were a sample or the total population (i.e., a complete enumeration of all planted plantations under the project). Further investigations revealed that it was a sample, and not a complete survey of all plantations. Given that it was a sample there was no description of how the samples were selected (sampling methods), and how the results were extrapolated to estimate survival for the whole afforestation project. Some journalists and the first author of this book independently observed that some of the plantations had completely failed as there were no living trees on them (either because they were burnt down, or the seedlings did not survive). It is expected that if a randomised sampling technique were used, some of these failed plantations would have been included in the sample, and the conclusions could have been different.

Irrespective of where one stands in the debate about the success or failure of the SADA afforestation project, there are clearly serious lessons to be learnt on how not to manage a forestry project. The Forestry Commission, would have been the appropriate body to manage this project, given its expertise and established infrastructure across the country. Having said that, we also recognise that the intent was to promote private sector delivery of the project, which, if well executed, has clear benefits. Before the audit report was released in April, 2014, the President of Ghana on January 31, 2014, had directed the "SADA Board to hold consultations with the Forestry Commission and the Ministry of Local Government to work out a strategy for the proper implementation of the afforestation and tree growing project, on a decentralised basis"(Peace FM 2014). At least, this was recognition that the FC is better placed to manage such a project and is indeed the right organisation to lead this effort.

1.8.3. Governance and Legislative Framework

Forest plantation development in Ghana is currently led by the Plantations Department (PD) of the Forest Services Division (FSD) of the Forestry Commission (FC), which is responsible for the implementation, coordination and management of the NFPDP. The programme is currently being implemented under three main strategies and five components: the Modified Taungya System (MTS), Government Forest Plantation Development Programme, and Private Plantation Development. In 2007, a purely research-based Model Plantation component was added to offer the FC plantation managers the opportunity to undertake mixed species trials, experiment various planting designs and tree spacing trials (Ghana Forestry Commission 2007b).

The Modified Taungya System (MTS) involves the establishment of plantations by the FSD in partnership with farmers. The farmers, in addition to the food crops they harvest, have a 40% share in the returns from the investment. The Government also has a 40% share while the landowner and community have 15% and 5% shares, respectively (Ghana Forestry Commission, 2007b).The MTS is funded by the African Development Bank (AfDB) Community Forest Management Project (CFMP) and the government funded Modified Taungya Component. The CFMP, unlike the other components under the NFPDP, is being executed by the Ministry of Lands and Natural Resources (MLNR) through the FSD and the Ministry of Food and Agriculture field staff in four Forest Districts in three Regions of the country.

The Government Forest Plantation Development Programme (HIPC): This second strategy utilises hired labour and contract supervisors to establish industrial plantations. Plantation workers are hired and paid a monthly allowance to establish and maintain plantations while plantation supervisors are given one-year renewable contract employment to supervise and offer technical direction (Ghana Forestry Commission, 2006b). This strategy is under the Government Plantation Development Programme (GPDP) which is funded by the Highly Indebted Poor Countries (HIPC) benefits. Under this scheme, the plantations developed are owned by the government and the respective landowners who are entitled to royalty payments. The Private Plantation Development component of the strategy involves the release of degraded forest reserve lands by the FC to private entities after vetting and endorsing their reforestation and business plans (Ghana Forestry Commission 2006b).

Policy and legislative support for plantations are provided by the 2012 Forest and Wildlife Policy and the *Forest Plantation Development Fund Act* (Act 583). The 2012 Forest Policy encourages the establishment of forest plantations and calls for the provision of incentives for their establishment and management. The main legislation governing forestry plantation development in Ghana is the *Forest Plantation Development Fund Act* (Act 583), which was passed in 2000 to offer financial support and other incentives to plantation developers in the private sector. The Act was subsequently amended in 2002 into the *Forest Plantation Development Fund (Amendment) Act* (Act 623), to cover plantation growers, both in the *public* and private sectors.

1.8.4. The Ghana Forest Plantation Strategy 2015-2040

In October 2013, a draft of the Ghana Forest Plantation Strategy 2015 – 2040 was made public. If implemented according to the 25-year implementation plan that was developed, it will cost an estimated US$3.5 billion. The document recognises the role of plantation forestry in Ghana given the current state of our forest resource base and the level of deforestation and forest degradation from other natural resource extraction activities. The strategy outlines the goals, the purpose and strategic objectives as follows:

The goal of the strategy is to achieve a sustainable supply of planted forest goods and services to deliver a range of economic, social and environmental benefits. The purpose of the strategy is to optimise the productivity of planted forests by identifying suitable tree species and improving their propagation, management, utilisation and marketing.

The Strategy has five main strategic objectives, described as crucial for success. They form the key levers for change, and proposed actions must reflect them.

Strategic Objective 1

- To establish and manage 500,000 ha of forest plantations and undertake enrichment planting of 100,000 ha through the application of best practice principles, by year 2040.
- To undertake maintenance and rehabilitation of an estimated 235,000 ha of existing forest plantations through the application of best practice principles

Strategic Objective 2

- To promote large scale and small-holder forest plantation investments

Strategic Objective 3

- To create employment opportunities and sustainable livelihoods in rural communities through forest plantation development

Strategic Objective 4

- To increase investments in research and development, extension, training and capacity building for forest plantation development and timber utilisation

Strategic Objective 5

- To improve governance in the regulation and management of forest plantations.

The Strategy is consistent with the Ghana Forest and Wildlife Policy (2012), and provides details of plans by the government and private sector to reforest degraded forest lands by developing commercial forest plantations of recommended exotic and indigenous tree species at an annual rate of 20,000 ha (i.e., 10,000 ha: public/public-private partnerships; 10,000 ha: private sector) over the next 25 years.

The strategy identifies challenges to past efforts and consequently outlines the strategic direction, actions and resources required to promote productively and sustainable forest plantations. It indicates the technical and financial resources required and performance measures necessary to track progress over the period (2015 to 2040).

This Strategy is probably the most comprehensive and thought-out plan for plantation forestry development in Ghana yet. It clearly identifies where plantation forestry can play a major role, and also identifies what type of planting is required and how much. The projections of the demand for forest plantation products fall into five categories as described below (Forestry Commission 2013):

The Timber Industry

Presently, the timber industry relies on a dwindling supply of large hardwood logs from the natural forest which can only sustain an annual allowable cut (AAC) of one million m^3 compared to the industry installed capacity of about 3.7 million m^3. Consequently, there is a serious supply-demand gap of over two million m^3 that threatens the survival of both the industry and its resource base.

Woodfuel: It is estimated that about 85% of the population - mainly in the rural areas - depend on fuelwood (charcoal, firewood etc.) for cooking. It is sourced mainly from the natural forest or savannah woodlands. The annual per capita fuelwood consumption is estimated to be 1.0 m^3 round wood equivalent (FAO 2010). This means the estimated current annual fuelwood demand is 24 million m^3. However, based on the following premises the annual fuelwood demand is calculated as 16.8 million m^3:

- The rural population is 65% of the total national figure (i.e., 15.6 million) and consume 1 m^3/capita of fuelwood
- The urban poor constitute 20% of the national population. They use a mix of fuels including gas with a fuelwood consumption of 0.25m^3/capita
- The urban rich/middle class make up 15% of the population and that their consumption of fuelwood is insignificant(i.e., zero)

Table 5. Projected planting requirements for identified plantation end-uses

End-Use	Estimated Annual Demand (m^3RWE)	Supply From Existing Forests (m^3 RWE)	Balance (to be met from Plantations) (m^3RWE)	Total Plantation Area Required[5] (ha)	Annual Planting rate (Rotation=25 years)(ha)
Timber Industry (mostly export + some local) [1]	2,000,000	800,000	1,200,000	120,000	4,800
Local Timber (mostly chainsawn)	2,500,000[2]	0	2,500,000	250,000	10,000
Fuelwood	16,800,000	15,960,000[3]	840,000	56,000	2,240
Poles[3]	20,000[4]	15,000	5,000	500	20
Bamboo & Rattan				500	20
Environmental uses (watersheds, degraded mine sites, carbon, biodiversity etc.)				37,000	1,480
Total	**21,330,000**	**16,775,000**	**4,555,000**	**464,000**	**18,560**

[1]Sawmill, bush mill, veneer & plymill, excluding chainsaw
[2]Marfo 2010
[3]95% assumed from forests.
[4]100, 000/poles/yr @0.2m3/pole
[5]MAIs (*m3/ha/yr*) used: *15 for fuelwood; 12 for poles; 10 for timber and poles*
Source: Ghana Forest Plantation Strategy 2015-2040

Poles: Thinnings from the FC teak plantations have been the main source of raw material for the wood pole treatment plants in Ghana. The current demand, however, exceeds the supply and some of the pole treatment plants in the country have been importing softwood poles to supplement the local supplies. It is estimated that about 100,000 wooden poles per year would be required for the national electrification programme for the next 30 years (Odoom 1998). Assuming an average pole volume of 0.2 m^3, this amounts to some 20,000 m^3 per year.

Bamboo and Rattan: The areas for these were assumed to be the same as those for poles.

Environmental Functions: A total area of 1,480ha will be planted annually for environmental purposes. This area is equivalent to 10% of the annual planting target for the timber industry requirements of 14,800ha.

A recent review by the Forests Service Division of the Forestry Commission also identified an estimated total area of 175,000 ha of potential areas suitable for plantation development within forest reserves in the high forest zone. An estimated 300,000 ha of potential sites, representing 75% of total forest reserve area in the Northern Savannah, were also identified (Ghana Forestry Commission 2013).

1.8.5. Extent of Forest Plantations in Ghana

Table 6 presents the areas planted under the National Plantation Development Programme that was launched in 2001. The implementation of the programme however, began in 2002. The planting target of 20,000 per annum was reduced to 16,250 ha in 2004 and further to 10,000 in 2005. The annual targets were reduced to allow the FSD field staff to cope with the large annual target while at the same time maintaining the large areas already established (Ghana Forestry Commission, 2006b). A wide range of tree species is planted in the plantations, including both indigenous and exotic economic tree species.

Table 6. Areas planted under the National Forest Plantation Development Programme from 2002-2012

Year	Area planted (ha)							
	Modified Taungya system (MTS)	Community Forest Management Project	HIPC Funded	Private Developers	Model Plantations	Expanded program	FC/ Industry Plantation fund	Total area
2002	17,341	-	-	1609	-			18950
2003	17541	-	-	1609	-			19150
2004	16090	-	5510	1609	-			23209
2005	9105	1136	3342	1609	-			15192
2006	9401	2298	2709	1609	-			16017
2007	8711	2731	2948	1613	79			16081
2008	111	2930	1807	6107	160			10698
2009	2427	4293	903	4015	140			11361
2010				4733		14,009	107	18,602

Table 6. (Continued)

Year	Area planted (ha)							
	Modified Taungya system (MTS)	Community Forest Management Project	HIPC Funded	Private Developers	Model Plantations	Expanded program	FC/ Industry Plantation fund	Total area
2011				4560		6,720	271	11,248
2012				3670		5,854	95	8400
2013				2747	2858		181	
Total	80,727	13,388	17,219	35,491	3237	26,582	654	177,298

Data Source: Ghana Forestry Commission (2014). Note: Some arithmetic errors were detected in the original table published by the Forestry Commission, and so totals here differ from theirs.

The indigenous species include *Mansonia altissima* (Oprono), *Terminalia superba* (Ofram), *T. ivorensis* (Emire), Khaya spp. (Mahogany), *Ceiba pentandra* (Onyina), *Heritiera utilis* (Nyankom), *Entandrophragma angolense* (Edinam), and *Triplochiton scleroxylon* (Wawa) and the exotics are predominantly *Tectona grandis* (Teak), *Cedrela odorata* (Cedrela) and *Eucalyptus camaldulensis* (Eucalyptus). The exotic species form about 95% of the areas planted (Ghana Forestry Commissio, 2008b).

The NFPDP has achieved some successes since 2002, but clearly the planting of 168,812 ha is insufficient to reduce the current burden on the natural forests. If it is assumed that all the trees planted in Table 6 survived, and assuming a mean annual increment of $14m^3$/ha/year (based on the average MAI in Chapter 2, Table 1), these plantations would be producing only 2.36 million m^3 of wood annually, which is more than the official annual allowable cut (AAC) of 1.0 million m^3, but probably still less than the unofficial quantities harvested in Ghana in a year.

1.9. Conclusion

Ghana's forest sector is important for its environmental, social and economic contributions to the lives of Ghanaians. Analyses of the current state of the natural forest estate in this Chapter have shown that this natural resource base is not sustainable without a serious effort to stop the high rates of deforestation. Plantation forestry development is a key component of the effort to arrest deforestation and ensure the availability of forest resources to society.

The government of Ghana has an important role to play in this regard. The government needs to implement serious measures to reduce deforestation. The forest industry and its supporters like to argue that reductions in harvesting levels will result in timber companies laying-off employees. Politicians like to use this argument to rationalise their inaction. It seems clear that this focus on short-term economic gain for long-term pain is unwise; precisely because when the forest is completely depleted everyone in the industry will then be unemployed. The most sensible approach is to manage the resources sustainably, even if it will cost some jobs in the short term.

Chapter 2

CONSIDERATIONS IN FOREST PLANTATION DEVELOPMENT

Chapter 1 provided an overview of Ghana's Forest sector and revealed that the sector is already in a crisis. In Chapter 2, we will examine the reasons why Ghana should intensify the development of forest plantations and explain why the current state and management of the natural forest resources would be unable to meet the future wood and non-timber needs of Ghanaians and the export market. Several models of plantation development are examined and the policies as well as the technical and economic factors affecting plantation establishment and management are discussed.

2.1. DEFINING A FOREST PLANTATION

Definitions are by nature difficult and controversial, and defining a forest plantation is no different. Over several decades, there have been considerable attempts to define many concepts related to forestry and forest plantations (Carle and Holmgren 2003). According to Ford-Robertson (1971), a forest plantation is "a forest crop or stand raised artificially, either by sowing or planting." This definition has been accepted and used for several decades until the FAO started work to harmonise forest definitions in order to ensure consistency, effective communication and data exchange on forest-related issues between countries and organisations. In 2003, the FAO proposed a comprehensive definition for plantation forests as *"planted forests that have been established and are (intensively) managed for commercial production of wood and non-wood forest products, or to provide a specific environmental service (e.g., erosion control, landslide stabilisation, windbreaks, etc.)."* Though intensive management is an important component of the definition, it should be noted that planted forests established for conservation, watershed or soil protection may be subject to little human intervention after their establishment (Carle and Holmgren 2003).

Plantation forestry is usually associated with afforestation or reforestation activities. Both afforestation and reforestation involve tree planting; the only difference between them is the time span over which there were no trees on the land being replanted. Afforestation is usually used to describe tree planting in areas that have not had a forest for more than 50 years.

Reforestation on the other hand, refers to land cover change into a forest of an area that has not carried a forest within the last 50 years.

2.2. The Need for Forest Plantations in Ghana

The reasons for pursuing plantation forestry differ across the various countries of the world. For Ghana, the overarching strategic reasons for promoting forest plantations are to ensure that the country can meet the demand for forest products for its growing population, reduce the pressure on the natural forests and hence contribute to their sustainability. Forest plantations, usually of fast-growing exotic species[1], also provide forest products to communities that do not have natural forests. Plantation forestry in Ghana is one arm of the three axes of policy tools to reduce deforestation, protect the natural forest resource base thereby arresting and /or reversing deforestation, and ensure the availability of forest products to meet social, environmental and economic objectives. Other reasons for pursuing forest plantations include: high productivity of plantations compared to natural forests, flexibility to site plantations, rural and economic development and for carbon sequestration to obtain carbon credits. These reasons are discussed in detail below.

2.2.1. Decreasing Supply of Forest Products from Natural Forests

Ghana's population has traditionally relied on the natural forests to meet their needs for timber and non-timber forest products. However, analyses of demand and supply of the forest resources suggest that in future, demand will outstrip supply and hence new resources need to be created. The decreasing supply of forest products is a result of increasing destruction of the natural forest due to natural and anthropogenic factors, unsatisfactory and unreliable natural regeneration, poor management of existing forests, and exclusion of some natural forests from wood production.

A) Deforestation and Forest Degradation

Ghana has one of the highest rates of forest loss in Africa, raising concerns about the adverse consequences of diminishing forest cover and woodland areas in the country. The accelerating rate of deforestation shows that Ghana's reliance on its natural forests to supply its raw material needs cannot be sustained in the long term. Table 1 shows the decline in forest area of Ghana from 1990 to 2010.

The total forest area in Ghana was estimated at 4.94 million ha in 2010 (FAO 2011). Detailed distribution of this forest area by function is given in Table 1. There are 204 forest reserves in the high-forest zone covering an area of 1.58 million ha and 62 forest reserves in the savannah zone covering 600,000 ha.

---

[1] The term exotic is used to describe a species introduced into a country from outside, as opposed to indigenous species that grow naturally within a country.

Table 1. Primary designated functions of forests in Ghana

*FRA 2010 Categories	Forest Area ('000 ha)			
	1990	2000	2005	2010
Production	1 694	1 386	1 255	1124
Protection of Soil and water	353	353	353	353
Conservation of biodiversity	43	43	43	43
Social services	89	73	66	59
Multiple Use	0	0	0	0
No/Unknown	5 269	4239	3800	3361
Total	**7 448**	**6 094**	**5 517**	**4 940**

Source: FAO: Fifth Assessment Report (2010). FRA is the Forest Resource Assessment by the FAO

Deforestation averaged about 1.8% per year between 1990 and 2000 but increased to 1.9% per year between 2000 and 2010[2], according to data from the Global Forest Resource Assessment (FAO 2011). Against this backdrop of a declining resource, roundwood production has been rising steadily over the years.

Figure 1 illustrates the continuous decline in the forest area vis-à-vis the drastic increases in the production of roundwood for wood fuel, saw logs and veneer logs, and for other industrial purposes.

Roundwood production increased steadily from 1961 until 1990 but after that rose drastically. In 1961, the forest area was about 9.6 million ha and declined to 4.94 million ha by 2010. This reduction represents a 49% of the original forest area in only 50 years. Based on the estimated remaining forest area and the deforestation rate in 2010, it is obvious that with no effort at maintaining and/or increasing the forest area, Ghana's forests could be completely depleted by 2058 or degraded to the point where no commercial logging can take place (see Section 1.7).

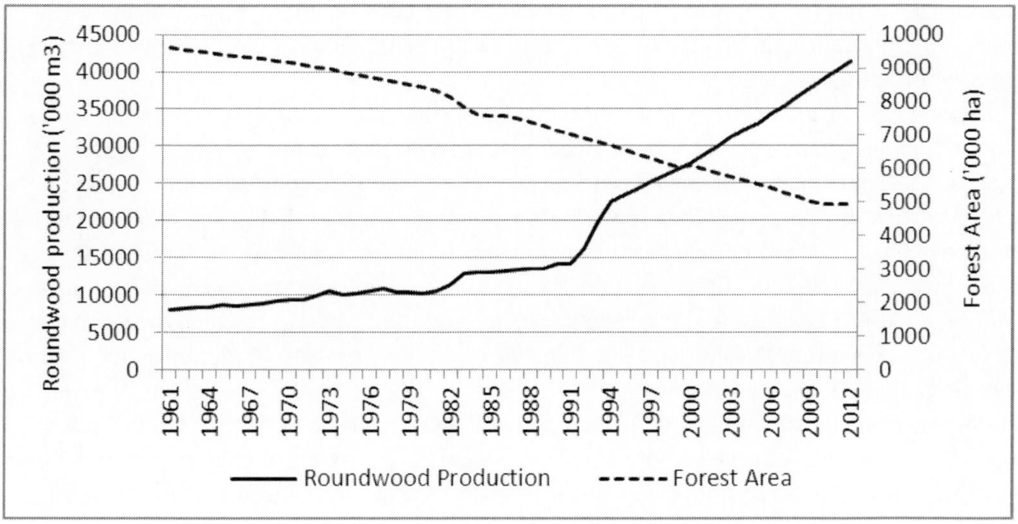

Figure 1. Forest area and roundwood production in Ghana from 1961 to 2012.

[2] The forest area reduced from 7.448 million ha in 1990 to 6.094 million ha in 2000 and to 4.94 million ha by 2010.

As a result of concerns over the sustainability of the timber resource, the commercial species were classified into 'scarlet', 'red', and pink' groupings based on their sustainability. 'Scarlet' species were being over harvested at a rate greater than 200% the estimated sustainable yield and were thought to be under threat of economic extinction, 'red' species were being cut at a rate of 50–200% the sustainable yield, and 'pink' species were being harvested at less than 50% the sustained yield (Ministry of Lands and Forestry 2004)

There is little consensus on the roles and, in particular, the actual causes of deforestation (Lugo and Brown 1982), partly because some studies mistake causes for effects (Deacon 1994), and because interactions among the causal factors vary from one country or region to another (Allen and Barnes, 1985; Saxena et al., 1997). The few rudimentary, primarily qualitative forestry studies for Ghana have generally concentrated on community-specific ethnobotanical and anthropological issues (the only exceptions are Benhin and Barbier (2001), Benhin and Barbier (2004), and Owusu (1998), but all focused on effects of the structural adjustment programme on deforestation). For example, Dei (1992) and Dei (1990) qualitatively analysed the process of deforestation for a specific rural Ghanaian community, while Townson (1995) assessed various economic activities generated from forest products. Appiah et al. (2009) through a survey in southern Ghana, identified four most highly ranked causes of deforestation as poverty-driven agriculture, lack of alternative rural wage employment other than farming, household population levels, and conflict in traditional land practices. These studies for particular communities help provide more focused solutions to local or specific deforestation problems but are inadequate in providing strategic policy direction for reducing deforestation at the national level.

There is potentially a long list of factors that cause deforestation in Ghana. The most common ones include: population growth, forest products exploitation, over-reliance on wood fuel for energy, uncontrolled forest fires, agricultural and mining expansion into forested areas, corruption, lack of enforceable property rights, etc. However, broad generalisations and qualitative statements about the factors that cause deforestation in Ghana do not contribute much to the search for specific policy prescriptions. Hence, quantitative assessments of the causes of deforestation in Ghana are required to analyse the complex relationships between deforestation and its hypothesised causes.

Nanang and Yiridoe (2010) provided the most comprehensive analysis of the causes of deforestation in Ghana yet. The study modelled deforestation using two-stage regression methods: as an interaction of interlinked key sectors in the Ghanaian economy (including forest products exports, fuelwood energy consumption, cocoa production, infrastructure development and food crop production), which compete for forest land use or forest products. In the first stage, the four direct (or first-level) causes of deforestation were regressed on various second-level causes. In the second stage, deforestation was regressed on the estimated first-level causes of deforestation. The impacts of various causes of deforestation were quantified using elasticity estimates.

Policies aimed at minimising deforestation in Ghana should be classified and prioritised, based on whether the effects on forest and woodland area loss are direct or indirect. The causes of deforestation are not linked to the forestry sector alone but are also affected by agricultural, economic, demographic and political factors. The results of the study showed that fuelwood consumption and food crop production are two of the leading direct causes of forest and woodland area loss, with the elasticity of deforestation being highest with respect to fuelwood consumption (3.5), and then with respect to deforestation in previous years (0.7) (Nanang and Yiridoe 2010).

B) Natural Disturbances in Natural Forests

Natural disturbances such as fires, pests and diseases are part of the dynamics of every natural forest ecosystem, and Ghana's forests are no exception. These natural disturbances act together to reduce the amount of wood available in the forests.

Fire is an important part of the Ghanaian culture; as it is frequently used in land clearing for agricultural purposes, hunting, honey collecting, palm wine processing and ceremonial celebrations. Fire activity in forests is strongly influenced by weather and climate, fuels, ignition agents and human activity. There are social and ecological benefits of fire in terms of forest ecosystem renewal, composition, species diversity and importantly, carbon balance. However, wildfire in forests and savannah woodlands cause irreversible environmental damage by reducing the productive capacity of forests, damage water supplies, impacts on water quality, reduces soil fertility and hence agricultural productivity, and kills wildlife and biodiversity. Forest fires occur regularly and cause severe damage; the annual financial loss due to wildfires is estimated to be about US$24 million (Ministry of Lands and Forestry 2004). FAO (2011) estimates that up to 80% of the forest area in Ghana were affected by wildland fire in 2005. The most devastating wildfire in recent memory was in 1983, which resulted in massive damage to forests, woodlands and cocoa farms across the country. It is estimated that about 30% of the forest in the moist semi-deciduous zone were either destroyed or degraded by these fires (Hawthorne and Abu-Juam 1995) and the loss of about 4 million m^3 of high-quality timber. Due to the high relative humidity and short dry season experienced in the HFZ, the frequency and extent of wildfires are less than in the savannah zone (FAO, 2002a). In the savannah zone, wildfires are an annual occurrence, and hence the trees are adapted not only to survive recurrent fires, but also to withstand drought conditions as well.

All forest tree species in Ghana have at least some pests that can cause serious damage under some circumstances (Wagner et al., 2008). Some of the most valuable timber species in Ghana are susceptible to various kinds of pests and diseases. For example, the shoot borer-*Hypsipyla robusta,* has been the major obstacle to the propagation of the *khaya* spp.; while the main constraint to *Milicia excelsa* is the gall *Phytolama lata,* which attacks the leaves of seedlings (FAO, 2002a). In savannah ecosystems, the *analeptes trifasciata*, which is a longicorn beetle, attacks several species in the family *Bombaceae* such as *Adansonia digitata* (baobab), C*eiba pentandra* (silk cotton) and *Eucalyptus alba.*

There is no formal monitoring of forests to detect pests and diseases, and hence there are no statistics to quantify the economic damage to forests though these costs are considered to be substantial. There is little if any, information on forest diseases in Ghana. Pests and diseases tend to reduce natural forest productivity and, therefore, impact on their sustainability. Efforts to control forest and forest product pests in Ghana date back to the colonial times when the West African Timber Borer Research Institute was established in Kumasi in 1953 (Nair 2007). Attention was only paid to the entomology of other pests after the establishment of the Forest Products Research Institute in 1964 (now the Forest Research Institute of Ghana (FORIG) (Nair 2007).

Plantation forests of exotic species are less vulnerable to indigenous pests and diseases of their new environment, at least for the first rotation. Large amounts of wood residue from felling debris and the presence of stumps are favourable for colonisation by insect pests and as sources of infection (Evans 2001). This can usually be reduced by modifying the silviculture or application of specific protection measures (Evans 2001).

C) Unsatisfactory and Unreliable Natural Regeneration

Ghana has, for many decades, failed to obtain satisfactory and reliable natural regeneration from its natural forests and woodland areas. There could be several reasons for this: poor management, little understanding of the ecology of desired species, institutional and regulatory failures, the focus on promoting regeneration of only economic species, and the inability and /or high cost of controlling weeds. The ecology of tropical forests is more complex than that of temperate forests or plantations, and understanding of this ecology is still limited for many tree species. In addition, laws, regulations, policies and standards for timber harvesting that could promote natural regeneration are often not effectively enforced by the forestry authorities.

Since 1946, several attempts at the development of a silvicultural system for the indigenous forests of Ghana have been made, beginning with the tropical shelterwood system (TSS), which was based on the Malaysian uniform system (MUS). This was followed by the modified selection system (MSS), implemented between 1956 and 1970. Thirdly, enrichment plantings were carried out to improve the stocking of the poorly stocked wet evergreen forest reserves as well as to sustain the supply of the then desirable species (FAO 2002a). All the silvicultural systems failed to achieve satisfactory natural regeneration and were abandoned.

Despite these failures, the *polycyclic felling system* (PFS) used in Ghana today evolved from the modified selection system in the 1950s. The PFS is basically selective logging with diameter limits and uses advanced regeneration (i.e., residual stands below the cutting limit at the time of initial harvesting) to develop into the next crop on a 40-year cutting cycle. The key considerations in implementing this silvicultural system are to minimise damage to advanced regeneration during harvesting and to define an appropriate length of the cutting cycle (Buschbacher 1990). A necessary condition for successful regeneration under the PFS is to ensure that logging operations have little to no impact on the residual trees through the use of proper felling directions, the right equipment and skilled workers (Buschbacher 1990). Though no statistics exist on the level of damage to residual stands during harvesting, there are widespread observations of considerable damage to residual stands and soil disturbance that are jeopardising any hopes of achieving sufficient natural regeneration under the current silvicultural system.

The *Timber Resource Management Act*, 1997 (Act 547) prescribes that a timber utilisation contract (TUC) holder shall submit an undertaking to execute a reforestation plan during the period of the contract to the satisfaction of the Chief Conservator of Forests (currently CEO of the FC) (section 8(d)). TUC holders are also required under Section 11 (d) and (e) of LI 1649 for the reforestation or afforestation in any area that the CEO of the Forestry Commission may approve and show evidence of capability to undertake reduced impact logging. These reforestation requirements, together with natural regeneration were expected to ensure the renewability of the forest resource. However, two main problems have been encountered in the implementation of these requirements over the years. First, forestry companies are reluctant to reforest because of the uncertainties of whether they would benefit from the reforestation activities, given the long rotation ages of the trees planted and the short duration and uncertainty of their harvesting rights to the contract area. Secondly, the forestry authorities have been unable to enforce strictly these regulations against the often powerful international forestry companies that operate in these concessions for fear that they might withdraw their investments. The best way to ensure reforestation occurs after logging is to levy the forest companies, and then use the revenue to hire a private company to undertake the reforestation.

From an ecological perspective, it is possible to achieve natural regeneration in Ghana's forests. Hawthorne (1995) showed that on a national level, the majority of the forest species in Ghana have sufficient natural regeneration. However, the appropriate silvicultural system and forest management plans have to be developed and implemented to harness this potential for sustainable forestry practices to be realised. Compared to plantations, natural forest management is less intensive, with low yields and low capital investments (Buschbacher 1990). Hence natural regeneration and growth in natural forest ecosystems continue to be poor and puts into question the ability of Ghana's natural forest estate to meet the forest products needs of its growing population.

D) Exclusion of Some Natural Forest Resources

Some of the natural forest areas in Ghana are inaccessible to harvesting. These areas include forests that perform ecological functions such as erosion control and watershed protection. About 5% of Ghana's land area is under legal reservation. This includes 352,500 ha representing 22% of the permanent forest estate under permanent protection (areas in hills and swamps, sanctuaries, areas of significant biodiversity, fire protection, etc.) that are excluded from commercial timber harvesting. Approximately, another 397,000 ha of the permanent forest estate is excluded from timber harvesting due to the state of degradation and require rehabilitation. These are found mainly in the savannah-forest transition zone. Other factors that render forests inaccessible include forests with difficult terrain (hills or swamps); timber located in faraway locations where the cost of harvesting is exorbitant compared to the benefits, low stocking, unavailability of roads, etc. These factors together reduce the amount of forest area that is potentially available for timber production to only 762,400 ha.

2.2.2. Flexibility to Locate Plantations

Unlike plantations that can be located by choice, natural forests in Ghana are endowed by nature and located in areas where we find them. This represents a major advantage of forest plantations over natural forests. Based on need, forest plantations can be located to serve environmental needs such as prevent soil erosion or protect watersheds and other vulnerable sites or close to where the products are in demand. Forest plantations can be used to promote and increase fuelwood supplies in areas that do not have natural forests or where the forests have been depleted or degraded. Also, to the choice of location, plantations offer the flexibility to choose the species to be planted to achieve the social, economic or environmental objectives of the plantation.

2.2.3. High Productivity of Plantations

The natural growth rate of Ghana's natural forests is estimated at 4.6 million m^3 with an annual increment rate of $4m^3$/ha (FAO 2011). This mean annual increment (MAI) falls well within the range of 0.4 to $7m^3$/ha for tropical forests, but still far below the 25-45 m^3/ha for tropical hardwood plantations reported by Evans and Turnbull (2004). Reasons for the high productivity of plantations include the ability to control inputs, match species to site characteristics, and intensive management. Unlike natural tropical forests that are complex in their species composition and structure, forest plantations are simpler, and the ecology and silviculture of the species are better understood and hence they are easier to manage. With plantations, the initial spacing can be chosen to ensure full stocking of the site and pursue subsequent stand density management that optimise tree growth.

If exotic species are used in plantation development, they are often immune from the pests and diseases associated with natural forest species that would have otherwise impeded their productivity. The high productivity of plantations permits less land to be used in producing the same quantity of timber compared to natural forests, freeing more land for other uses. However, it has to be noted that plantations may not be able to replicate the full range of goods and services provided by natural forests. Table 1 presents the MAIs of selected plantation species grown in Ghana by vegetation zone.

2.2.4. Rural and Economic Development Tool

The ability to develop plantations in a location of the planter's choosing allows governments, non-governmental organisations (NGOs), and the private sector to use forest plantation initiatives as rural or economic development tools and to improve food security and alleviate poverty through payment for ecosystem and environmental services (e.g., developing eco-tourism). To achieve poverty reduction objectives, it is necessary to incorporate poverty alleviation as a major objective of both governments- and private sector-led plantation development efforts. The benefits accruing from plantations must reach rural communities for their livelihoods to improve. Additional discussion on the role of plantation forestry in rural development is found in Section 10.2.

2.2.5. Environmental Benefits

Given the multi-purpose functions of forest plantations, they can play critical roles in protecting the environment, such as soil conservation, watershed protection, flood reduction, and shelter from wind and rainstorms. This is especially true in the savannah areas of Ghana that is more prone to these environmental problems due to lack of adequate tree cover. In degraded landscapes, plantation forestry can be a useful tool in restoring, rehabilitating or reclaiming these areas. In recent years, flooding has become an annual occurrence in the Guinea and Sudan savannah areas of Ghana. Such areas would benefit significantly from increased tree cover to reverse the degraded lands and reduce the impacts of these environmental hazards.

Table 1. Estimated mean annual increments (MAI) for selected species in Ghana

Species	Good site		Poor site	
	Zone	MAI (m^3/ha/yr	Zone	MAI (m^3/ha/yr
Tectona grandis	Moist semi-deciduous	12	Transition zone	6
Gmelina arborea	Moist evergreen	20	Moist semi-deciduous	12
Cedrela odorata	Moist semi-deciduous	18	Moist semi-deciduous	12
Azadirachta indica	Guinea savannah	12	Guinea savannah	4
Ceiba Pentandra	Moist semi-deciduous	18	Guinea savannah	12
Triplochiton scleroxylon	Moist evergreen	20	Moist semi-deciduous	12

Data sources: Nanang (1996), Nunifu (1997) and FAO (2002a).

Increasing concerns about global warming and the role forests play in mitigating the concentrations of greenhouse gases in the atmosphere through carbon sequestration have provided an additional impetus for plantation development. For example, deforestation and forest degradation are estimated to account for almost 20% of the global greenhouse gas emissions (IPCC 2007). Forest sector mitigation measures are considered to be cost effective in combating global warming while providing a range of social, environmental and economic benefits to society. A growing tree can store as much as 45% of its stem dry weight as carbon (Evans 1992), and this makes forest plantations particularly attractive in achieving climate change mitigation because of their fast growth. One of the flexible mechanisms of the Kyoto Protocol is the Clean Development Mechanism (CDM), which allows Annex 1 countries (those countries that are signatories to the Protocol and have binding emission targets) to develop afforestation or reforestation projects in developing countries and use the resulting carbon credits to meet their Kyoto targets. This flexibility provides an incentive for Annex 1 countries to meet their greenhouse gas obligations more cost effectively, and an opportunity for developing countries to attract the kind of resources required to establish plantations. It is important to point out that the benefits that plantations provide in reducing greenhouse gases is highest where trees are planted on degraded or grasslands, not when plantations replace natural forests.

2.3. Policy Considerations in Plantation Development

Developing forest plantations for any purpose involves several long-term decisions that have economic, environmental and social implications, and, therefore, requires that careful planning and due diligence be undertaken before embarking on the project. In this Chapter, we focus on the important considerations in plantation forestry development. There are two levels to this: the national level policy decisions; and the plantation level considerations. At the national level, policy decisions are based on the overall national requirements for forest resources and environmental protection and the role of plantations to meet these needs. From the analyses in Chapter 1 and the reasons given in Section 2.2, it is clear that at the national level, Ghana needs to implement forest plantation development initiatives in order to avoid acute wood shortages and mitigate environmental degradation.

Once a decision is made to pursue forest plantations, there is a secondary question of what approach (model) or models of plantation development should be used. Approaches in the literature include government-sponsored plantation projects, community-based approaches, and private industrial plantation development initiatives (Enters et al. 2004). Each of these has their advantages and disadvantages and each country needs to decide which combination of these approaches would best suit its needs. At the plantation level, there are factors that investors have to take into consideration when establishing forest plantations. At the planning stage of the plantation project, these factors should be considered to ensure that the investor is aware of the requirements surrounding the investment and also identify where technical help may be required by the investor.

2.3.1. Survey of Plantation Models

In an FAO study in the Asia-Pacific Region reported by Enters et al. (2004), there were notably three distinct models commonly applied for the establishment of large-scale forest plantations:

i plantation establishment led by a central or state-government planting programmes;
ii plantation establishment under community-based development; and
iii private sector-led plantation establishment.

The first model is popular in China, Vietnam and Malaysia. For example, the Chinese Government had planned to increase the country's forest cover from 14% to about 20% by 2010 (Enters et al. 2004). In fact, China achieved a 20.36% forest cover by 2009, two years ahead of schedule! The Chinese government attaches great importance to forestry development through various government-supported planting programmes, including the promotion of private-sector participation with funding through foreign investment and joint ventures. Ghana has implemented a few state-led plantation development programmes in the past including the Rural Afforestation Programme (1989-1995) and the National Forest Plantation Programme that was started in 2001.

In the second model, participatory activities in plantation establishment at all levels form the key element for the successful implementation of the project. This model is adopted in a number of developing countries including India, Nepal, the Philippines, Bangladesh, Laos, Myanmar and Sri Lanka (Enters et al. 2004). The basic concept of this model recognises the importance of local people in protecting, managing and developing forests and envisages mobilising the communities through community forestry programmes. In this context, communities are empowered to manage degraded forests, including undertaking reforestation activities on a benefit sharing basis (Enters et al. 2004). This system is identical to the modified taungya system being applied in the HFZ of Ghana under the National Forest Plantation Development Programme. The RAP also emphasised community participation in plantation forestry, and hence about 90% of all seedlings planted during the implementation of that project were in community plantations (FAO 2002a). Recently, the FC has taken community involvement in forest plantation development more seriously and as a result, a Community Forestry Unit has been created within the FC.

The third model is through the involvement of the private sector. In this model, the government initially establishes a critical mass of plantations, but ownership and responsibility for the management of the plantation is gradually handed over to the private sector (Enters et al. 2004). Countries that have adopted this model are New Zealand and Australia. Plantation forestry is regarded as a business and afforestation investments are primarily assessed based on their potential profitability (Enters et al. 2004).

2.3.2. Plantation Models Used in Ghana

The development of forest plantations in Ghana has relied on different types of models. Generally, forest plantations are either undertaken by individuals at their expense, or sponsored by the central government, the private sector, or non-governmental organisations. Private plantation forestry is limited to a few companies, whiles almost all plantation forestry initiatives since the 1950s, were sponsored by the central government. Historically, the most widely used models in Ghana are the smallholder individual and community plantations. The modified taungya system and agroforestry systems have also been applied on limited scales across the country. Most industrial plantations are owned by private companies to produce poles for electrification, sawn wood and pulpwood.

A) Individual Farm Plantations

Smallholder plantations in Ghana include tree planting on individual household farm properties and around homes. These are either private or government sponsored. Small-scale plantations on farmlands often reflect land tenure constraints that limit land ownership sizes to small parcels of land per person. Farmers have known for centuries that incorporating trees among agricultural crops can produce a valuable cover for shade-tolerant subsistence crops, recycle nutrients from deep in the soil and improve crop yields. Household trees can provide individual families with their own supply of nearby and protected fuelwood, thereby eliminating or reducing lengthy travel time to collection areas.

Traditionally, trees have been planted for shade, protection against the wind, and for medicinal purposes. However, tree planting in plantations or on farmlands for poles and fuelwood has not been practised by rural and urban communities, until recently. For example, mature neem trees around homes in northern Ghana are usually managed by the coppice system for the continuous production of poles and rafters for building construction. With the advent of the government- and World Bank-sponsored Rural Afforestation Programme (RAP) in 1989, neem was grown for poles and fuelwood in small to medium plantations (0.5 - 3.0 ha) adjacent to villages where they are easily maintained and readily accessible for harvesting (Nanang 1996).

Under the NFPDP, about 37,000 ha of smallholder plantations were developed in the first three years (2001-2003). To ensure success of smallholder tree planting it is important to identify the target group in need of this service, assess their needs, use local knowledge, solicit local participant's input into project constraints and potential, and recognise and accommodate the socio-cultural dynamics.

B) Community Plantations/Woodlots

Community plantations have the potential to provide large amounts of wood products for a community and were the most popular plantation model under the RAP. Communities are heterogeneous in character, encompassing a variety of conflicting goals and expectations, and communities are embedded within a larger regional population (Skoupy 1991). When community members can participate in all phases of the project planning, implementation and evaluation they are afforded a sense of ownership and control and will, as a result, be more committed to the project (Skoupy 1991).

Although they may produce large quantities of wood, community plantations may be ineffective in areas where land tenure is uncertain or where land shortages occur. Moreover in such systems, the dietary subsistence needs of smallholder farmers are often ignored at the expense of wood production. Crush and Namasasu (1985) studied a woodlot project in Lesotho and concluded that its failure was due to uninterested communities because (a) the project did not recognise the realities of rural existence and (b) it did not meet the immediate needs of the people. The project was successful in growing trees and after 12 years of operation across 5000 ha in numerous localities, there was still limited farmer participation in the establishment and operation of the woodlots (Crush and Namasasu 1985). Because community woodlots have relatively long rotation lengths they cannot provide immediate benefits. Furthermore, equity problems often arise in the distribution of benefits to poorer community members, those with lower status, or where benefits cannot be transferred across generations. Such people usually receive a significantly smaller proportion of the benefits (Crush and Namasasu 1985).

In regions where community plantations may be an appropriate fuelwood conservation strategy, they may succeed in promoting self-sufficiency by allowing more wood to be produced, sold and used by a target group. Communal fuelwood plantations can be successful in locales where there is a strong sense of community and a history of community action (Munslow et al. 1988). Additionally, community woodlot initiatives are likely more appropriate in areas where secure communal land ownership is possible (Crush and Namasasu 1985). With increasing population, it is likely that land that is truly communal, both in title and practice, will become scarce in Ghana and pose serious challenges to community forestry projects.

C) Agroforestry Strategies

Agroforestry is a collective name for land-use systems and practices in which woody perennials are deliberately integrated with crops and/or animals on the same land management unit (World Agroforestry Centre (ICRAF), 1993). The integration can be either a spatial mixture or in a temporal sequence. There are normally both ecological and economic interactions between woody and non-woody components in agroforestry (World Agroforestry Centre (ICRAF), 1993). The initial concept of agroforestry was based on its usefulness at the farm level, while more recent thinking conceives agroforestry within a landscape level, which is capable of providing wood products and environmental services to sustain the livelihoods of communities.

Agroforestry often provides significant benefits to communities in terms of wood production and subsistence measures. If start-up costs are not available through outside donors or well-funded government programmes, agroforestry systems may be the best choice for peasant farmers. Agroforestry technologies are usually lower in cost because they require no new technology and are simply an extension of present land use systems where trees are incorporated into existing agricultural crop systems (Manshard 1992). Such systems may be among the best suited for communities across Ghana where the fiscal resources to implement large-scale initiatives like state forest department strategies often rely on outside donor funding.

Furthermore, land tenure constraints that restrict individual land holdings to a few hectares (ha) of land, and the need to produce enough food on the same piece of land means that for most farmers, it is not possible to acquire additional land solely for growing trees. For these reasons, agroforestry would be more attractive than conventional plantation forestry. It is well known that in Ghana, farmers traditionally practised agroforestry on their farmlands by protecting some indigenous tree species as they cleared land for crop production and plant trees for shade on cocoa farms.

The Ministry of Food and Agriculture created an agroforestry unit within the ministry to promote agroforestry practices within the country. Research in agroforestry had also increased, with local universities establishing departments to teach and research into agroforestry. Ghana developed a National Agroforestry Policy in 1986 to promote agroforestry practices for sustainable land-use (MOFA/AFU 1986). The National Agroforestry Policy recognised the fact that an organised and coordinated approach was required if agroforestry was to play a role in the promotion of sustainable agricultural development (Asare 2004). In this light, the Government of Ghana, with assistance from the UNDP and FAO, initiated a national programme to support agroforestry. The three main areas for implementing this policy was research, training, and extension education in agroforestry.

Several NGOs such as the Ghana Rural Reconstruction Movement (GhRRM), Adventist Development and Relief Agency (ADRA), CARE-Denmark, and Conservation International have been influential in supporting government's effort in empowering farmers to engage in sustainable agriculture through agroforestry (Asare 2004). For example, ADRA supported the government's effort in 1989 by launching the Collaborative Community Forestry Initiative (CCFI) programme that established nurseries and supported households with seedlings. Under this programme, 20 nurseries were established within ten years producing more than 4 million assorted tree seedlings including fruit trees like mangoes, cashew, orange and guava. Woody tree species including teak, eucalyptus spp., neem, and *Albizia lebbeck* were also produced (Djarbeng and Ameyaw 2002).

D) Modified Taungya System

The Taungya system, which was developed in Myanmar (formerly called Burma) is a plantation development model in which farmers are given parcels of degraded forest reserves to produce food crops and to help establish and maintain timber trees (Agyemang et al. 2003). Ghana adopted this system in the 1930s to help produce a mature crop of commercial timber in a relatively short time, while also addressing the shortage of farmland in communities bordering forest reserves (Agyemang et al. 2003). This approach has merit because cocoa, the dominant cash crop grown by farmers in the reserve, is a shade tolerant plant, especially during its initial stages of growth. The problem with the traditional taungya operations was that timber seedlings were simply ignored by the farmers. Farmers had no incentive to protect the forest trees.

The modified taungya system (MTS), introduced in 2002, is a framework that allows farmers and their families to keep portions of the revenue from harvested trees on their farms as an incentive for them to tend the trees. This system allows degraded forest reserves to be forested with selected tree species, intercropped with food crops and designed to entitle all stakeholders to the plantations' benefit and give them a long-term interest in maintaining tress (Agyeman et al. 2003). The MTS in Ghana relies on both indigenous and exotic tree species. The system offers 40% of proceeds from sales on the standing tree value to participating farmers. The farmers contribute their labour for their share of the benefits from trees and food crops. Additionally, the contribution of the taungya system to ensuring food security in the country since 2002 has been remarkable. For example, food production from the plantation coupes under the MTS in 2008 was estimated at 471,875 tonnes (Ghana Forestry Commission, 2008b).

E) Private Industrial Plantations

Apart from the Government and NGO-initiated programmes, several private companies and individuals have established plantations at various scales and for various purposes using various partnerships, programmes and strategies in the past decades (FAO 2002a). However, these private industrial plantations are still limited in terms of the number of companies participating and the areas planted. Notable among these are the Pioneer Tobacco Company Ltd. (PTC) which owns about 5,000 ha of teak plantations mainly in the Brong Ahafo Region, the Ashanti Goldfield Company Ltd., with over 1,400, 50, and 42 ha of teak, gmelina and eucalyptus species, respectively. The Bonsu Vonberg Farms Ltd., owns about 500 ha of teak (FAO 2002a).

The Samreboi Timber and Plywood Company Ltd. (Samartex) has initiated a forest plantation programme at Oda Kotoamso near Asankrangwa in the Western Region. The area is an abandoned cocoa farm outside the reserved forests, while the Ghana Primewoods Products Ltd., Takoradi (GAP) has initiated a "Joint Forest Management Project" outside the reserved forests in late 1995 at Gwira Banso in the Western Region (FAO 2002a). In addition, some other timber companies have also established plantations on limited scales, e.g., Sunstex Ltd., Western Hardwoods, WLVC, Ghana Prime Woods Products Ltd., Specialised Timber Products, John Bitar and Co., Asuo Bomosadu Timbers and Samartex Ltd (FAO, 2002a). The recent rural electrification programme in Ghana has created a market for electric transmission poles, which hitherto were supplied from thinnings of the Government owned plantations. This market has generated unprecedented interest in teak plantations by individuals and organisations.

F) Timber Companies

As part of their timber utilisation contract (TUC) obligations and also to meet their future demands, timber companies are expected to develop and implement afforestation plans in the areas in which they operate. These companies usually involve farmers in reforestation and often provide them with training in nursery, plantation establishment and agroforestry techniques. Farmers are also provided with basic materials such as tree seedlings and nursery equipment. Examples of timber companies undertaking afforestation/reforestation under the TUC obligations include Suhuma Timber Co. Ltd, Swiss Lumber Co. Ltd, Samartex Timber and Plywood Co. Ltd, Logs and Lumber Co Ltd, Bibiani Logs and Lumber Co Ltd, and AG Timbers.

G) Other Potential Models

FAO (2002a) has identified three other models used in rubber and oil palm plantations in Ghana, which may be applied to forestry plantations. These models normally operate between companies and individual farmers under the headings: smallholder scheme; out-grower scheme; and the lease-back system. Under the smallholder scheme, a company arranges for credit and provides extension services and planting material for the establishment and maintenance of the tree crops to an individual planter or farmer who does not own the land. The company also serves as the market for the produce. The out-grower scheme is identical to the smallholder scheme, except that the planter owns the land or has a lease, freehold or share-cropping arrangement. In the lease-back scheme, the farmer owns the land and leases it to a company for plantation development. The farmer usually takes a share of the final value of the harvest (FAO 2002a).

2.4. Technical and Economic Considerations in Plantation Establishment

Establishing a forest plantation involves several technical steps and decisions. There are three distinct management phases that can be distinguished: seed collection and handling; nursery practices and plantation establishment; and plantation management. The initial plantation establishment phase is divided into the following activities: species selection, site preparation and planting operation (Camirand 2002). The remaining decisions include where to plant trees, acquiring seeds, producing the planting stock, preparing the site, planting, weeding, pruning, thinning, final harvest, and regeneration. Depending on the purpose of the plantation, the frequency and timing of these operations would vary. The following factors need to be considered in making decisions related to the plantation enterprise.

2.4.1. Purpose

The first consideration in establishing a plantation is to define clearly the purpose for which the plantation is being established. The investor will ask questions such as: "what are the outputs required and within what period?" The answers to these questions will have tremendous impacts on many other decisions that will have to be made in respect of the plantation enterprise. The purpose of the plantation will determine the species to be planted, where the plantation should be sited, the kinds of management regime to be used, etc.

2.4.2. Site Selection

A second important consideration is selecting the site for the plantation. Good site selection is one of the most important decisions that can lead to improved yields, reduced rotation lengths, and increased economic returns for plantations (FAO 2002b). In this case, the most important factor is to match species to the site in order to optimise productivity. Furthermore, sites are selected to minimise the risk of plantation failure due to poor soil drainage, drought or inability to control weed competition. The important components of site quality are: soil depth and drainage; soil physical and chemical composition (including pH); amount and pattern of yearly soil moisture availability; frequency and nature of common and occasional winds, storms and fires; and the general climate of the area (FAO 2002b). A general approach to assessing site quality is to examine the presence and importance of competing vegetation, and of populations of animals, insects and microorganisms that are either damaging or beneficial to trees, which can affect management of forest plantation sites. The size and health of trees already present on the sites are good but imperfect indicators of site quality (FAO 2002b).

Site selection may be limited by the availability of land for the investor. Most smallholder farmers are usually restricted to planting trees on their farmlands or family lands which may not be the most suitable for forest plantations. This could reduce the productivity of such plantations, thereby requiring more land to achieve the same level of output. Community plantations rely on communal lands, while industrial plantation investors would acquire large parcels of land through purchasing or leasing. Industrial investors have the most flexibility in choosing the site since they would be purchasing or leasing the land. Large-scale investors would also need to consider the location of the plantations vis-à-vis processing facilities, markets for their products, the cost of land preparation, labour force availability and available infrastructure. The ability to match species to the site is limited by the lack of technical information on many plantation species and the sites that optimise their productivity. Matching species to sites is complicated by the wide range of potential plantation species and infinite combinations of sites available to choose from. Consequently, a spatial map that shows the suitability of various plantation species for the different vegetation zones would be a useful tool for plantation forestry investors in Ghana.

2.4.3. Choice of Tree Species

In addition to choosing tree species to match their most productive sites, species are also chosen to fulfil the purpose of establishing the plantation (end-uses). There is also the question of whether the plantation should be a single species (monoculture) or multiple species (mixed). Each of these has its advantages and disadvantages. If the forest plantation is to provide only wood products, then most often, a single species suitable to the site may be sufficient. However, if the forest plantation is to provide aesthetic, wildlife habitat, biological diversity, and other services besides wood production, multi-species forest plantations will often be favoured (FAO 2002b). Extensive planting of one species, whether indigenous or exotic, inevitably results in some areas where trees are ill-suited to the site and suffer stress. This may occur where large mono-specific blocks are planted or where exotics are used extensively before sufficient experience has been gained over a whole rotation (Evans 2001).

Species choice is also related to whether an indigenous species or an exotic species should be used. In Ghana, most exotic species outperform indigenous ones in terms of productivity (Section 2.2.3), and would be more favoured in plantation projects. In addition, exotic species offer a wider range of choices, are often free from local diseases and pests, and usually their silviculture is better understood than indigenous species (Evans and Turnbull 2004). However, indigenous species offer advantages in terms of acclimatisation and would be preferred in re-stocking degraded forest reserves, for example. If two species are identical except that one is indigenous and the other is exotic, the former should be preferred. In most cases, however, the choice may be to use a mixture, rather than a monoculture.

Species are also chosen based on the outputs desired. If fuelwood is the major need, then species that produce large amounts of biomass within short periods preferably with known desirable fuelwood characteristics such as low ash content and high calorific value, and also coppice well would be the appropriate choice. However, if the need is for building construction and electrification, then tree species that have straight boles and are naturally durable would be more suitable. Tree species that are water intensive should neither be planted in the drier areas nor used as ecological restoration around rivers or streams. Species for restoring degraded lands and watershed protection would have characteristics that are consistent with their functions such as fast growth, provision of shade, litter production, deep-rooted, erosion resistance, ability to survive and grow on poor sites, etc. When it comes to ecological restoration, non-indigenous species can pose a major problem because they are often aggressive and can overwhelm native species, thus altering ecosystem structure (Berger 2006).

Because most plantations in Ghana and other tropical countries use exotic species, it is often assumed that exotic species are better suited for forest plantations than native species. The literature identifies several advantages and disadvantages of using either species. The main advantages of native species are (after Evans and Turnbull 2004):

- the growth of indigenous species in natural stands gives an indication of their potential, reducing the risk of complete failure when grown in plantations;
- indigenous species are naturally adapted to their environment and developed resistance to known diseases and pests;
- indigenous species play ecological roles, and may be critical habitats for local animals and plants; and
- plantations of native species conserve the native flora, especially in areas of deforested or degraded forests.

Advantages of exotic species include (Smith, 1986; Evans and Turnbull, 2004; Zobel and Talbert 1984):

- there is a wide variety of exotics to choose from, and hence increases the chance of finding one that would be suitable for the purpose of the plantation;
- tree improvement programmes have identified some exotic species with preferable genetic and phenotypic characteristics;
- exotic species lend themselves to mass propagation, as required in large plantation programmes;
- most exotic species grow more rapidly, providing higher yields of wood and other products and higher economic value;
- exotics may have resistance to local diseases and pests;
- experience with the exotic species in other parts of the world may provide dependable silvicultural methods for managing the species; and
- uniform stands of exotic species are easier to manage intensively over short rotations.

Despite these advantages, plantations of exotic species have been known to produce poor results in test plantations on sites unsuited for long-term plantation development, susceptibility to local pests and diseases, failure of single species stands due to slowly developing effects of sites, unacceptable stem form, undesirable wood qualities, poor adaptability to environmental conditions and development of wood characteristics, and unsuitable to local market's needs (Zobel and Talbert 1984; Zobel et al. 1987). In order to minimise the risk of plantations of exotic species failing and to optimise the benefits from these plantations, Zobel et al. (1987) recommend that species should be carefully chosen to match the site, use only resistant species and seed sources with attributes of local interest; use a variety of seeds or cuttings from several sources to ensure genetic diversity; ensure sanitation in raising nursery stock and transporting seedlings; and provide protection to the established trees.

2.4.4. Regeneration Method

The method of regenerating the plantation and subsequent management regime are critical to determining the success or otherwise of the plantation investment. Plantations can be regenerated from direct seeding, using a container or bare rooted-stock seedlings, or by using cuttings. The best method will depend on the species and site characteristics. Another method of regeneration from an existing plantation is the coppice method. For example, coppicing is an alternative reforestation tool for teak plantations in the tropics, substantially reducing regeneration time/costs and associated demands for labour and seed when available (Bailey and Harjanto 2005). Growth rates of coppiced material are rapid in most situations and make this method suited for managing woodlots in Ghana. Bailey and Harjanto (2005) compared coppiced teak plantations to paired seed-origin plantations, at ages 3, 8, and 13 years, on Forest State Corporation managed land located in Java, Indonesia. The plantations were evaluated for height, diameter, lower-bole straightness, and the presence of disease in both plantation types and at three ages.

Mean height and diameter of trees in coppiced plantations were both significantly greater than those in their paired seed-origin plantations at all three ages. Furthermore, heights and diameters in coppiced plantations were higher than expected based on established growth tables for Java. Coppiced plantations were less symptomatic of disease than seed-origin plantations, which promise better wood production and quality. Lower-bole stems in coppiced plantations developed less straight than those in seed-origin plantations, but these deviations faded with time and could likely become insignificant within a 60-year rotation. Bailey and Harjanto (2005) concluded that coppiced plantations in Java can make a major contribution to teak production in Indonesia.

2.4.5. Level of Investment

An investor in plantation forestry needs to consider the amount of investment required for the enterprise in terms of labour, capital, time and infrastructure. In industrial plantations, hired labour will be needed, and the investor would need to ensure that there is available labour force within a reasonable distance from the plantation site, as this will affect the cost of labour. In Ghana and other developing countries, labour costs are low and the labour force is usually available, especially in rural areas where unemployment rates are often high. For small-scale farmers, labour is usually not an issue, as they rely on their labour and that of friends and family members. Unless subsidies are granted, capital investments in plantations can be high, especially in the first few years. And hence investors need to consider the costs of planting and managing the plantation until the final harvest. There is also the investment of time in the plantation venture for managing the staff and ensuring that the plantation is protected from fires, insects and diseases.

Basic infrastructure such as roads, electricity and transport are required to get to a forest plantation site to plant, for later management and harvest. The relative availability of access does not directly affect productivity; however, it can indirectly affect it by its influence on the ability of management to exert leverage through such things as post-planting release, thinning and pruning, and other management activities (FAO 2002b). Furthermore, accessibility affects the energy it requires to deliver forest products to their users, and thus the economic feasibility of using them (FAO 2002b). If basic infrastructure such as roads, water, communications are not available at the chosen site, the investor may have to provide these at his expense (either fully or partly) and this will impact on the profitability of the enterprise.

2.4.6. Economic Factors

Economic factors relate to first and foremost whether the investor will turn a profit on the venture. Profitability is affected by several other economic, social, and technical aspects of the plantation development enterprise. For example, economic viability of plantations is affected by the technical aspects of the plantation enterprise discussed in the preceding sections (e.g., matching species to sites), as some of these may impose economic costs on the plantation investment.

Another important economic consideration is the ability to finance the project at reasonable interest rates. Even if the investor can finance from his resources, the issue of the rate of return on the plantation over the entire rotation will still need to be taken into account and compared with alternative investments. Benefits accruing to the investor are partly affected by the level of taxation imposed on the plantation and its outputs, and the property rights regime under which the investor operates. Therefore, investors need to be aware of the taxation regime in effect where the plantation is located. Furthermore, investors need to worry about whether there is a market for the outputs and the distance to those markets. Also, the opportunity cost of the plantation will be an important consideration. That is, alternative uses of the inputs (e.g., land) will need to be compared with the use of the land for forestry to ensure that all resources invested in the plantation are put to their best uses.

There is also a combination of taxes, levies and fees on forestry operations and forestry products in Ghana. From the legislative framework, there is still some confusion as to what levies and fees apply to privately established forest plantations. The potential fees and levies include stumpage fees, fees for management services, rent for contract areas, property marks fees, export levies for selected species, and timber rights fees. With the exception of export levies which apply to teak (10% under the *Trees and Timber (Amendment) Act* (1994)), it is not clear which of the other charges plantation owners are subject to.

Consideration of profitability is not restricted to industrial investors in forest plantations. At first glance it may not be entirely obvious that forest plantations at the community level for the provision of fuelwood and other basic needs such as fodder, construction wood, fruits, etc. should also consider economic principles in their decision-making. It is true that under these instances, the emphasis is not normally in the selling of commodities to maximise net returns based on commercial transactions, but communities try to maximise their well-being. Evans (1992) provides evidence that under certain conditions, smallholders do plant trees in anticipation of future markets.

2.4.7. Legal and Institutional Requirements

Plantation establishment is also affected by the legal and institutional framework within Ghana. Forestry practices are governed by various policies, laws and regulations and these have to be understood and respected when establishing plantations. Secondly, investors need to be aware of the different incentives available to forest plantation developers, and the kinds of restrictions that exist on exports of various forest products from their plantations. The *Trees and Timber Management (Amendment) Act*, 1994 (Act 493) and the *Timber Resources Management Regulations*, 1998 (L.I. 1649) also specify various export levies and stumpage rates for timber products, including plantation species such as teak (export levy 10%) and ceiba (export levy 30%). Investors need to consider these in assessing the economic feasibility of the plantation enterprise.

Property rights form the bedrock of private decisions in resource use and allocation, and hence the property rights regime within the location of the plantation and how benefits are shared among property rights holders will also affect the kinds of management activities that can be undertaken and the benefits to be derived. There may also be land-use zoning restrictions, water and electricity rights-of-way, etc., which could interfere with plantation location and management. These need to be carefully considered.

2.4.8. International Environmental Concerns

Another factor that is likely to impact on the marketability of forest products from plantations that has to be considered, especially for large-scale industrial plantations, is the emerging issue of environmental concerns regarding forest plantations. These environmental considerations have led to the introduction of forest certification programmes around the world. Forest certification is a system for identifying forests that are managed to maintain ecological, economic and social components of the ecosystem. Certification is a market-based mechanism to reward sustainable forest management (SFM), and in a way resembles a command-and-control approach as those who do not certify their forests or products could be penalised by the loss of market share for their products.

Forest products and forest certification grew out of environmental concerns about the sustainability of natural forests, and is seen as a tool for reaching consensus on forest management and a means of addressing market failures in forest management resulting from multiple benefits and information asymmetry. Through certification, producers can prove to consumers that environmental concerns have been addressed, and hence certification acts as a link between the consumer and producers of forest products.

Although importing countries have not yet made forest certification mandatory for tropical plantation products, it is clear that countries that wish to have unfettered market access around the world would be compelled to show that their wood is coming from sustainably managed forests. Ghana has been developing the processes needed to certify its forests and this is a step in the right direction. The Ghana Forest Management Certification System Project was initiated in 1997 with the assistance of the European Union and the Netherlands to develop standards for certification. Field tests on the resulting chain-of-custody and log-tracking systems have been carried out since 2002, as has the development of standards for sustainably managed forests (ITTO 2005a).

At the global level, the two competing certification schemes with different operating modalities are the Forest Stewardship Council (FSC), which provides all the necessary elements of certification through centralised decision-making on standards and accreditation and the Programme for the Endorsement of Forest Certification (PEFC), on the other hand, which operates as a system for mutual recognition of national certification systems. Almost two-thirds (65%) of the world's certified forests (in 22 countries) carry a PEFC certificate, while the FSC's share is 28% (in 78 countries); the remaining forests are certified solely under national systems. Most of the certified forests in the tropics are FSC-certified (ITTO 2008).

Although originally intended for natural forests, certification schemes have recently been extended to cover plantation forests. The FSC has ten principles and criteria that must be satisfied before a plantation is certified (after FSC 2010):

1. Compliance with all applicable laws and international treaties;
2. Demonstrated and uncontested, clearly defined, long–term land *tenure* and use rights;
3. Recognition and respect of indigenous peoples' rights;
4. Maintenance or enhancement of long-term social and economic well-being of forest workers and local communities and respect for worker's rights in *compliance* with International Labour Organisation (ILO) conventions;
5. Equitable use and sharing of benefits derived from the forest;
6. Reduction of environmental impact of logging activities and maintenance of the ecological functions and integrity of the forest;
7. Appropriate and continuously updated management plan;
8. Appropriate monitoring and assessment activities to assess the condition of the forest, management activities and their social and environmental impacts;
9. Maintenance of High Conservation Value Forests (HCVFs) defined as environmental and social values that are considered to be of outstanding significance or critical importance; and
10. In addition to *compliance* with all of the above, plantations must contribute to reducing the pressures on, and promote the restoration and conservation of, natural forests.

Despite the general acceptance of certification schemes, there are major problems associated with regulating markets with certification due to insufficient information for consumers in their decision-making; unpredictable changes in consumer and producer response to certification, whether the goal of forest management is only for consumer values or societal values, the lack of coordination between certification schemes and government regulations, and how to deal with goods and services that are not traded in the market system (Haener and Luckert 1998).

In spite of these problems, it is accurate to say that certification schemes in whatever form they take are here to stay, and large-scale plantation developers who may be competing internationally for market share for their products, will be obliged or persuaded to certify their plantations. Even if plantation investors in Ghana do not follow any nationally or internationally recognised certification schemes, there may still be some pressure to ensure that environmental concerns of forest plantation development are addressed in one way or the other. These could increase plantation development and management costs that may be non-trivial, but which may be compensated for by increased market share of marketed products.

2.5. Sustainability and Environmental Concerns in Plantation Forestry Development

2.5.1. Sustainability Considerations

The word sustainability means different things to different people. With plantations, there are two components to sustainability. The first component relates to the broad issues of whether using land and devoting resources to tree plantations is a sustainable activity from the economic, the environmental or the social sense (Evans 2001). The second component focuses on the narrow issue of whether tree plantations can be grown indefinitely from one rotation to the next; i.e., whether their long-term productivity can be assured (Evans 2001).

Economic and social sustainability have to be assessed on an individual basis, as they depend on several factors. Economic sustainability relates more to whether the enterprise is profitable enough to be self-sustaining, considering all the timber and non-timber benefits from the plantation as well as all direct and indirect costs. Establishing forest plantations has economic implications beyond the forestry sector. For example, forest plantations may increase local and regional land and property prices and increase local inflation as well. But there are also positive economic benefits and spin-offs from plantation forestry including the generation of local employment opportunities at all phases of the plantation project, development of infrastructure or the development of local housing and accommodation to meet the needs of migrating populations.

Social sustainability has to do with how the plantations fit into the social structure of the community, social acceptability and social benefits that are derived by communities. The establishment of forest plantations may impact on rural populations (e.g., displacing them from their homes or farmlands). People may move into, or away from, locations where plantations are established, hence the community and population dynamics are likely to change in response to plantation development in a region. The remaining discussion in this section focuses on the narrow definition of sustainability, which has attracted more attention.

Because of the intensive management often associated with plantations and the high nutrient demands of most plantation species, there is the question of sustainability of forest plantations as a result of depletion of the soils following continuous biomass removals. The issue of how much biomass can be removed without negatively affecting the sustainability of plantations has gained greater importance with current emphasis in international climate change discussions to use bioenergy as an environmentally friendly energy supply alternative to fossil fuels.

Hartemink (2003) has extensively studied nutrient losses from plantations in the tropics and notes that harvesting techniques for logs affect the soil carbon and nitrogen stocks in forest plantations. Forest plantations mimic the natural forest in which nutrient cycling is fairly closed, and unless thinning of trees has occurred during the first years after planting, the major drain of nutrients is at harvest (Hartemink 2003). Nutrient losses can be minimised if only the stem wood is removed from the field, but soil disturbance and complete removal of the above ground biomass is likely to induce considerable nutrient losses as well (Hartemink 2003). For example, Chacko (1995) observed site deterioration under teak in India with yields from plantations below expectation and a decline in site quality with age and attributes this to poor supervision of establishment, over-intensive taungya (intercropping) cultivation; delayed planting; and poor after-care. Chundamannii (1998) similarly reports declines in site quality of teak plantations over time and blames this on poor site management. Site deterioration has also been identified as a problem in Indonesia and according to Perhutani (1992), is caused by repeated planting of teak on the same sites.

Comparisons of the physio-chemical properties of soils under two distinct adjacent forest cover types (logged native forest and teak plantations) at three locations in the forest-savannah transition zone of Ghana by Salifu (1997) showed that in two of the locations, nitrogen (N) and magnesium (Mg) concentrations and organic matter (OM) contents in the soil horizons were significantly higher under logged forest than under teak plantations. In the third location, phosphorus (P) and potassium (K) concentrations were significantly higher in the logged forests. In general, total nutrients were higher in soils under adjacent logged forest compared to teak plantations. The higher nutrient concentrations and OM contents in soils under logged forest were due to more undergrowth, litter and organic matter under logged forest and a lesser demand for these nutrients by tree species in the logged forest. Lower soil macro-nutrient concentration and contents in soils under teak was due to lower organic matter content under teak cover and associated with higher nutrient demand and nutrient immobilisation by teak (Salifu 1997).

Evans (2001) acknowledges that plantations and plantation forestry operations do impact the sites on which they occur and under certain conditions nutrient export may threaten sustainability. However, site quality can be maintained through care with harvesting operations, conservation of organic matter, and management of the weed environment (Evans 2001). Evans (2001) concludes that plantation forestry appears entirely sustainable under conditions of good husbandry, but not where wasteful and damaging practices are permitted.

To help respond to some public concerns, including scientists, who were questioning the prospects of tropical plantations established in short-rotation forestry as a sustainable natural resource, a research project was initiated in 1995. It was an international partnership of public and private organisations coordinated by the Centre for International Forestry Research (CIFOR). The main aim was to examine critical effects of site management on productivity of successive rotations of plantations and soils. The project has 16 experimental sites (10 eucalypt, 4 acacia and 2 conifers) in Australia, Brazil, Congo, China, India, Indonesia, South Africa and Vietnam (Nambiar and Kallio 2008).

The study concluded that subtropical and tropical plantations can be managed to increase and sustain productivity. Conserving site resources (organic matter and nutrients) to maintain production is very important. No major risks to soils were identified that could not be managed by scientifically-based practices. The study, however, recognised that all questions of long-term sustainable production of plantations cannot be resolved from experiments of one rotation at a few sites so there is a compelling case to continue the research programme to support plantation forestry in the tropics (Nambiar and Kallio 2008).

There are several approaches proposed by Evans (2001) to help sustain the productivity of forest plantation sites through subsequent rotations. These include: (1) genetic improvement of planting stock, which includes change in species, seed origin, use of new clones, use of genetically improved seed and, in the future, genetically modified trees all offer the prospect of better yields in later rotations; (2) better understanding of the silviculture of plantation species through manipulation of stocking levels to achieve greater output of fibre or a particular product, matching rotation length to optimise yield, and use of mixed crops on a site to aid tree stability, lower pest and disease threats; (3) fertiliser application to compensate for nutrient losses on those sites where plantation forestry practice does cause net nutrient export to the detriment of plant growth; (4) site preparation and establishment practices to alleviate soil compaction after harvesting or weed control strategies to reduce competitive vegetation; and (5) conserve organic matter through preventing systematic litter raking or gathering during the rotation and conserving organic matter at harvesting.

2.5.2. Environmental Considerations

Despite the advantages of forest plantations over natural forests described in the preceding sections, it is important to note that due to their simpler structure and most often monoculture nature, forest plantations may not be able to provide the same kinds of environmental and social goods and services provided by natural forests. As a result, there are some environmental issues to keep in mind when considering, establishing and managing plantations.

Sites for plantation establishment have to be chosen carefully to exclude areas that are critical habitats for species needed to conserve biological diversity. In this case, replacing natural forests with a forest plantation is a bad idea. Forest plantations usually have less total biological diversity than do indigenous forests, and their associated biota are also different in composition from those of indigenous forests in the same area (FAO 2002b). It is also possible for exotic plantation species to introduce diseases and pests to their new environment, which could infest native tree species. Moreover, some exotic species are invasive by nature, and could become difficult to control once they establish themselves within a location. An example is of neem trees in some areas of the coastal savannah areas of Ghana. Therefore, tree species for introduction must be chosen carefully. Despite these potential problems with plantations, it is fair to say that the different suites of biota provided by species planted in forest plantations can contribute to regional biological diversity. Also, tree planting can be used to improve stocking of a natural forest where necessary.

Some forest plantations such as teak hardly support any undergrowth, often leading to excessive soil erosion especially when teak is planted on a slope. Some soil degradation under teak in Ghana has been observed in parts of the plantations at the Ho Hills Forest Reserve in the Volta Region and the Yendi Town Plantation (FAO 2002a). There are two schools of thought about the lack of undergrowth in pure plantations of teak. The first attributes the lack of undergrowth to allelopathic effects of leachates from teak that tend to limit the amount of seedlings that can survive and grow under teak plantations. This claim has, however, been more circumstantial with little or no empirical studies to support it (Healey and Gara 2003).

Although studies by Murugan and Kumar (1996) found significant concentrations of phenolic acids which have been implicated in regeneration failures in some forest types (e.g., Pelissier and Souto, 1999; Mitzutani 1999) in the foliage of teak, there has not been any strong documented evidence that directly links the release of these substances at concentrations high enough to impede the establishment of undergrowth in teak plantations. Studies by Jadhav and Gaynar (1994) and Tripathi et al. (1999) on the effects of teak leachates on rice, soybean and cowpea have demonstrated potential allelopathic effects.

Healey and Gara (2003) investigated the effects of teak plantation on the establishment of native species and observed significantly lower abundance, diversity and size of the native species growing under teak plantations as compared to adjacent land. Their conclusion on the potential allelopathic cause of these differences was speculative. Empirical investigations of potential allelopathic effects of teak are often confounded by other factors. As alluded to by some researchers (e.g., Pelissier and Souto 1999), allelopathy is not about the concentrations of allelopathic substances; other factors including climate, target plant physiology and the potential interaction with other soil compounds are also important (Healey and Gara 2003).

The more popular school of thought attributes lack of undergrowth to the excessive shading of the broad leaves of teak trees particularly after crown closure, which makes it impossible for the development of any competing vegetation in the understory. This, according to FAO (2002a), is made worse by fires in the dry season which burn the leaf litter, further reducing the possibility of any undergrowth and leaving the soil surface unprotected from heavy rains, particularly at the beginning of the rainy season before the overstory has developed new leaves. This issue appears to be much more legitimate and may relate well to concerns about the sustainable management of multiple rotations of teak. While teak is a high nutrient demander, erosion of top soils under teak plantations coupled with high concentration of nutrients in teak leaves (Weaver 1993) which are burnt annually may lead to a net nutrient loss between rotations. While this may be known from experience, there is a lack of information about the effect on yields of the second rotation (Hall et al. 1999). The use of mixtures of different species especially the nitrogen fixers with teak has been widely promoted to help prevent soil nutrient loss and improve productivity. An extensive literature review, carried out as part of a general study on mixed tree species plantations in the tropics (FAO 1992) showed many examples of teak grown in species mixtures in the 1930s largely from India but also Indonesia, Nigeria, Benin and Sri Lanka (Hall et al. 1999).

Some environmental organisations strongly oppose the use of exotic monoculture plantations, arguing that they are not environmentally sound. According to AFORNET (2008), these claims about the ecological fragility and damaging effects of such plantations are only partly justified because (AFORNET 2008):

- there are few cases where monitoring and investigation of soil decline affecting long-term production have shown that this actually is a problem;
- fast growing plantations, whether exotic or indigenous, will have an impact on water balances in direct proportion to their growth rates;

- biodiversity in monoculture plantations will inevitably be less than in a natural forest but not necessarily less than in agricultural lands on which these plantations are established. Also, biodiversity increases with increasing age of plantations;
- incidences of pest and disease damage occur, but there are only half a dozen or so examples where such attacks have acquired serious and permanent proportions (e.g., *Dothistroma pini* on *Pinus radiata*); and
- plantations can also play positive ecological roles in terms of enhancing the natural regeneration of indigenous species on reforested degraded sites.

2.6. CONCLUSION

The main reasons for promoting forest plantations in Ghana are to ensure the availability of forest products for present and future generations and reduce the pressure on the natural forest resource, thereby contributing to the sustainability of the latter. The concept of forest plantations is not new to Ghana; neither is the need to develop plantations a recent invention. In fact, the need for plantations to provide wood products to the savannah regions of Ghana has been recognised since 1956 (FAO 2002a).

The slow rate of plantation development has more to do with the lack of commitment to follow up and meet the targets set by previous programmes. There is opposition in certain NGO communities to plantation development, but a fair assessment of plantations has to take into account the actual land use prior to planting and the positive benefits plantations provide. Some environmental groups have argued that plantations are detrimental to biodiversity, but this is only the case if they replace natural ecosystems or traditionally managed forests. Even though plantations may not replicate all of the functions of a natural forest, they do play positive environmental roles in protecting sensitive areas and prevent or even to some extent reverse land degradation.

The benefits of plantations in most cases would outweigh their economic, social and environmental costs. There is a cost to doing nothing about the land and environmental degradation resulting from the lack of tree cover, which those who criticise plantations underestimate. The costs of environmental disasters that can be avoided by the presence of forest cover can be very high.

Plantation establishment and management involve a series of complicated processes, as they are affected by technical, social, economic, environmental, institutional and legal factors. Investors at the smallholder, community or industrial plantation levels would need to weigh these factors and make decisions that will determine the success or failure of the plantation venture. These decisions are further complicated by high levels of uncertainty in future costs and prices, legal and institutional operating environments. Feasibility of the plantation establishment would rely on sound and technical expertise from professionals in each of these aspects to minimise risk and optimise the benefits to be derived.

Historically, Ghana has relied more on promoting individual and community forestry plantations, often sponsored by the central government and international donors. However, it would serve the country well to look at other models that support industrial plantations more seriously than is being done presently.

Although certification programmes are already in place, there are some sticky issues that still need to be resolved. Despite these concerns, it is likely that forest plantations that are targeted at producing timber for the international markets will be obliged either by the national government or international environmental movements to certify their plantations.

Chapter 3

CHALLENGES OF, AND OPPORTUNITIES FOR, PLANTATION DEVELOPMENT IN GHANA

3.1. INTRODUCTION

Ghana has not been successful in developing large areas of forest plantations, probably because of large tracts of natural forests that were available. As a result, the rate of plantation development has lagged far behind the rate of deforestation. The limited success in plantation development may be due to barriers and lack of opportunities to promote plantation forestry.

This Chapter analyses the barriers to, and the opportunities and incentives for, plantation development in Ghana. In general, there are four kinds of barriers to plantation development: economic, technical, institutional and biophysical. Conversely, there are clearly some opportunities at the national and international levels that Ghana can seize to achieve its plantation development goals. These opportunities include Ghana's forest legal and policy framework, availability of land and a favourable climate, available labour force, the Clean Development Mechanism under the Kyoto Protocol, etc. The final section of this Chapter examines the issues around incentives to promote plantation development. It is recognised that the characteristics of forest plantations as an investment are unique, compared to other alternative land-use investments. The reasons for providing incentives are explored, followed by a description of the different kinds of incentives for plantation development in Ghana, and the problems associated with incentive programmes.

3.2. CONSTRAINTS OF PLANTATION DEVELOPMENT

Constraints of, or barriers to plantation development refer to the factors that prevent or discourage individuals, private firms, communities, NGOs and other public sector organisations from engaging in plantation establishment and management. It makes sense to assume that because Ghana is a tropical country that has a favourable climate that supports fast tree growth, it should do better than temperate countries in plantation development. However, growth rates are only one aspect of economic competitiveness in forestry. The fact that Ghana has not done well in this area, whilst countries such as Finland and Sweden continue to develop forest plantations and compete with tropical countries such as Brazil, suggests that there are other factors more important than a favourable climate that determine the success of plantation forestry. These other success factors may be lacking in Ghana, which together would constitute barriers or constraints. Even with a favourable climate it still takes considerable time for investors to recover investments from trees. Private and public individuals and firms will not invest in plantations unless the political, institutional and economic environments not only permit, but actually encourage, people to make money from the plantations. The constraints to plantation development have been categorised into economic, technical, institutional, and site and biophysical factors depending on their origin and impact.

3.2.1. Economic Constraints

The net financial returns from wood production is determined by a complex interaction of factors such as climate, soil type, land use regulation and those aspects of market structure that influence input and wood prices (Bhati et al. 1991). The main economic barriers to plantation development in Ghana are:

A. Uncertain Economic Profitability

A major economic barrier is the fact that plantations have long rotation periods (the time interval between planting and harvesting of the plantation) and may not be the most profitable investment option compared to other, more short-term, investments. Most native species in Ghana have very long rotations of up to 100 years or more while exotic species may have shorter rotations of about 20-50 years. Economic profitability is highly dependent on the length of the investment: the longer the rotation, the less profitable the investment is likely to be, all things being equal. Questions regarding economic profitability also reflect increased risk of fires, pests and diseases and accumulation of interest on the capital that is tied up in the plantation venture. Another issue is that entrepreneurs are usually interested in investments that would yield profits during their lifetime, and hence would tend to shy away from long-term investments as encountered in plantation forestry. Unless thinning is undertaken, most trees do not generate any revenues between planting and harvesting; therefore landowners who do not have other sources of income to live on while the trees reach maturity will be reluctant to invest in plantations compared with other land uses such as agricultural and cash crops.

Available published analyses of profitability of forest plantations in Ghana are limited. FAO (2002a) compared the benefit/cost ratios of several native and plantations of exotic species on good sites in the HFZ using a discount rate of 10%. The study concludes that among the exotic species, teak and *Cedrela odorata* were the most profitable, with benefit/cost ratios of 2.4 and 1.7 respectively; while *Ceiba pentandra* and wawa (*Triplochiton scleroxylon*) were the most profitable among the native species with 1.4 and 1.3 benefit/cost ratios, respectively. It is not possible to make a general case as to whether *all* forest plantations in Ghana are profitable investments or not. Several factors influencing plantation profitability are: choice of species and site, rotation length, prices of outputs, input costs, and the discount rate. Therefore, economic profitability has to be analysed on a case-by-case basis.

B. Lack of Financing

Timber plantation establishment requires huge investments of financial resources to be successful. Given the risky nature of such long-term investments, commercial banks are usually reluctant to provide credit for plantation investors. Furthermore, the land titling system in Ghana does not provide enough security for forestland to be accepted as collateral by the commercial banks. Lack of financing at reasonable interest rates, therefore, constitutes an economic barrier to plantation development. Interest is the fee for borrowing money, and usually this amount is paid to the lender over and above the amount borrowed (principal). In Ghana, commercial interest rates are habitually high (between 20-40%). With such high rates, it is almost impossible to return a profit from investments in forest plantations that span over several decades. The higher the interest rates, the higher the interest that a borrower would pay on the initial capital. For example, if an individual borrows Gh¢100, 000 to establish a plantation at a 5% interest rate over 50 years, the amount of interest at the end of the 50 years will be Gh¢1, 046,740. However, if the interest rate were 20%, the interest on the same capital will be Gh¢909, 943,815 over the same period.

The main goal of the Forest Plantation Development Fund (FPDF) was to support the development of plantations in both the public and private sectors in Ghana and thus, reduce the financial barrier for investors. Despite the good intentions of the FPDF, management of the fund has been weak and ineffective, with allegations of corruption and misapplication of funds by some beneficiaries. For example, in August 2009, the Minister of Lands and Natural Resources lamented that there was nothing to show for the Gh¢226 million that had accrued to the FPDF since 2000 (Business News 2009). The Minister indicated that the former Board of the FPDF had contravened the FPDF Act by investing a large part of the resources accruing to the fund in money market instruments for interest, rather than investing in actual plantations, which was the core objective of the fund. The Minister probably forgot that the problem arose from the ambiguity in Section 7(b) of the FPDF Act itself that empowers the Board to invest the funds. Consequently, members of the Board were replaced in August 2009 (Business News 2009). There are also institutional problems plaguing the FPDF. Since its inception, no loans have been given for plantations under three hectares and to farmers who intercropped with cocoa or palm oil (Boni 2006). The land rights documentation required to obtain a loan is costly and the bureaucratic procedure lengthy, tiresome and inaccessible to small-scale farmers (Boni 2006).

C. Marketing Challenges

The installed milling capacity of the timber industry in Ghana over the years has been tailored to processing large-diameter trees that are harvested from natural forests. Plantation-grown trees tend to have smaller diameters, either because they are harvested much sooner than their natural forest counterparts, or because some exotic trees are smaller in size than native species. The local milling sector is not equipped to process small-diameter plantation logs and may, therefore, be unable to utilise some of the plantation-grown wood in their mills. This could reduce the demand for plantation grown logs. While there may be alternative markets for this wood, eliminating the milling sector from this market has a potential to depress log prices of plantation wood, and hence have a negative impact on investments in plantation development. Depressed log prices are exacerbated by the complete ban on log exports in 1995, which has already depressed the prices of logs sold in the domestic market (FAO 2002a). As a result of this ban, private companies without timber processing plants located in Ghana will be obliged to sell their logs in alternative markets or to the few millers that can utilise the products at depressed log prices. Another issue is related to marketing of lesser-known plantation species. While teak is well-known internationally for its superior wood qualities and properties, the properties and uses of other popular plantation species in Ghana such as gmelina and cedrela are still relatively unknown and have therefore only been marketed on a limited scale in Africa (FAO, 2002a). These marketing challenges continue to persist despite the fact that some of these plantation species have properties that are identical to some widely marketed native species.

Getting forest products to markets requires a certain minimum level of infrastructural development, such as roads that can be used to transport inputs to, and outputs from, the plantation and processing sites. Most areas in Ghana still lack basic feeder and trunk roads, technology and skilled labour to support industrial plantation development. Therefore, investors in these areas may be required to develop the infrastructure themselves, which will ultimately decrease the profitability of their investments. The lack of infrastructure that can support integrated wood-processing industries is, therefore, a barrier to forest plantations, given that integrated wood production, processing and marketing offer more advantages than non-integrated production systems. For example, integrated forest companies (i.e., those that own both forests and processing facilities) can ensure constant supply of raw material at predictable prices and have lower transaction costs compared to those who own only one of these. Integrated forest companies are therefore able to reduce their production costs and increase their competitiveness.

3.2.2. Technical Constraints

Technical barriers are related to the lack of materials, limited information on matching species to sites, limited fire management capacity, limited knowledge and technology required for successful plantation development, and the transfer of that information to those who need it most.

A. Limited Technical Information and Dissemination

Successful plantation development is underpinned by a good understanding of the silviculture and ecology of the tree species used. Before using a tree species in plantations, it is important to understand the site and climatic requirements of the species, the pests and diseases that affect the species, growth characteristics, management requirements, uses of the species, and market potential. In Ghana, this technical information is available for a limited number of tree species. Hence, this has limited the number of species that have been planted in plantations so far to only a few exotic and native species such as *Tectona grandis, Gmelina arborea, Cedrela ordorata, Eucalyptus spp.*, *Terminalia superba* and *Triplochiton scleroxylon.*

A critical piece of information for those interested in plantation development is to know what species to plant on what type of land. At the national level, there seems to be no scarcity of potential land for forest plantations. However, by determining and providing information to potential clients that help them to match correctly species to their most productive sites, the profitability of plantations can be greatly enhanced. The Forestry Department provided some guidance on species to be used in the HFZ for plantations and improvement planting of degraded natural forests (line planting and enrichment planting) (FAO 2002a). This information should be expanded to include the savannah and transition zones, and cover more tree species.

Research is needed to develop plantation-related technical information over time and provide these to the end-users. Although some work has been done by local research organisations such as the Forestry Research Institute of Ghana (FORIG), much still needs to be done. There has been a lack of a consistent long-term research into plantation forestry issues in Ghana, which is the main reason for the current state of insufficient information. Research should be a key component of any national plantation development strategy. Such information needs to get to the target investors through effective extension services in the form of technical packages. Historically, extension services are well developed for the agricultural sector; the forestry sector can benefit from a similar model.

B. Limited Fire and Pest Management

Fire is a major part of Ghana's forest ecosystems, and hence fire management should be incorporated into forest plantation planning and development. The high relative humidity and short dry season in the high forest zone combine to reduce the amount of combustible material (fuel load) in the forest and hence reduce the risk of forest fires. However, the savannah zones are subjected to annual recurring fires, which put all plantations at risk of burning. Unfortunately, there are limited forest fire management strategies in place across Ghana to eliminate or minimise the impacts of these fires on forest plantations. Even more serious than that, unlike other countries, there is no significant active research into forest fire management in Ghana. In the savannah zone, it is common for trees planted in the rainy season to be completely burnt down the next dry season due to the lack of fire management.

Research that develops management techniques to protect plantations at the individual, community or industrial levels will be indispensable to successful plantation development. The current approaches to managing fires in plantations by the Forest Services Division of the FC include wildfire awareness campaigns, fire ride construction, ground patrol, fire suppression and enforcement of legislation relating to forest fires (Ghana Forestry Commission 2008b). The FSD has acquired 12 fire tenders for its Wildfire Management Project to help in firefighting operations in the fire-prone zones of Ghana, especially in forest plantations. Despite these efforts, there is still a need for additional resources, training, research and capacity in fire management.

In addition to fire, there is also limited information on how to manage the pests and diseases that affect forest plantations, especially, the exotic species. Additional research will be needed in disease and pest management to ensure the survival and sustainability of forest plantations.

C. Lack of Materials

Another technical barrier to plantation development is the lack of materials required for raising seedlings, transporting, transplanting, and managing the plantations over the rotation. These technical barriers particularly affect individual plantation owners, who lack the financial capacity to purchase these materials on their own. To optimise the benefits from forest plantations, it is critical that seedlings be of superior phenotypes, or of genetic quality that are resistant to common local diseases and pests, be able to survive fire and drought conditions, have fast growth and high- quality wood. These characteristics suggest that a national organisation takes responsibility for providing such seedlings to end-users. During the implementation of the Rural Afforestation Programme (RAP), the then Forestry Department was responsible for providing all planting materials and technical assistance to individuals, communities and other organisations.

3.2.3. Institutional Constraints

A) Policy and Regulatory Framework

As discussed in Chapter 1, the forest sector is governed by a hierarchy of policies and regulations that have been developed and modified over time. The overarching document that provides the strategic direction for forest sector activities is the Forest and Wildlife Policy adopted in 2012. These policies and regulatory frameworks determine the operating environment, provide incentives, and in some cases, act as barriers to plantation development.

The 1992 Constitution, the 2012 Forest and Wildlife policy, the *Timber Resources Management Regulations*, 1998 (LI 1649) and *Timber Resources Management (Amendment) Regulations*, 2003 (LI 1721), the *Trees and Timber Management (Amendment) Act*, 1994 (Act 493), the *Administration of Stool Lands Act*, 1962 (Act 123) and the *Forest Plantation Development Fund (Amendment) Act*, 2002 (Act 623) contain provisions that potentially pose barriers to developing forest plantations. The barriers posed by the policy and regulatory framework relate to ambiguities in some provisions of the policies/regulations, lack of coordination of the policies, and poor implementation of existing policies and regulations.

The *Timber Resources Management (Amendment) Regulations*, 2003 (LI 1721) impose stumpage fees on natural forest timber including some native species that are planted in plantations. According to these regulations, the stumpage rate shall be determined by the Minister responsible for forestry in consultation with the FC and the Administrator of Stool Lands, having regard to the market demand and inventory levels of timber species. This provision, in particular, is very problematic as it provides no clarity and could become a major disincentive for investment in plantations. Who would want to invest in a system when the investor does not know how much stumpage fees they will be charged? It should be pointed out that the previous LI 1649 of 1998 included a formula for calculating the stumpage fees for the various species, which was eliminated when the Regulations were amended in 2003 under LI 1721. In fact, this power for the Minister to determine stumpage rates has resulted in a court action by the Ghana Timber Association (GTA) against the Forestry Commission.

Ayine (2008) narrates the lawsuit against the FC by the GTA about the determination of stumpage fees. In April 2005, the GTA sued the FC before the High Court, challenging its formula for computing stumpage fees. As noted above, under the LI 1721 (2003), the Minister responsible for forestry has discretionary power to fix stumpage fees, in consultation with the FC and the Administrator of Stool Lands. The GTA challenged the stumpage fees determined by the Minister on several grounds. It claimed that the FC's determination was illegal, as no guidelines regulating the exercise of the Minister's discretion had been formulated or published; and that the Minister did not consult the FC and the Administrator of Stool Lands as required. In addition, as the fees were also charged on trees that were not commercial logs, GTA claimed that its members had to pay higher-than-acceptable stumpage fees, which significantly affected their business operations. Therefore, the GTA sought a court declaration that the FC's system for the calculation of stumpage fees is defective and has resulted in the GTA's members being overcharged. The FC denied that the stumpage fees were illegal and threatened to publish in the media the names of the timber companies that were not paying stumpage fees as required. In retaliation, the GTA also threatened to expose the collusion between forestry officials and its members. Eventually, the FC settled the case out of court (Ayine 2008).

Another ambiguity in the legislation is whether export levies apply to timber harvested from plantations and if so, how much. There are export levies on processed and unprocessed wood from natural forests under the *Trees and Timber Management Act*, 1994 (Act 493). The list of species for which these levies apply includes only one exotic plantation species (teak) and leaves unanswered the question as to whether all timber harvested from plantations will be subject to similar export levies and royalties applicable to natural forests.

The *Timber Resources Management (Amendment) Act*, 2002 (Act 617) specifies that timber harvesting rights are transferred from the State to companies through a timber utilisation contract (TUC).The Act further specifies that no timber rights shall be granted in respect of land with private forest plantation or land with any timber grown or owned by any individual or group of individuals. Although it is clear that the government has no interest in the ownership of the trees planted by individuals or in private plantations, and will not allocate such trees to timber firms for harvesting, it is still unclear whether the grower of the trees is accorded ownership rights regardless of their rights to the land on which such trees are planted. It can be interpreted from Act 617 that a private plantation owner does not require a TUC to harvest his/her trees for private and commercial purposes. However, there are still some ambiguities regarding whether a plantation owner needs a registered property mark to harvest his/her own trees for individual and commercial purposes as required under the *Timber Resources Management Act*, 1997 and *Trees and Timber Management Act*, 1994 (Act 493) for natural forests.

With the competitive bidding system used to allocate forest harvesting rights, companies who win bids are expected to pay a fee for the timber rights. Given that plantations do not qualify to be given out under a TUC, it is safe to assume that no such timber rights fees are applicable, although this is not explicitly spelled out in the legislative framework. These legal and regulatory ambiguities in Ghana's forestry sector show the lack of strategic direction and coordination, which tend to confuse private investors. It is not obvious that potential investors in plantations clearly understand the nuances and technicalities involved in navigating through these legal and policy confusion.

Another institutional constraint is the ineffective implementation of existing policies and regulations. Under the *Timber Resources Management Act*, 1997 (Act 547) and the *Timber Resource Management Regulations*, 1998 (LI 1649), TUC holders are required to develop and implement reforestation, afforestation and plantations in any area that the Chief Conservator of Forests[1] may approve. TUC holders make undertakings (including a performance bond) to execute the reforestation plan during the period of the contract to the satisfaction of the Chief Conservator of Forests. The FC and its predecessor Forestry Department have lacked the capacity to enforce these regeneration and plantation establishment regulations. This inability

[1] Although Ghana's legislations still refer to the Chief Conservator of Forests, this position no longer exists. The equivalent position is the Chief Executive Officer of the Forestry Commission.

to enforce legislated plantation initiatives contributes to ineffective plantation development in Ghana.

A potential cause for concern with the FPDF is Section 7(b) of the *Forest Plantation Development Fund Act* (2000), which specifies one of the functions of the FPDF Board as "*attracting contributions into the Fund and investing the moneys of the Fund.*" This legal provision of "investing" the money has resulted in a situation where the money for plantation development was being invested in money markets for interest, rather than in plantations (Business News 2009). Therefore, the FPDF has not achieved much since 2001, and the financial difficulties faced by investors in plantation forestry persist. This Act needs to be amended to eliminate this problematic clause.

B) Customary Practices and Tree Tenure

Apart from the legal and policy barriers, there are cultural practices that also pose challenges to forest plantation development in Ghana. Traditionally, most subsistent farmers have planted, tended and conserved only fruit trees with an economic value around their homes and farmlands. In the northern regions, these trees include shea, dawadawa and mango (*Mangifera indica*), which can be observed on most farmlands and around homes. In southern Ghana, the most common fruit trees include citrus, banana, plantain and cash crops such as cocoa. However, there is no traditional culture of planting and tending trees for purposes of fuelwood or timber production in any part of Ghana. The concept of forest plantations to provide wood products is, therefore, alien to the culture and traditional practices of most farmers in Ghana.

Insecure tree tenure, resulting from the different configurations of ownership of planted trees, and the potential benefits that accrue to farmers/tenants can influence incentives for plantation development (Owubah et al. 2001). The forms of tenure that have longer terms, are more clearly defined, provide more of the economic benefits to their holders, are likely to stimulate tree planting (Zhang and Pearse 1996). In most cultures, a tenant cannot exercise ownership rights over trees on the land for which he was only granted farming rights. Under the traditional land-use system, the planting of trees by tenant farmers is generally considered as an attempt to perpetuate their stay, which may in turn indirectly imply ownership of the land (FAO, 2002a). Except in the Upper West Region, a tenant farmer does not own the natural or planted trees on the land issued to him (FAO 2002a). These cultural barriers work together to impede the development of forest plantations.

C) Land Tenure

The importance of land tenure to forestry is related to the fact that tenures define the types of property rights held by users of the land. Property rights define the extent to which the holder can enjoy the benefits accruing from the asset; hence property does not have to be a tangible good, but rather a defined set of rights over something (Pearse 1992). Therefore, efficient land tenure systems that offer appropriate property rights to holders contribute to effective resource management and socio-economic development.

The configuration of land ownerships across Ghana is complex. This is a result of a mixture of cultural practices, traditions and legal and constitutional frameworks, which produces a complex series of rights and interests in land. There are two main categories of land ownership in Ghana: public and customary lands (Stool/Skin lands, clan lands, family lands, etc.). Under the 1992 Constitution, all public lands in Ghana are vested in the President on behalf of, and in trust for, the people of Ghana. The constitution also recognises stool/skin lands, which are vested in the appropriate stool/skin on behalf of, and in trust for, the subjects of the stool/skin in accordance with customary law and usage. In Ghana, customary lands constitutes about 78% of land ownership, State lands constitute 20%, while vested lands (split ownership) is 2%. These complex land ownership arrangements lead to the inadequate security of tenure, conflicts, difficult accessibility to land, and confusion over ownership of some lands and tension between the state and customary authorities. These tenures confer different kinds of ownership and benefits and hence may imply incentives or barriers to plantation development.

Generally, three kinds of tenure arrangements are available: leasehold, freehold interest, and communal or family land-use rights. Under indigenous land-use rights, the user has no discretionary land transfer rights because the land belongs to a corporate body (Benneh 1989). Individuals can sometimes enhance their rights in such holdings by making some long-term investments (Aidoo 1996). Freehold interests in lands confer an absolute and secure land rights to the user (Benneh, 1989; Aido 1996). Leaseholds take two forms: sharecropping and annual land rental payment. Sometimes land rental payment is small or merely symbiotic (Zhang and Owiredu 2007). Leaseholds are restrictive and do not offer any security of land tenure to the farmers (Benneh 1989). The period of leasehold ranges from a year for the cultivation of annuals to 50 years for the establishment of tree plantations (Zhang and Owiredu 2007).

Several studies in Ghana have examined land tenure systems, methods of land acquisition, and management models for forest plantations (Odoom 1999); farmers' willingness to establish plantations (Owubah et al. 2001) and the impact of land tenure and market incentives on farmers' actual forest plantation activities (Zhang and Owiredu 2007). These studies conclude that the type of ownership is an important determinant of the willingness of people to invest in plantations. In particular, farmers who own lands outright are more likely to invest in plantations (Zhang and Owiredu 2007) compared to those who do not have such security to a title.

The key constraint imposed by the land tenure arrangements in Ghana related to forest plantation development are summarised by FAO (2002a) as: a) multiplicity of interests and rights in land which may vary in different parts of the country leading to conflicting claims to ownership; b) lack of reliable maps indicating stool/skin land boundaries which can give rise to disputes; c) cumbersome land disposal and documentation procedures; and d) potential conflict concerning tenurial and management arrangements for plantations within forest reserves where parties other than the FC are involved.

In addition, communal and family ownership of lands leads to fragmentation of the land base into small parcels, which poses a challenge for large-scale plantation projects that require lands that will cross ownership boundaries. Conflicts over land ownership and land fragmentation together act as serious barriers to plantation development in Ghana. An investor will have to negotiate land titles with numerous owners to acquire land to meet his/her needs. These negotiations can be long, frustrating and expensive for investors either at the negotiation stage or after the project has begun.

3.2.4. Site and Biophysical Constraints

In general, most vegetation, climatic and soil zones in Ghana can support tree growth; however, not all soils can support economically profitable forest plantations. The risk to the economic sustainability of plantation forestry depends on the ecological capability of the site to support the planted trees. Site characteristics such as low nutrient reserves, poor nutrient retention ability and susceptibility to drought are therefore major limiting factors to using tropical soils for short rotation tree crops (Tiarks et al. 1998). Because of the importance of the litter layer to nutrient supply and soil structure, disturbance of the surface should be kept to a minimum to maintain productivity (Tiarks et al. 1998). Maintenance of the litter layer or another vegetative cover is necessary for limiting erosion and keeping a suitable soil moisture balance (Spaargaren and Deckers 1998). Despite the high rainfall received in parts of the tropics, soil water deficit can be a recurring constraint on productivity on many sites in subtropics and tropics (Gonçalves et al. 1997; Landsberg 1997). In the semiarid and arid tropics, water stress may limit rates of growth below commercially viable levels; soil management for conserving available water is a critical consideration (Tiarks et al. 1998).

Forest and savannah fires contribute to dry conditions and may limit tree growth as well. The soils of the savannah zone tend to be eroded, have only a small capacity to keep water and are deficient in nutrients, notably phosphorous and nitrogen (Boateng 1966). The soils in the HFZ, due to the high rainfall are highly leached and acidic (pH 4.0 -5.5), with low cation exchange capacity, available phosphorus, nitrogen and organic matter (FAO 2002a). These differences in soil characteristics mostly account for the low productivity of the savannah zones compared to the high forest zones in Ghana.

3.3. Opportunities for Plantation Development

3.3.1. Legal and Policy Framework

Even with the imperfections of the regulatory and policy framework governing Ghana's forestry sector identified above, these frameworks do provide some opportunities for developing forest plantations. The major opportunities are contained in the 2012 Forest and Wildlife Policy, the *Timber Resource Management Regulations*, 1998 (LI 1649), the *Forest Plantation Development Fund Act* (2000), the *Forest Plantation Development Fund (Amendment) Act* (2002), and the National Forest Plantation Development Programme (NFPDP).

The second strategic objective of the Forest and Wildlife Policy (2012) is to promote the rehabilitation and restoration of degraded landscapes through forest plantation development, enrichment planting, and community forestry informed by appropriate land-use practices to enhance environmental quality and sustain the supply of raw materials for domestic and industrial consumption and environmental protection. The support to plantations provided by this Policy is significant, given that this is the document that provides strategic direction for Ghana's forestry sector. It is a sign of commitment from the highest levels of government to promote plantation development and provides the enabling policy for subsidiary policies and regulatory initiatives to be developed to promote plantation forestry.

The *Timber Resource Management Act*, (1997) provides for a TUC holder to establish and manage forest plantations in contract areas and reforest or afforest in any area that the Chief Conservator of Forests may approve. These provisions provide opportunities for the Chief Executive of the FC (which is equivalent to the Chief Conservator of Forests under the previous Forestry Department) to direct plantation development through the timber utilisation contract process.

Since 2000, governments of Ghana have taken steps towards promoting forest plantation development with the passing of the *Forest Plantation Development Fund Act* (Act 583). In 2002, this *Act* was amended to cover plantation growers, both in the *public* and private sectors. If managed effectively, the funding provided by the Fund would help to reduce the financial barriers faced by plantation developers. It also offers opportunities for the Fund to support meaningfully research into plantation development, given the limited research funding from alternative sources.

In 2001, the government launched the National Forest Plantation Development Programme (NFPDP). Under the programme, both indigenous and exotic economic tree species have been planted. These efforts should be continued and intensified especially at the community level. In the savannah regions, the greatest need is for fuelwood and other wood products for building construction. Therefore, the financial assistance under the FPDF should be provided to individuals and communities who show the willingness and capacity to develop community plantations.

3.3.2. Land Availability and a Favourable Climate

Data from the Ministry of Lands and Natural Resources on land use patterns in Ghana show that there are about 10.7 million ha of savannah woodlands and unimproved pasture, comprising 45% of the total land area of Ghana (FAO 2002a). These lands are potentially available for forest plantation development. This area is in addition to degraded reserved and unreserved forest areas that can be replanted into plantations or improvement planting. The actual area that is available for forest plantations will need to be determined precisely, taking into consideration other land use needs and whether each parcel of land is suitable for plantations.

Ghana is also blessed with a favourable climate without extremes of temperatures, rainstorms and other weather-related hazards. Rainfall is adequate to support tree growth in almost all areas of Ghana. Not only that, as shown in Table 1, Section 2.3, some plantation species can reach MAIs of up to $20m^3$/ha/yr on good sites in the HFZ. Even on poor sites in the savannah zone, plantation growth is as good as that of natural forests on good sites in the HFZ (section 2.3). As with the case of land, there is a need to develop maps that match tree species to soils and climate for each ecozone.

3.3.3. Labour Force Availability

Reliable official statistics on unemployment are difficult to find, and so estimates of the unemployment rate vary significantly among agencies reporting these statistics. Currently, the unemployment rate is estimated to be between 11 and 20%. Unemployment in Ghana is a function of over-reliance on the government as the main employer in the country and limited job opportunities in the private sector. The agricultural sector employs about 50% of Ghana's population. There are many junior and high school drop-outs who are walking the streets in cities, selling everything from dog chains to candies to survive. Many others are under-employed, especially in the savannah areas of Ghana, who have little to do during the long, dry season. Plantation development initiatives will have access to this large pool of labour at reasonable costs to perform skilled and unskilled tasks from tree nurseries to harvesting and processing wood products. Available labour at low costs will serve as opportunities for investors to reduce their investment costs, and will also be an economic development tool for rural areas with few alternative employment opportunities. The similarity between the operations in the agricultural and forestry sectors is an advantage because many job seekers are already familiar with the basic tasks associated with managing food crops.

3.3.4. Available Market Demand

Current and future shortages of wood products from the natural forests for domestic use and export provide an opportunity for forest plantations to fill the gap between demand and supply. As shown in Chapter 1, the demand for wood products will continue to rise and hence there is a ready demand for plantation-grown wood. For example, with increasing rural electrification, the demand for teak for use as electric transmission poles will continue to increase. Secondly, increased demand for teak products by India offers an exciting market opportunity that Ghana is already exploiting. The majority of Ghanaians depend on wood as their main energy source (Benhin and Barbier 2001), and hence the demand for roundwood in Ghana is driven mainly by the demand for fuelwood. Although it is fair to note that not all plantation species will have as much demand as teak, species that yield large quantities of biomass within short rotations will have ready markets in bioenergy applications.

3.3.5. International Climate Change Initiatives

The Kyoto Protocol (KP), which was negotiated in December 1997, requires developed countries as a whole to reduce their greenhouse gas (GHG) emissions by 5.2% compared to 1990 levels, during the first commitment period from 2008 to 2012. The KP and subsequent international climate change initiatives recognised that forests, forest soils and forest products all play important roles in mitigating climate change. The Clean Development Mechanism (CDM) is one of the flexible mechanisms that was included in the Kyoto Protocol to help developed countries meet their reduction targets in a cost-effective way. Specifically, the KP recognises afforestation and reforestation as the only eligible land uses under the CDM. Since CDM projects can only be undertaken in developing countries, this offers interesting opportunities for the establishment of plantation forests for sequestering carbon in Ghana.

While investors in developed countries will benefit from carbon credits, Ghana will benefit from forest products and environmental services (e.g., reduced levels of GHG in the atmosphere) provided by such plantations. These climate change frameworks present opportunities for Ghana to incorporate CDM projects into its plantation development agenda, and draw foreign investors and institutions into the country.

3.3.6. Environmental Consciousness and Wood Shortages

Ghana's population is increasingly educated in all spheres of life including environmental issues. Both governmental (e.g., the Environmental Protection Agency (EPA)) and local and international non-governmental organisations have been carrying out environmental campaigns across the country on the need to protect the environment. These environmental agencies continue to use their existing networks to provide further education to people on the benefits of increasing tree cover through plantation development. The need for additional forest resources to meet the needs of society is widely recognised, especially in the savannah areas where tree cover is disappearing at a very fast rate. Rural dwellers now travel longer distances to gather fuelwood and other forest products; while city dwellers now pay more for charcoal because it has to be transported over longer distances to markets than previously. Soil fertility is declining, and rainfall patterns have become more erratic and unreliable as well. In short, many people are aware of the need to increase tree cover in many parts of the country. This general awareness of the populace about the effects of deforestation and environmental degradation and the increasing scarcity of forest resources within their lifetimes will reduce their resistance to plantation establishment.

3.4. Incentives for Plantation Development

3.4.1. Definitions of Incentives

Several definitions of incentives abound in the literature. For example, Enters (2001) refers to incentives as the "incitement and inducement of action." Gregersen (1984) defines incentives as "public subsidies given to private sector to encourage socially desirable actions by private entities." For the purposes of this chapter, incentives refer to anything that motivates people to do something they would otherwise not have done. In line with this definition, incentives for forest plantations include economic and policy instruments that motivate investments in forest plantations.

3.4.2. Reasons for Incentives in Plantation Forestry

Most of the world's forest plantations have been established with incentives in the form of a subsidy of one sort or another, either directly or indirectly (Whiteman, 2003; Cossalter and Pye-Smith 2003). For example, in many parts of Latin America, Oceania and Asia, plantation programmes paid more than 75% of the establishment cost with additional allowances made for land, maintenance and many other costs (Brown 2000). Given the pervasive nature of incentives in plantation forestry, the natural question is, why do we need incentives?

The long-term nature of growing trees, whereby establishment expenditures are incurred early in the life of the plantation, and benefits accrue many years down the road, is the most important reason that makes promoting plantation forestry difficult. The long gestation period for trees makes the investment riskier than many other alternatives. The question is: why would anyone want to invest in such a risky enterprise? Incentives are appropriate when the private net returns are lower, but the returns including externalities are greater than the returns from alternative land uses. In this situation, incentives could be effective by creating a more socially desirable land-use pattern (Haltia and Keipi 1997; Williams 2001). If not, incentives will represent a misallocation of public-sector resources, merely helping investors to earn higher returns (Enters et al. 2003).

Incentives may not be needed if the private returns from forestry exceed those from other land uses or if the addition of incentives will still not provide an attractive private return. In selecting an appropriate level of assistance, rates of return not only have to be compared with those from alternate land uses but also with investments in other sectors (Williams 2001). Incentives may also be used to help investors overcome barriers such as the high capital cost of establishment and the relatively long waiting period for a profit (Williams 2001). If incentives are implemented effectively, they could provide the initial impetus to develop national forest industries for either foreign exchange or ecological and environmental benefits. Furthermore, incentive programmes can be used to encourage economic development and generate employment in specific, less developed regions and diversify the economy away from areas with limited economic potential, such as agriculture (Williams 2001).

Most governments realise that they cannot undertake plantation development programmes successfully without the participation of the private sector, hence incentives are meant to reduce barriers to investments and remove structural impediments and operational constraints to maintain private sector interest and investment in plantations (Williams 2001).. Incentive programmes in Ghana should be targeted at reducing or removing entirely some of the barriers to plantation development identified earlier in this Chapter.

3.4.3. Types of Incentives

In a comprehensive study of incentives for forest plantation development in the Asia-Pacific region, Enters et al. (2004) categorise the kinds of incentives into direct and indirect incentives. Direct incentives are provided directly by governments, development agencies, non-governmental organisations and the private sector and include the following: goods and materials (e.g., seedlings, fertilisers, etc.); specific provision of local infrastructure; grants; tax relief or concessions; differential fees and access to resources; subsidised loans; cost-sharing arrangements and price guarantees. For example, under the RAP in the late 1980s in Ghana, the government and World Bank-funded project provided direct incentives to individuals and communities interested in plantation establishment. These were mainly in the form of free seedlings, transportation and technical assistance in the planting and managing the plantations.

Enters et al. (2004) further divide indirect incentives into *variable incentives* and *enabling incentives*. Variable incentives are economic factors that affect the net returns that producers earn from plantation activities such as input and output prices, exchange rates, general taxes, specific taxes, interest rates, trade restrictions (e.g., tariffs) and fiscal and monetary measures (Enters et al. 2004). Enabling incentives, on the other hand, mediate an investor's potential response to variable incentives and help to determine land use and management (FAO 1999). These include: land tenure and resource security, accessibility and availability of basic infrastructure (ports, roads, electricity, etc.), producer support services, market development, credit facilities, political and macroeconomic stability, national security, research and development, and extension services. Enabling incentives can also be viewed as elements in the investment environment that affect decision making (Enters et al. 2004).

3.3.4. Incentives for Plantation Forestry in Ghana

Incentives for plantation development in Ghana can be broadly classified into two categories: institutional and policy incentives, and economic incentives. Institutional and policy incentives provide the long-term enabling environment for investments in plantations. The most notable reform in this direction is provided within the context of land management under the National Land Policy reform. Economic incentives are more wide ranging and are intended to influence the benefit/cost structure of the plantation enterprise to make it more worthwhile for investors and/or support rural farmers with basic inputs into plantation establishment. The main policy and legislative frameworks that provide economic incentives are contained in the *Timber Resources Management Act 617 (Amendment) Act,* (2002), the *Forest Plantation Development Fund Act,* 2000 and the *Forest Plantation Development Fund (Amendment) Act,* (2002).

A) Institutional Incentives

One of the main constraints to plantation forestry development in Ghana identified above is the land tenure system. The National Land Policy (NLP) was the first comprehensive approach to dealing with the land tenure constraints in Ghana. It was formulated from 1994-1999 through participatory processes that included consultations with traditional authorities, farmers' organisations, academia, public sector institutions, researchers, government authorities, etc. According to the World Bank (2003), the overall long-term goal of the NLP is to stimulate economic development, reduce poverty and promote social stability by improving security of land tenure, simplifying the process for accessing land and making it fair, transparent and efficient, developing the land market and fostering prudent land management practices.

The specific objectives were to facilitate equitable access to land, protect land owners and their descendants from becoming landless, ensure prompt payment of compensation for compulsorily acquired lands, minimise and where possible eliminate land boundary disputes, create and maintain effective institutional capacity at national, regional, district and community levels, and promote community participation and public awareness in sustainable land management (World Bank 2003). The NLP is expected to provide the appropriate institutional structures, and clear, coherent, and consistent polices to support land development activities. The NLP is implemented under the Land Administration Programme in five-year phases over 15-25 years (starting from 2004). When these institutional changes are completed, they will minimise the barriers posed by the hitherto antiquated and confusing land administration system in Ghana and provide incentives for investments in forestry and other land-related projects (World Bank 2003).

B) Economic Incentives

The *Timber Resources Management (Amendment) Act, 2002* (Act 617) provides the framework for a wide range of incentives for forestry and wildlife sectors, including forest plantation developers. Section 14A prescribes that an investor in any forestry or wildlife enterprise is entitled to such benefits and incentives as are applicable to its enterprise under the *Internal Revenue Act*, 2000 (Act 592) and under Chapters 82, 84, 85 and 98 of the Customs Harmonised Commodity and Tariff Code scheduled to the *Customs, Excise and Preventive Service Law*, 1993 (PNDCL 330). The chapters of the Customs Harmonised Commodity and Tariff Code specify items that have been zero-rated under the law.

In addition, an investor whose plant, machinery, equipment or parts of machinery are not zero-rated under the Customs Harmonised Commodity and Tariff Code scheduled to the *Customs Excise and Preventive Service Law*, 1993 (PNDCL 330), may submit an application for exemption from import duties, Value Added Tax (VAT) or excise duties on the plant, machinery, equipment or parts thereof to the Forestry Commission which shall submit it to the appropriate tax authority. For the purpose of promoting strategic or major investments in the forestry and wildlife sector, the Minister responsible for forestry may determine or negotiate specific incentives in addition to the incentives provided above for such period as may be specified in the relevant timber utilisation contract.

Act 617 also provides for investment guarantees, transfer of capital, profit and dividends. This *Act* gives investors guaranteed unconditional transferability through any authorised dealer bank in freely convertible currency of dividends or net profits attributable to the investment; the remittance of proceeds, net of all taxes and other obligations, in the event of sales or liquidation of the operations of the investor or any interest attributable to the investment. There are also guarantees against expropriation. The operations of an investor shall not be nationalised or expropriated by Government; and no person who owns, whether wholly or in part, the capital of any forestry or wildlife investment shall be compelled by law to cede the interest in the capital to any other person; and there shall not be any acquisition of the operations of an investor by the State unless the acquisition is in the national interest or for a public purpose and under a law that makes provision for payment of fair and adequate compensation.

The *Forest Plantation Development Fund (Amendment) Act,* 2002 provides for financial assistance for the development of forest plantations and research and technical advice to persons involved in plantation forestry on specified conditions. The Act encourages investment in forest plantation development through incentives and other benefits to be determined by the Fund Board. Incentives under the Act are defined as loans, rebates, grants and insurance to qualified beneficiaries. A beneficiary under this Act is entitled to tax rebates and such other benefits that apply to it under the *Ghana Investment Promotion Centre Act*, 1994 (Act 493). The Forest Plantation Development Fund is the main funding source for the National Forest Plantation Development Programme (NFPDP), which was launched by the government in September 2001. The programme was aimed at encouraging the development of a sustainable forest resource base that will satisfy future demand for industrial timber and enhance environmental quality. The programme was also expected to generate jobs and significantly increase food production in the country thereby contributing to wealth creation and reduction in rural poverty.

The NFPDP provides direct incentives to farmers participating in the programme in the form of paying for hired labour for seed collection, peg cutting, ploughing and lopping. The government also provides free seeds and seedlings, technical supervision, technical information on matching species with the site, the market for products, and the benefits of the food crops planted under the system.

3.4.5. The Problems with Incentives

Despite the great success of forest plantation incentives in countries such as Brazil, Chile, New Zealand and Australia, direct subsidies to industrial plantations have been widely criticised as inefficient and inequitable because of market distortions that can lead to economically incorrect allocation of productive factors (Evans and Turnbull, 2004; Enters et al., 2004). Inefficiency results from two reasons: a) the difficulty of proving a linkage between the incentive and environmental benefits resulting from the investment; and b) when incentives are provided to plantation growers who would have planted trees without them or when a higher rate of incentive is paid than would have been necessary to induce a grower to plant trees (Enters et al. 2004). Indirect incentives provided by policy and land tenure reforms are considered essential to industrial and socio-economic development. Most countries are re-directing their incentives at individual small-scale plantations and community forestry programmes (Evans and Turnbull 2004).

Direct incentives offered in the early stages of a forestry project may simply buy participation and may not lead to long-term interest in the project (Enters et al. 2004). Subsidies have often succeeded in stimulating the adoption of conservation measures that were abandoned or even actively destroyed once payments ceased (Lutz et al. 1994). The same has been observed for plantations (Sawyer 1993). In fact, during the RAP in Ghana, the first author observed that some farmers did not bother to protect their seedlings from fire in the dry season following planting, especially those who received food aid from the Adventist Development and Relief Agency (ADRA) as incentives. These same farmers requested for seedlings the following year to replant the same sites that were planted in the previous year! If an incentive is a primary cause of behavioural change, the discontinuation of that incentive is likely to become a cause for reversal (Enters et al. 2004).

3.5. CONCLUSION

Plantation development in Ghana continues to face economic, technical, institutional and biophysical constraints. The importance of these barriers differs across the various categories of plantation developers. The needs of individual and community small-scale farmers are different from those of large-scale industrial plantation investors. Clearly, there are opportunities as well, offered by local, national, and international demands for goods and services produced by forest plantations. The implementation of the Land Administration Programme is critical to streamlining land administration and minimising land-related barriers to plantation development in Ghana. However, such institutional changes should not be limited to legislation, judicial decisions, records management, titling, community-based land use planning, monitoring and evaluation, but should be expanded to include much needed reforms of the land tenure system itself as well.

Except the incentives provided directly to farmers under the NFPDP, the most extensive economic incentive programmes provided by legislation are intended to create an enabling environment for commercial timber producers and foreign investors. There is no published information on the success of the incentives provided to industrial plantation developers in Ghana, and so it is difficult to assess their effectiveness. As noted in Chapter 2, response to the incentives provided to farmers under the MTS is encouraging.

Given the land tenure constraints discussed above, it seems that majority of the forest plantations in Ghana will continue to be small-scale individual and community plantations and hence incentives targeted at these groups should be encouraged and improved, in addition to those for industrial plantations. In the meantime, the debate as to whether incentives for forest plantations work or not and their impact on plantation development and national economies will continue to rage on.

PART II. FOREST GROWTH DYNAMICS AND ASSESSMENTS

Chapter 4

FOREST GROWTH DYNAMICS AND PRODUCTIVITY

Understanding and predicting the behaviour of forests through time are fundamental to forest management. In this Chapter, we provide the basic concepts of stand growth dynamics and productivity. The purpose is to describe the concepts used in forest stand dynamics and provide a theoretical basis for understanding forest stand growth dynamics and the principles that underpin forest management. In the later sections of the Chapter, we discuss the environmental factors that influence tree growth, methods of improving site productivity and the impacts of climate change on forest plantations.

4.1. FOREST STAND DYNAMICS AND GROWTH

4.1.1. Forest Stand Dynamics

Forest stand dynamics are the changes in forest structure, function and composition through time, including stand behaviour before and after disturbances (Oliver and Larson 1996). Knowledge of forest stand dynamics is applied in many areas of forest management, including plantations for timber production (e.g., Cannell and Last 1976), semi-natural woodlands where conservation objectives are a priority (e.g., Smith et al. 1997) and multi-functional woodlands managed for both timber and biodiversity values (e.g., Kerr 1999). The usefulness of understanding forest dynamics is to be able to make reliable predictions about how forests will change over time and in the face of natural and anthropogenic factors.

Most descriptions of stand development characterise it as a progression through stages toward an older forest, possibly an old forest, in the absence of disturbance (O'Hara 2004). Disturbances, from human or other causes, can move stand development backward or forward in the process, depending on their type, severity, and timing (O'Hara 2004). Oliver and Larson (1996) describe the four generally recognised phases of forest stand dynamics. The first is the stand initiation phase, which corresponds to the phase of recruitment of stems to the stand. In plantations, planting ensures that the stand achieves full stocking of the site in one or more planting operations. In this phase, the trees are still young and growing; hence full site occupancy will not have occurred. The second stage is the stem exclusion phase, where the stand develops a closed canopy and the deep shade in the understory prevents further recruitment of trees to the stand while competition, site factors and genetic differences lead to differentiation of crown dimensions and stem diameters (Wilson and Leslie 2008). The stand then continues on to an understory re-initiation stage, where herbs, shrubs and advanced regeneration of trees appear and survive as a result of gradual thinning of crowns or occasional gaps that allow increased levels of solar radiation to reach the forest floor (Oliver and Larson 1996). Finally, the stand reaches an old-growth stage where the overstory trees die in an irregular pattern, either from natural causes or disturbance, creating space for recruitment into the canopy of trees from lower strata (Oliver and Larson 1996). The transition from one stage to the next generally holds true for plantations, if retained for long enough (Kerr 1999), and those natural forests that regenerate after a large-scale, stand-replacing disturbance such as a forest fire (Kimmins 2003).

However, each stage of the sequence may be disrupted as a result of managed interventions or natural disturbance events (Johnson and Miyanishi, 2007; Kimmins 2003). Because plantations are usually made up of one tree species, even-aged and most often of uniform spacing, the trees are often clear-felled before reaching the understory re-initiation stage in order to maximise the financial return on investment costs (Savill et al., 1997; Matthews, 1989; Evans and Turnbull 2004).

Figure 1 shows the growth patterns of neem plantations in Northern Ghana in terms of basal area and gross volume. The data were based on temporary sample plot inventory of neem plantations planted at 2 m x 2 m spacing. This graph represents the natural progression of a stand from planting until age nine years, without thinning. The volume available for harvesting in year 9 is approximately $100m^3/ha$. The basal area growth follows identical patterns to that of volume. The figure provides a baseline for how the stand would evolve without any disturbance, and could be the basis for predicting the impacts of management decisions on the development of identical stands in the savannah zone.

4.1.2. The Concept of Forest Growth

Growth is the increase in a particular stand characteristic over a period of time. Growth includes increases in height, diameter, volume, crown or any other forest attribute of interest. In simple terms, if a forest stand was measured in 2010 and it contained $100m^3/ha$ of wood and a subsequent re-measurement in 2015 showed a volume of $150m^3/ha$, then the growth can be said to be 50 m^3/ha over the five years or on average, $10m^3/ha/year$.

Individual trees grow using photosynthesis through adding woody biomass to all parts of the tree - stem, branches, roots, etc. Most of these growth components are difficult if not impossible to measure in the field (Davis et al., 2001). Growth is affected by how many trees die between the measurement periods for which growth is being assessed, i.e., mortality. In single species plantations, mortality is easier to estimate than in natural forests with high species diversity. For example, if a plantation is planted at an initial density of 2500 trees/ha, and after five years there are 2000 trees, then the mortality is 500 trees over the 5-year period.

Depending on the purpose of measurement, growth can be categorised into total growth, potentially usable growth, and growth that is actually removed and used (Davis et al. 2001). Total growth refers to all biomass produced by the trees, including roots, stump, bole, branches, leaves, etc. Potentially usable growth refers to woody components that could be used by a manufacturer, given current technology, while the amount of the growth that is actually removed and utilised constitutes the third kind of growth (Davis et al. 2001).

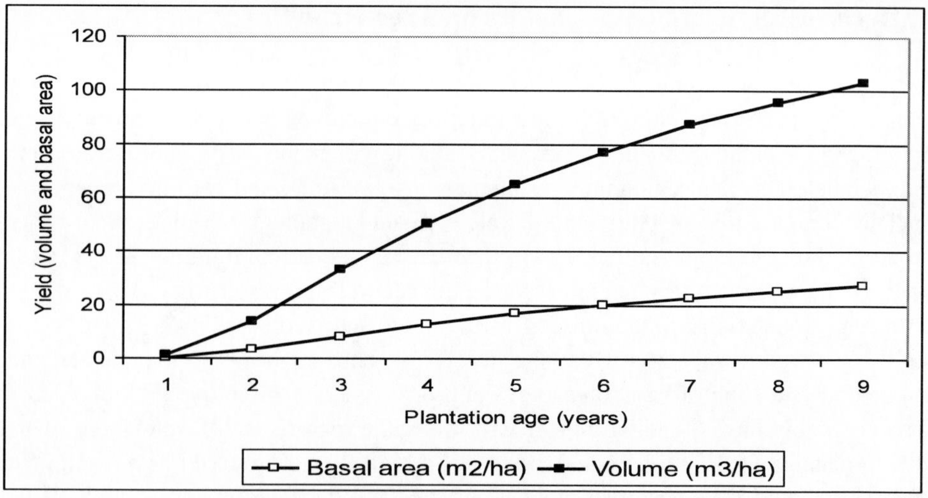

Figure 1. Growth patterns of neem plantations in Northern Ghana.

4.1.3. Measuring Forest Growth

Growth measurement is complicated by mortality. Because the increase in volume, for example, between any two measurement periods has to consider that some trees might have died in that time, while within that interval some trees might have grown into the measurable size as well (Davis et al. 2001). Following Beers (1962) the following framework is defined for estimating volume growth within forest stands:

V_1= volume of living trees at the beginning of the measurement period
V_2= volume of living trees at the end of the measurement period
M = volume of mortality over the measurement period
C = volume removed during the measurement period
I = volume of in-growth over the period

Depending on the type of growth required, the following five components of growth are defined (after Davis et al. 2001).

1. Gross increment including in-growth = $V_2 + M + C - V_1$
2. Gross increment of initial volume = $V_2 + M + C - I - V_1$
3. Net increment including in-growth = $V_2 + C - V_1$
4. Net increment of initial volume = $V_2 + C - I - V_1$
5. Net change in growing stock= $V_2 - V_1$

To understand and predict growth, stand growth models are used. A stand growth model represents an abstraction of the natural dynamics of a forest stand and depicts growth, mortality and other changes in stand composition and structure (BC Ministry of Forests and Range 2009). Chapters 5 and 6 of this book will discuss in detail the measurement, development and use of growth and yield models.

4.1.3. Effects of Environmental Factors on Tree Growth

Trees grow in environments consisting of a complex integration of physical and biological elements. Climate, soil, light, precipitation, the movement and composition of air, and topography constitute the main physical (abiotic) elements. The biological (biotic) elements include other plant communities, animals, insects, fungi and the microorganisms in the soil. This complex mix of biotic and abiotic elements together determines plant growth, productivity and distribution. As these environmental factors can be limiting factors in plant growth, it is important to understand their overall impact on forest productivity.

To manage plantations to optimise growth and benefits derived from them without reducing the productive capacity of the site, the forest manager needs to manipulate stands based on an understanding of basic physiological processes involved in tree growth. Also, the manager must know how the environment will affect the growth and development of trees they wish to manage. It should be noted, however, that if trees grow faster as a result of one of the factors, it could have negative implications for the other factors. For example, if trees grow faster because of an increase in temperature, trees will use the stocks of nutrients in the soil more quickly, and may eventually deplete them. Lack of nutrients will then become a limiting factor in tree growth in this case. The main factors discussed below are: light, temperature, water and nutrition.

Light

Light regulates growth and development of a tree through some incompletely understood reactions. Light directly affects tree growth by its quantity (intensity), quality, and duration. Light quantity refers to the intensity or concentration of sunlight and varies with the season of the year (OSU, undated). In most tropical countries, the annual variation in light quantity is less compared to temperate climates. In general the more sunlight a plant receives (up to a point), the better capacity it has for photosynthesis (OSU, undated). As the sunlight quantity decreases the photosynthetic process decreases as well.

Light quality refers to the colour or wavelength reaching the plant surface. Red and blue light have the greatest effect on plant growth (OSU, undated).. While the forester has no control over the quality of light, he can manipulate the intensity of light by cutting and thus can control species dominance, differentiation into crown classes, live crown ratio, and overall crown dimensions (OSU, undated).

The amount of time that a plant is exposed to sunlight is referred to as light duration or photoperiod. As sunlight is essential to photosynthesis, in general, therefore, the longer the duration of light, the longer photosynthesis can take place. The duration of light is also known to influence the flowering of plants in different ways.

Temperature

Temperature directly affects the day-to-day physiological processes of plants and indirectly influences their seasonal or cyclic development (OSU, undated).. Irrespective of where a plant is grown, it is subject to daily and periodic changes in temperature. The daily temperature change is referred to as the thermo period (OSU, undated).. The effect of temperature on plant growth depends on the type of plant and whether the plants are adapted to the temperatures or not. When temperatures are high, they lead to increased respiration sometimes above the rate of photosynthesis. Increased respiration implies that the products of photosynthesis are being used more rapidly than they are being produced which will impact growth because growth occurs only when photosynthesis is greater than respiration.

On the other hand, low temperatures can also result in poor growth because of the slowing down of photosynthesis. If photosynthesis is slowed, then the growth of trees is slowed, and this results in lower stand productivity. Given the different tolerance levels for the temperature of various plants, plants are usually classified as either hardy or non-hardy depending on their ability to withstand cold temperatures. Extremes of temperature can lead to stress in plants.

The effect of temperature on tree growth and productivity has to be evaluated with caution, depending on whether the trees are in the tropics, boreal or temperate regions of the world. For example, Way and Oren (2010) found that increased temperature generally increases tree growth, except for tropical trees. They suggest that this probably occurs because temperate and boreal trees currently grow below their temperature optimum, while tropical trees are at theirs already. In the tropics, increased temperature would more likely lead to increased heat stress than to increased productivity.

Moisture (water)

Adequate water supply to trees is one of the basic conditions needed for survival and optimisation of tree growth and productivity. Water (moisture) is available to trees primarily through the soil, and so the moisture-holding properties of the soil are of major importance to tree growth. The source of water in the soil and hence the water supply to plants comes from atmospheric precipitation, which is mainly rain in the case of Ghana, but can be snow in countries where it snows.

Water plays several important roles in plant growth and productivity. First, it is a primary component of photosynthesis in plants. Secondly, it is the main vehicle by which nutrients are transported throughout the tree and maintains the turgor pressure or firmness of plant tissues. Third, transpiration in trees is regulated through the opening and closing of the stomata, which is the result of turgor pressure and other changes in the cell. Tree roots can move through the soil due to the pressure in them provided by water. One of the most critical roles of water in tree growth and productivity is to act as a solvent for minerals moving into the plant and for carbohydrates that are transported through the tree for use or storage. Finally, the temperature of trees is regulated by the gradual evaporation of water from the surface of the leaves, through the stomata.

As a result of their deep rooting systems, trees can survive droughts better than annual food crops. However, depending on the duration and severity of the drought, it can have major impacts on tree health and survival by effectively slowing and reducing growth and in extreme cases lead to death of portions of the tree or the entire tree. A more common effect of drought on trees is the increase in susceptibility to insects and diseases (Purcell 2012). A water-deprived tree is unable to produce its usual levels of carbohydrates, significantly lowering its energy reserves, which are needed for a tree to produce chemicals that protect it from pathogens (Purcell 2012).

Trees can also be stressed by too much water, i.e., flooding, which is the other extreme end of a drought. Many tree species can survive long periods of flooding as long as their canopies remain above the water level. However, when canopies are completely submerged, the trees can die in a matter of months. As with drought, flooding causes stress in trees and reduces their ability to withstand insect and disease attacks (University of Arizona 1998).

Nutrition

Trees require adequate nutrition to survive, grow and produce. Plant nutrition refers to the steps by which a plant assimilates nutrients and uses them for growth and replacement of tissue. Nutrition is in contrast to fertility, which refers to the inherent capacity of a soil to supply nutrients to plants in adequate amounts and suitable proportions (University of Arizona 1998). In general, plants need about 18 elements for normal growth. Some of these, carbon, hydrogen, and oxygen are found in air and water. The following are found in the soil: nitrogen, phosphorus, potassium, magnesium, calcium, and sulphur. As discussed later in this Chapter, the latter six are called macronutrients because they are used in relatively large amounts by the plant. The remaining nine nutrients are referred to as the micronutrients, as they are used in much smaller quantities.

To optimise plant growth, adequate water and oxygen must be available in the soil. The role of water in plant growth was discussed above. The main role of oxygen is for soil respiration to occur to produce energy for growth and the movement of mineral ions into the root cells across their membranes (University of Arizona 1998). Oxygen from the soil is essential to producing energy for nutrient absorption.

4.1.4. Potential Effects of Climate Change on Forest Growth and Productivity

There is increasing scientific evidence, and consensus that increased emissions of greenhouse gases into the atmosphere are influencing the world's climate, with a potential to cause a range of impacts on forests. Climate change refers to a statistically significant change in climate characteristics over a long period of time (such as from one century to another), whilst *climate variability* means variations (ups and downs) in climatic conditions on time scales of months, years, decades, centuries, and millennia. Climate change refers to the long-term change in average weather conditions at the global level. Within any given day, month or year, however, variations in the weather occur at regional and local levels. Thus, climate variability can take the form of natural hazards, such as floods, drought, cold- and heat waves, cyclones and storms.

The main effects of climate change on forests are related to changes that will occur in some of the main environmental factors affecting tree growth and productivity discussed above. These include:

Temperature rise: Climate change can lead to a rise in temperature, particularly at the extremes. Higher temperatures are also expected to increase evaporation and reduce relative humidity. A report by the Inter-governmental Panel on Climate Change (IPCC 2007) suggests that global surface temperatures for the period 1995-2006 were the warmest years since 1850, giving rise to lengthy droughts and episodes of intense rainfall. Historical climate data observed by the Ghana Meteorological Agency across the country over a forty-year period between 1960 and 2000 showed a progressive and discernible rise in temperature and a concomitant decrease in rainfall in all agro-ecological zones of the country (Kuuzegh 2009). Based on the forty-year observed data, projections of future climate scenarios indicate that temperature in Ghana will continue to rise on average of about 0.6°C, 2.0°C and 3.9°C by the year 2020, 2050 and 2080, respectively, in all agro-ecological zones in Ghana (Kuuzegh 2009). Increases in temperature will only increase tree growth and productivity in places where low temperature is a limiting factor otherwise temperature rises may actually reduce productivity as a result of potential drought conditions and reduced relative humidity.

Changes in rainfall: Changes in total rainfall and seasonal rainfall distribution as a result of climate change will be shown as changes in soil water availability. Rainfall is predicted to decrease on average by 2.8%, 10.9% and 18.6% for 2020, 2050 and 2080, respectively, in all agro-ecological zones of Ghana (Kuuzegh 2009). This decrease in rainfall will lead to a decrease in water availability for trees and hence will negatively impact on plantation growth and productivity.

Extreme weather: The incidence of extreme weather events is also expected to change, with increased frequency and intensity of heavy rainfall events, drought and extreme fire days. Already, growing incidence of floods and dry spells in the northern savannah zone of Ghana is exposing natural resource-based livelihoods to the threats of degradation and ultimately increasing the incidence of food insecurity and malnutrition (Yaro et al. 2010). As discussed above, floods and droughts have the potential to reduce growth and in extreme cases kill trees. Increased incidence of fire and storms will increase chances of forest destruction and hence reduced productivity.

Increased atmospheric concentrations of CO_2: Climate change is expected to lead to changes in atmospheric carbon dioxide concentration. Based on international research carried out over the past 20 years, a certain amount is understood about the direct effects of increasing CO_2 concentrations on plant growth and function. Doubling of CO_2 concentration generally leads to increases in plant growth of 10%–25% (Nowak et al. 2004; Luo et al. 2005b; Norby et al. 2005). Most studies suggest carbon fertilisation will increase tree growth though this growth will be limited by competition, disturbance, and nutrient limitations (Shugart et al. 2003).

Change in distribution of trees and growing season: In the longer term, climate change will lead to changes in the distribution of trees. For example, trees that are currently adapted to high temperatures in the savannah areas of Ghana may now grow well in southern Ghana as a result of high temperatures down south. Where tree growth is not limited by other factors (e.g., water), climate warming is likely to expand the growing season depending on the location.

There is a high degree of interaction between these factors and responses to climate change will vary depending on the particulars of the local site and climate (Battaglia et al. 2009). Without appropriate human interventions, it is possible that the effects of climate change - compounded by more direct sources of anthropogenic stress - will prove devastating to the world's forests (Broadhead et al. 2009). In general, management to help forests adapt to climate change will involve maintaining forest health and ecosystem diversity and resilience. There is a particular need to develop responsive management systems and to improve ecosystem resilience. Other steps are needed as well (Broadhead et al. 2009):

- forest monitoring to quickly detect and tackle outbreaks of pests and diseases;
- effective fire management;

- restoration of forest functions after disturbances;
- reduced impact logging;
- increases in the number of locations where specific habitats are managed; and
- efforts to connect habitats and landscapes

4.2. Site Quality

4.2.1. The Concepts of Site, Site Quality and Site Productivity

According to Skovsgaard and Vanclay (2007) the term *site* refers to a geographic location that is considered homogeneous in terms of its physical and biological environment; and in forestry, site is usually defined by the location's potential to sustain tree growth, often with a view to site-specific silviculture. Sites are therefore classified into *site types* according to their similarity regarding climate, topography, soils and vegetation. *Site quality* refers to the combination of physical and biological factors characterising a particular geographic location or site, and may involve a descriptive classification (Skovsgaard and Vanclay 2007). The properties that determine site quality are generally inherent to the site, but may be influenced by management. *Site productivity* in forestry is usually evaluated in terms of how much wood biomass (or any other desired material) the site can produce. However, the history of the stand such as when the stand originated, management regimes, pests and diseases, weather etc., all affect the productivity of a site (Davis et al. 2001).

4.2.2. Methods of Site Productivity Assessment

Clutter et al. (1983) describe both direct and indirect methods of site quality evaluation. Estimation of site quality based on historical yield records, stand volume data and stand height data constitute the direct methods. The indirect methods include the use of overstory interspecies relationship, lesser vegetation characteristics and topographic, climatic and edaphic factors using a) site index, b) non-tree-vegetation, and c) basic environmental and land attributes (Jones 1969). Indirect methods are used because of the difficulty of measuring the productivity of a site growing at its maximum potential through volume production. Non-tree vegetation typing as an indicator of site quality has been used extensively in Canada and Scandinavia. This method is based on the knowledge that in certain ecosystems, certain plants or plant communities are associated with certain forest types and to some extent, with site quality (Davis et al. 2001). The method of using physical environmental factors as measures of site quality evolved from the difficulties of correlating understory vegetation with site quality (Coilie 1938). The most widely used land attribute is to link soil type with site productivity, as soil has a large and often controlling influence on tree growth, especially as it relates to moisture and nutrients (Davis et al. 2001).

In the remaining sections, we will focus on the use of site index as a measure of site quality, because, of all the indirect methods of site quality that have been investigated, the use of tree height growth appears to be the most practical, consistent and useful indicator of site quality (Davis et al. 2001). Consequently, site quality estimation from stand height data is the most popular method and usually involves the use of site index curves. Any set of site index curves is simply a family of site development patterns with qualitative symbols or numbers associated with the curves for referencing purposes (Clutter et al. 1983). Site index curves are derived either graphically or by statistical curve-fitting procedures (Alder, 1980; Clutter et al. 1983). Data for the development of site index curves are obtained from three main sources: temporary sample plots (TSPs); permanent sample plots (PSPs); or stem analysis. TSPs provide the most inexpensive and quickest data, but the use of such data involves the assumption that the full range of site indices is well represented in all age classes within the sample (Alder, 1980; Clutter et al. 1983; Avery and Burkhart 1994).

Although site index is still the most widely used index of site productivity, it is important to point out the inherent assumptions in this method and concerns that have been raised by previous researchers. The main assumption in using site index is that height growth correlates well with stand volume growth (Skovsgaard and Vanclay 2007) and is least affected by stand density. Some empirical studies by Assmann (1955, 1959) have shown that this assumption does not always hold (e.g., repressed stands), which has led to the development of the concept of *yield level*, which is the stand volume growth per unit of height growth. To apply this concept, a stand density index based on the combination of stem number and quadratic mean diameter is used to provide an indication of the yield level, which may be used to adjust height-age – based estimates of site productivity (Skovsgaard and Vanclay 2007).

4.2.3. Developing Site Index Curves

The statistical methods used for site index curves are of three general types: the guide curve, the difference equation and the parameter prediction methods (Clutter et al. 1983). The most common equation forms used to fit site index curves are the Schumacher (1939) and Chapman-Richards equations (Richards, 1959; Chapman 1961).

To develop site index curves using the proportional curves method described by Alder (1980), a single equation is fitted to the plot level top height[1]/age data, using the logarithmic transformation of Schumacher's (1939) equation given as:

$$\ln(H_{top}) = \ln(H_{top0}) + \beta A^{-k} \tag{1}$$

where H_{top} is the mean top height, H_{top0} is the maximum top height the species could reach on the site (the asymptotic top height), A is the age of the stand, β is a regression coefficient and k is a constant. Using nonlinear regressions, we can estimate the value for k iteratively. Nanang and Nunifu (1999) have found that the value of k for teak and neem plantations in Northern Ghana was approximately 0.5. The site index (SI) for each plantation is determined by the algebraic re-arrangement of Equation (1) as follows:

1. A reference or base age of for the species is fixed. In this example, a base age of 20 years is used.
2. SI of each stand is then defined as the top height of the stand at the base age (20).

[1] The average height of the largest 100 trees per hectare in terms of DBH.In some cases height (100 tallest trees) are used, but this is usually referred to as dominant height.

3. By substituting H_{top} = SI and A = 20 into Equation (1), eliminating $\ln(H_{top0})$ and re-arranging terms, the expression for SI is obtained as follows:

$$\ln(SI) = \ln(H_{top}) + \beta\left[\left(\frac{1}{A}\right)^{0.5} - \left(\frac{1}{20}\right)^{0.5}\right] \quad (2)$$

Equation (2) can then be used to calculate the site index of each stand.

4.2.4. Choosing a Functional Form to Fit Site Index Curves

Foresters have been interested in the specification and estimation of site index curves for a long time now. The proportional or guide curve method of site index curve development described above has become the most popular method for even-aged single-species plantations. The reason has been the lack of re-measured permanent plot databases and the inability of most species to show annual or seasonal growth rings for height growth to be estimated. To address urgent needs for these curves for management, the common practice has been to collect data from temporary plots from stands of different ages to cover, at least, a major part of the rotation of the species. The basic requirement is to assume that the distribution of plots with respect to site quality is identical across all ages (Alder 1980). Failure of this assumption (when certain site/age combinations are unavailable) will bias the fitted guide curve away from the true average height growth trend (Walters et al. 1989). Other factors that can lead to biases in the site index curves when temporary sample plot data are used and alternative explanations of these biasing effects are discussed by Beck and Trousdell (1973), Smith (1984) and Walters et al. (1989). However, biases that arise due to the use of an inappropriate functional form were only addressed in a comprehensive way by Nanang and Nunifu (1999). The main conclusion is that the mathematical properties of the models provide some guidance for the selection of the most appropriate functional form using both the data-related criteria and the characteristics of the models themselves (see Nanang and Nunifu, 1999 for details).

4.3. Methods of Improving Site Productivity

4.3.1 Essential Plant Nutrients

There are 18 essential chemical elements for plant growth. The main plant nutrients obtained from the soil and required by all plants are: nitrogen (N), phosphorus (P), potassium (K), calcium (Ca), magnesium (Mg), sulphur (S), iron (Fe), manganese (Mn), zinc (Zn), copper (Cu), boron (B), molybdenum (Mo), cobalt (Co), nickel (Ni) and chlorine (Cl). Other elements are essential only for some plants and have positive impacts on plant growth, such as sodium, silicon, nickel, cobalt, selenium and iodine. All of the essential nutrients are equally important for healthy plant growth. However, these nutrients are generally divided into macronutrients and micronutrients depending on the amounts required (see Section 4.1.3).

Mineral nutrients in the soil are the main source of plant nutrition. These nutrients come from decomposition of plant residues, animal remains, and soil microorganisms; weathering of soil minerals; fertiliser applications; manures, composts, biosolids (sewage sludge), kelp (seaweed), and other organic amendments such as food processing by-products; N-fixation by legumes; ground rock products including lime, rock phosphate, and greensand; inorganic industrial by-products such as wood ash or coal ash; atmospheric deposition, such as N and S from acid rain or N-fixation by lightning discharges and deposition of nutrient-rich sediment from erosion and flooding.

These mineral nutrients are available to plants as long as they can be accessed by plant roots for uptake. However, these nutrients can be lost from the soil system through surface runoff; erosion; leaching; gaseous losses; and biomass removal through harvesting forest products. If these nutrient losses reach water bodies, they can become a source of contamination.

4.3.2. Impacts of Successive Harvesting of Plantations on Nutrient Reserves

Forest plantations of fast-growing species managed in successive crops accumulate large amounts of nutrients in short periods of time. This can lead to a drain on the nutrient reserves of the soils on which these plantations are grown and threaten the long-term sustainability of the plantation enterprise. Most soils on which trees are planted in Ghana are marginally fertile at best and cannot sustain commercial forest plantation productivity for successive rotations unless fertiliser application and/or other sources of nutrients are used to maintain productivity.

In forest plantations, nutrient removal through the harvested tree components represents a considerable drain on the nutrient capital of the ecosystem, since a large proportion of the nutrients are in the biomass (Gonçalves and Barros 1999). Losses may also occur indirectly through processes such as leaching, erosion, and volatilisation. This loss is enhanced by a faster rate of organic matter decomposition at the exposed sites.

Slash burning and intensive soil preparation for planting are also potential factors that influence nutrient loss. The amount of nutrients removed by harvesting depends on the amount of harvested tree components, species, stand development, and site quality (Gonçalves and Barros 1999). Even if only the stem is harvested, the amount of nutrients removed is very high, especially if the forest is managed in short rotations (Gonçalves and Barros 1999).

To maintain sustainable yields of wood and non-wood products, the land manager needs to employ strategies that ensure the continuous supply of nutrients to (or fertility of) the soil. Some of these are discussed in the sections that follow.

4.3.3. Planting Mixed Species

Planting trees of different species in a single plantation has many benefits including increasing biodiversity, providing multiple forest products to communities, reduced risk of all trees dying due to a forest fire and stress or diseases. Planting mixed species also increases the likelihood that at least one species will be well suited to the site conditions and be very productive, even if the other species are not doing well, especially with a changing climate. It provides a form of insurance against a complete failure of the plantation.

In addition to the above benefits, planting mixed species has the potential to increase plantation productivity, either through increased nutrient content or exploiting the different productive capacities of the different species planted on the site. Trees of different species and rooting patterns take nutrients from different soil horizons and distances from the tree. Consequently, planting such species mixtures reduces the intra-specific competition for water and soil nutrients from the soil, compared to a single species. The different mix of species may also result in different growth rates, crown closure times and reduction in the competition for sunlight (above ground competition). Also, leaves of different species may decay at different rates to recycle nutrients at different times of the year.

A key concept for designing highly productive mixed-species stands is the need to combine species that differ in characteristics such as shade tolerance, height growth rate, crown structure (particularly leaf area density), foliar phenology (particularly deciduous versus evergreen habit), and root depth and phenology (Kelty 1992). If the species differ substantially in these characteristics, they may capture site resources more effectively and/or use the resources more efficiently in producing biomass, resulting in greater total stand biomass production than would occur in monocultures of the component species (Kelty 2006).

Several studies have tested the question of whether biomass productivity in mixed plantations with complementary characteristics exceeds that in monocultures of the component species. Early studies involved plantations established in the late 19th century in Germany and Switzerland (Assmann 1970). Other studies of mixed plantations in the Czech Republic (Poleno 1981) and Russia (Prokopev 1976) also showed that mixed stands had greater yields than monocultures of either species.

Two species with similar growth characteristics have interspecific competition in a mixture that is equal to that of intraspecific competition in a monoculture. Species with complementary characteristics have interspecific competition that is much lower than intraspecific. For that reason, the phenomenon of higher production from this type of interaction is also called the competitive production principle (Vandermeer 1989).

The proportions of the species in mixtures can have important effects on stand development. When two species are combined in a fine-grained mixture in a 1:1 proportion, the upper canopy species may suppress the lower canopy species to the extent that the stand will shift toward the production levels of a monoculture of the upper canopy (Kelty 2006). It may be necessary with many species combinations to reduce the proportion of the taller species to achieve the highest production (Kelty 2006).

Despite these advantages, it should be noted that mixed species are more difficult to plant and manage compared to a single species mainly because the process of gathering information on several species and their interactions may be more difficult; a large processing plant may not have an adequate supply of timber from mixed stands; and certain tree species may compete too strongly with each other, or with crops, to be suited to mixtures (FAO 1995).

4.3.4. Minimum Soil Cultivation

Site preparation seeks to improve growing conditions for trees and not impair sustainability or productivity (Evans 2000). Longer-term benefits of site preparation include a reduction in bulk density, increased infiltration capacity and aeration, improvement in moisture storage and enhanced mineralisation rates of accumulated organic matter (Evans 2000). However, the type and intensity of soil preparation used before planting trees have a major influence on the soil and water properties. For example, a very significant amount of nutrients present in a forest plantation is found in the crown (leaves and branches), bark and litter (main components of the harvest residues) and therefore if burning is used to prepare the land, then a large amount of nutrients that were present on the site are lost. Though the ash left on the ground may lead to a temporary increase in nutrient availability to planted crops (Gonçalves and Barros 1999).

During site preparation, harvesting residues can be burnt or mixed with surface soil (e.g., conventional ploughing) or retained on the soil surface with restricted soil disturbance (e.g., 'minimum cultivation' in lines or spots) (Gonçalves et al. 2008). Minimum soil cultivation is a technique that is used to reduce nutrient loss from a harvested plantation by avoiding burning and avoid the detrimental effects of intensive soil preparation especially in areas to be replanted (Disperati et al. 1995). The method that has been applied in Brazil consists of not burning the slash of the previous harvest and laying it in rows without removing the top soil. Seedlings are then planted in hand or machine-made pits or in furrows (about 20 cm wide and 40 cm deep) opened by a one-tine ripper (Disperati et al. 1995; Gonçalves and Barros 1999).

Many managers have accepted minimum cultivation as a good alternative to maintaining or increasing soil fertility in the long term as it reduces nutrient and organic matter losses and preserves crucial soil physical properties (Gonçalves et al. 2008).

4.3.5. Slash from Harvesting Operations

Slash from harvesting operations forms an important source of organic materials that can be added back into the soil to improve soil nutrient availability. The amount of slash available depends on the harvesting methods used. Whole-tree harvesting methods tend to take away more slash from the site than tree-length methods. It is important to point out that continuous removal of all harvesting slash from the site will lead to rapid declines in soil fertility unless measures are taken to increase soil fertility in other ways.

Deleporte et al. (2008) studied the effects of slash and litter management practices on soil fertility and eucalypt growth over a full rotation in the Congo. The experiment demonstrated the great positive influence of (1) debarking stems in-field and spreading the bark; (2) retaining stem tops on the site; (3) avoiding slash burning; and, (4) reducing the delay between stand harvesting and crop planting.

Goncalves, et al. (2008) showed that silvicultural operations such as soil preparation, logging residue management and application of fertilisers can influence soil fertility, hence nutrient uptake and tree growth in Brazil. The experiment assessed the effects of complete harvest residue removal, residue retention and residue burning. Highest productivities were obtained where residues were retained or burned and the lowest where all residues (slash, litter and bark) were removed. Results highlight the temporary, but large, nutrient release due to burning and the effect of forest residues on tree growth.

4.3.6. Fertilisation

Fertilisers may be needed for most soils that are not able to supply all the nutrients required for suitable growth of planted trees. Fertilisation is an expensive option for rural communities that may not be able to afford the cost. Even for commercial plantations, the use of fertiliser increases the cost of production and has implications for the profitability of the plantation enterprise. Despite these costs concerns, if fertilisers have to be used, then the type of fertiliser has to depend on the nutrients lacking in the particular environment. The timing of fertiliser application also affects its effectiveness. It is recommended that fertilisers be applied at planting and also just before canopy closure. This strategy reduces nutrient loss by volatilisation, leaching, immobilisation and erosion (Gonçalves and Barros 1999).

As Evans (2000) observes, regular application of mineral fertiliser is not presently a feature of plantation forestry and most forest use of fertiliser is to correct known deficiencies in micronutrients such as boron in much of tropics, and macronutrients such as phosphorus on impoverished sites in many parts of both the tropical and temperate regions of the world.

4.3.7. Manure and Other Organic Materials

Many investigations have shown that treatment of organic matter both over the rotation and during felling and replanting is as critical to sustainability as coping with the weed environment (Evans 2000). It is important, therefore, to use strategies that maintain organic matter in the soil during the rotation, such as weeding, pruning and thinning practices. Also, at harvesting, methods that deplete the site of organic matter should be avoided. It may be possible in some cases to transport and deposit on the site, organic materials from outside of the site, such as woody materials.

4.3.8. Industrial and Urban Residues

In addition to mineral fertilisers and organic materials, several other types of industrial residues have been used and/or tested in eucalyptus and pine plantations. These include *iron slag* resulting from iron processing, *dregs and grits* residues resulting from the " Kraft" process used in the pulp and paper industry, *organic sludge,* which is an effluent produced by the pulp and paper industry, *forest biomass ash* obtained from the burning of wood and bark for the production of thermal energy by steam generation and o*rganic compost* from municipal garbage (Gonçalves and Barros 1999).

4.3.9. Weed Control

Most tree species planted in forest plantations are sensitive to or impacted by weed competition. A reduction in plant survival and growth may result from competition for water and nutrients because weeds use up larger volumes of soil than the young tree seedlings (Nambiar 1990). However, it has been observed that complete weed removal may not be necessary in every case. Lowery et al. (1993) compiled part of the information on weed control in tropical forest plantations, and reported that although in most cases, complete weeding resulted in better tree growth and survival, partial weeding in strips within the tree rows could represent a good compromise between tree soil resource use and nutrient conservation on the site. It is very important to define how long tree seedlings and weeds can grow together before the survival and the growth of the trees are reduced (Gonçalves and Barros 1999).

The main purpose of weed control is to reduce or eliminate competition from vegetation growing within forest plantations. Competition occurs in various forms, both below ground and above ground. By eliminating competing vegetation, the survival and productivity of the site are enhanced. Weeds that are removed and incorporated into the soil also add organic matter to it and hence increase nutrient availability and productivity of the site.

4.4. Conclusion

Understanding forest growth and dynamics and the factors that contribute to productive plantations is fundamental to allowing forest managers manipulate planted trees for optimum productivity. The complex environments within which trees grow implies that the environmental factors that affect tree growth must be considered in any management decisions. Though forest plantations can be grown successfully in many parts of Ghana, conscious efforts must be made to maintain site productivity using some of the techniques described in this Chapter.

Chapter 5

FOREST PLANTATION MEASUREMENTS

5.1. INTRODUCTION

Fundamentally, the subject of forest management has been mostly about measurements. The measurements that are taken to collect tree and forest stand information are referred to as *mensuration*. Forest mensuration covers all activities and techniques used for obtaining accurate measurements from trees and forests. Mensuration helps to answer questions such as how many trees to plant per unit area, at what spacing to plant, when and how much to thin or prune the trees, when and how much of the forest to harvest at a given time, etc.

Measurements in forestry could be distance or length, angles, areas, positions of objects in relation to others, rating of tree growth, the condition of trees or the number of trees per unit area. Measurements can be taken of the geographic and topographic variables that describe the plantation or the forest stand or on the site and soil characteristics of the forest stand. But most measurements are taken on the individual trees themselves and can be used to describe the production of wood products, the structure and diversity of the forest, or the condition of the forest.

The overall goal of measurements in forestry is to provide a detailed understanding of the forest (cost permitted) so that managers can make well-informed management decisions. This Chapter will focus on measuring individual trees and forests. Other categories of measurements will be described, but only to the extent that they help our understanding of the significance of measurements to forest management.

5.2. IMPORTANCE OF MEASUREMENTS IN FOREST RESOURCE MANAGEMENT

Forest resources management deals with the planning, administration, and management of a wide variety of forest resources to provide social, environmental and economic benefits to society. These benefits include wood, wildlife, special forest products, water, and recreation opportunities and may vary depending on the forest type. In a plantation forestry setting, wood products may be the primary or sole benefit of forest resource management although recreation and watershed maintenance benefits may also be derived.

The planning component of plantation forest resource management may require data to make planning decisions such as what species to plant, how to plant and on what site. The data needed to make such decisions may include the growth rates of the species and the soil properties of the site, all of which will require some measurements. Throughout the growth of the plantation and until final harvest, several intermediate activities like thinning may have to be made which require the measurement of some characteristics of the plantation such as basal area and average tree size. The final decision on when to harvest may be made based on the growth characteristics of the species on the site type on which the species is planted.

In multiple plantations settings, where the objective is to sustainably supply raw materials such as sawlogs or pulpwood, rotation age determination and harvest planning are important concepts that require enormous amount of data on the growth characteristics of the forest tree species and the site productivity. At the national level, policies to regulate the extraction and the use of the forest resource also require good quality data to develop, implement and evaluate those policies and measurements provide data for all aspects of forest resources management.

5.3. Measuring a Single Tree

Measurements in forestry are typically carried out on individual trees. Therefore, measurement techniques, tools and methods have been developed and refined over time, to accurately measure individual tree dimensions such as: age, diameter, height, weight, crown size, etc., to provide data and critical information required for management decisions. In the sections that follow, we describe the procedures and technology used to measure individual tree variables.

5.3.1. Age

Without additional information, determining the age of a tree will be considerably difficult.

The easiest way is to determine the age from the date of planting of the tree (from seed) if it is known. However if the tree is planted from a seedling, the time taken to raise the seedling in the nursery may have to be taken into consideration, particularly in regions where it takes multiple seasons to raise the seedlings. In tropical regions where seedlings are raised and planted within the same year, no adjustments are needed.

The second way to estimate tree age is to count the annual growth rings of a tree in temperate regions of the world. But this is possible only in trees that show annual growth rings. In most tropical trees, this method is impossible to implement particularly in the evergreen trees. Annual rings can be counted using two different methods. Tree growth rings can be counted from an increment core taken from the tree using an increment borer.

In most cases, however, the increment borer is either not available or may cause injury to the tree. If trees are dead and have been cut down, age can be determined by counting the rings on the stump.

Another method, though less accurate is to divide the circumference of the tree by 2.5 cm. This method is based on the assumption that trees grow by an average of one inch (2.5 cm) a year. In this case, you will measure the circumference of the tree up to 1.5 m from the ground and divide the circumference of the tree at this height by 2.5 cm.

5.3.2. Diameter

Tree stem diameter is one of the most important and the easiest measurements that can be taken on a tree. Tree stem diameter is often used as an indicator of the wood product(s) that can be obtained from the tree as larger trees produce larger logs from which larger volumes of timber can be produced. Another important reason for measuring tree diameter is that, it often correlates very well with other tree characteristics that are more difficult to measure, like height or the wood volume in the stem of a tree or the biomass (usually the dry weight) of the tree. In most tropical countries where plantations of exotic species are raised for such products as transmission poles, tree diameter may reflect the monetary worth of the tree, given that larger trees produce larger and/or longer transmission poles. Tree size, measured by tree diameter (and sometimes height) may reflect the competitiveness of a tree in the stand and, hence, how well it is likely to grow in relation to the other trees. Diameter is commonly used as the basis for selecting the timing for thinning and what tree to thin out when the objective is to remove smaller and less competitive trees in favour of the more competitive ones.

Stem diameter is not uniform but declines progressively from the base of the stem to the top. This phenomenon is called stem taper. The difficulty introduced by stem taper is the question of at what point along the stem the diameter should be measured. The consensus is to measure tree diameter at breast height and although there are local variations, the generally acceptable point for measuring tree diameter is 1.3 m vertical height above ground from the base of the tree. This is known in the forestry field as diameter at breast height (DBH). If the tree is growing on a sloping ground, breast height is measured from the highest ground level at the base of the tree. The base of the tree should be cleared of any loose litter and debris before making the measurement at breast height.

For younger trees less than 1.3 m tall, the point of measurement of the diameter is 0.1 or 0.3 metres above ground level, and the diameter is referred to as root collar diameter. Typically, root collar diameter is used in regeneration performance surveys to assess seedling survival and performance. However, the need to make root collar diameter measurement is increasing in forestry, as some products are now being harvested from very young forests. For example, plantation forests are being grown for only 3–5 years to produce wood for bioenergy uses.

The tools commonly used for measuring tree diameter include diameter or girth tapes and callipers. Typically made from fiberglass, a diameter or girth tape (also called logger's tape) is usually marked in both normal measuring units for the direct measurements of tree circumference and marks that are direct conversions of circumference to diameter (i.e., the ratio of tree circumference to π=3.142). To measure tree diameter in the forest, we start by determining and marking the point of measurement (1.3 m above ground) on the tree. The practical way of doing this in a ground inventory setting is to use a stick (wood, metal or plastic) slightly longer than 1.3m (preferably 1.5m) and place a permanent mark at 1.3 m from one end. The stick is then placed against the stem of the tree to be measured on the side of the tree where the ground level is the highest. If the diameter or girth tape is used, the tape is wrapped around the stem of the tree at the exact point of measurement as horizontally as possible and as tight as possible before the measurement is read. It is important to use a stick that is only *slightly* longer than 1.3 m and not much longer than that to avoid the error of wrongly using the short end of the stick and measuring diameter below breast height.

The diameter measurement can also be taken using callipers by clumping the calliper on the tree stem at breast height, just tight enough to hold the callipers in place without denting the stem and taking the measurement. Since tree stems are seldom perfectly circular in cross-section, calliper measurement is best taken twice at right angles to one another, and the average of the two measurements taken as the diameter of the tree.

The above discussion assumed that a tree has a single stem with normal stem form at breast height. In some cases, a tree may have a fork below breast height or misshape of the tree trunk such as abnormal swelling or decay at breast height. Under these circumstances, it may not be possible to measure accurately tree diameter at breast height, and some adjustments have to be made. The general practice in the case of a fork below breast height is to:

i) Measure the diameter of each stem (d_i for i =1,2..,n),
ii) Calculate the cross-sectional areas using the diameter measurements (i.e., $A_i = \frac{\pi \times d_i^2}{4}$),
iii) Sum the cross-sectional areas to obtain total cross-sectional area ($A = A_1 + A_2 ......A_i$),
iv) Calculate the diameter *d* of the summed cross-sectional area, A (i.e., $d = \sqrt{\frac{4 \times A}{\pi}}$),
v) Record *d* as the average diameter of the stem.

There are several methods of dealing with swelling at breast height. The most common approach is to take the diameter measurement either immediately below or above the swell as the DBH measurement. Another approach involves taking both the diameter measurements above and below the swell and the averaging the two measurements. No single method is universally favoured, and the choice is often left to the practitioner. Regardless of which method is used, it is important to ensure a consistent application of the chosen method for the entire inventory exercise.

5.3.3. Height

Another important parameter in forest measurements is tree height. Tree height is usually measured from the ground level to the tip of the terminal bud. Tree height measurement has practical and ecological significance. Tree height is the second most important variable for predicting tree stem volume. The ratio of tree height to diameter (usually called slenderness ratio or coefficient) is used as an indicator for the presence of such ecological phenomena as competition and relative crowding in the stand, presence of wind sway or the social rank of the tree in the stand. A high slenderness ratio, for instance, indicates the tree is experiencing relatively more competition or is in a very crowded stand. Also, younger trees tend to have higher slenderness coefficients.

For practical purposes, however, the main use of tree height is to estimate the quantity (volume) of wood in the stem. Due to the high correlation between tree height with stem wood volume and the fact that tree height growth is least affected by competition, tree height is also used in the concept of site index as a measure of site productivity.

Tree height measured to the tip of the terminal bud is called total height. Tree height may also be measured from ground to the base of the crown, i.e., to the first crown-forming branch. This height is known as the bole height or height-to-live crown. The height-to-live crown is usually a reflection of the vigour of the tree; a tree with short height-to-live crown typically carries more foliage than a tree of comparable height with a longer height-to-live crown. And more foliage typically implies higher primary productivity and hence vigorous growth. Merchantable tree height is a more subjective measure determined by the type of products expected to be harvested from the tree. Merchantable height is the distance from the point where the tree can be felled (stump height – usually 0.3 m above ground) to a point on the upper part of the stem with specified minimum stem diameter. For sawlogs where the emphasis is on bole size, the top diameter to which merchantable height may be measured may be bigger (up to 25 cm or more) while pulpwood may require a smaller top diameter.

There are several ways of measuring tree height in forestry. In younger forest stands where trees are relatively shorter (10 m or less) a height measuring pole may be used to measure directly tree height. Trees in older stands may be much taller than what the height measuring pole can reach. For such tall trees, heights are typically estimated based on trigonometric relationships and most instruments that are used for measuring tree heights are calibrated using these relationships. The basic principle is that an observer, standing at a chosen distance *L*, measures two angles relative to the horizontal line of sight; one to the tip of the terminal bud of the main branch of the tree (angle of elevation A) and the other angle to the bottom of the tree (angle of depression B). Then using the trigonometric relationship among the distance (*L*), the two angles (A and B) and the height of the tree (H) an estimate of the tree height is obtained. This is illustrated in Figure 1 below.

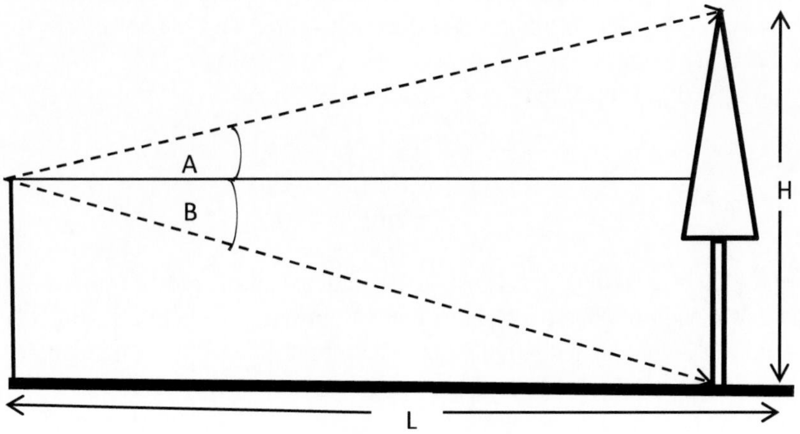

Figure 1. Illustration of the process of measuring tree height.

From Figure 1, it can be shown that the tree height H is given by (West 2009);

$$H = L[\tan(A) + \tan(-B)] \quad (1)$$

Under certain circumstances such as the presence of a relatively steep slope, the horizontal line of sight may actually be below the base of the tree and the angle B will have to be measured above the horizontal line of sight if the observer's position is down slope. In such a case, -B in Equation (1) is replaced by +B. Common tree height measuring instruments that use the principle described above are clinometer and the hypsometer. Tree heights can also be measured using other instruments such as Haga altimeter, Abney level, and Spiegel relaskope.

Some precautions and/or adjustments have to be made under the following situations:

i) The distance from the tree to the observer must be a true horizontal distance;
ii) In the forest where trees are closer together, it may be difficult to see the top of the tree to measure. Where possible physically shaking the tree may help identify the target tree. Or the observer may have to adjust the distance or change the direction in order to get a good view of the tree top; and
iii) If the tree has a lean, the height measurement will have to be adjusted using the angle of the lean (from the true vertical position).

5.3.4. Tree Basal Area

Tree basal area is the cross-sectional area in m^2 (measured over the bark) at breast height (1.3 m above the ground). Tree basal area is strongly correlated with and can be used to estimate tree volumes. At the stand or plantation level, tree basal area is a measure of stand competition. Basal area is not measured directly but derived from the DBH. Basal area is calculated using an equation that is based on the assumption that the cross-section of the tree is a circle. Therefore, basal area is calculated using the formula for calculating the area of a circle (i.e.,$BA = \pi r^2$ where r = radius of the tree and π= 3.142). First the measure of tree DBH is converted to the radius in metres and then substituted into the basal area formula to determine the basal area in m^2.

In many forest mensuration texts, tree basal area is given as:

$$BA = \frac{\pi \times D^2}{40000} \quad (2)$$

The factor 40000 is used to convert DBH to the radius and the radius from centimetres to metres. Given that π and 40000 are constants, there are instances where Equation (2) may be written as $BA = k \times D^2$, where $k = \pi \div 40000$. For example, if the DBH of a tree measured from a teak plantation is 40 cm, the basal area (in m^2) is given as:

$$BA = \frac{3.142 \times 40^2}{40000} = 0.126\, m^2$$

5.3.5. Tree Form and Taper

In general tree form is used to describe the rate at which the stem diameter decreases from the bottom (stump) to the top. A related term, called taper is the decrease in diameter of a stem of a tree or a log from the base upwards. Therefore, the form is the rate of taper. Both taper and form are important variables that affect the volume of a tree stem.

Tree form is expressed in two ways: as form factors or form quotients. A form factor is the ratio of the volume of the tree bole and the volume of a cylinder with the same diameter and height. Two types of form factors are commonly recognised in the forestry literature: the absolute form factor and the artificial form factor. The term absolute form factor is used if the tree diameter at the stump (0.3 m above ground) is used to calculate the form factor while artificial form factor is used if diameter at breast height is used (Philip 1994). In some literature, a form factor determined using DBH is referred to as the false form factor. The Hohenadl's form factor is less popular and uses diameter at 1/10 of the tree height.

Tree form quotient is an important variable in determining tree volume next to tree height and DBH. Form quotient is an expression of the ratio of an upper stem diameter to the diameter at breast height. Two common form quotients have been used in forestry literature; the normal form quotient (Schiffel 1899) uses diameter at one-half the total height of the tree, while absolute form quotient (Johnson 1912) uses diameter at one-half the height above breast height; that is half way between breast height and the tip of the tree. Form quotient decreases with height and is always greater than or equal to 0. The most accurate way of measuring form quotient is to use diameter inside bark since tree bark thickness varies along the stem.

Tree taper is usually represented by a mathematical equation, which expresses tree diameter along the stem as a function of the height above ground from which the diameter is measured. These mathematical equations are called tree taper functions and are typically developed for individual species or group of species preferably with similar growth characteristics. Taper equations are constructed by destructive sampling of trees. A tree is cut down and a series of diameters (usually inside the tree bark) are measured along the stem from the base to the top and recorded together with the heights from the base of the tree at which the diameter measurements are taken. Prior to cutting each tree, the height and DBH are measured and recorded. There are quite a number of ways by which the diameters are measured along the stem. While some researchers prefer to measure the diameter at fixed intervals along the stem (e.g., 0.5m or 1 m), others prefer to measure a fixed number of diameters along the stem regardless of the tree height. It is argued that the first option of taking measurements at fixed intervals will lead to severely unbalanced records in cases where tree heights vary greatly. In the second approach, a fixed number of diameters to be measured are decided on and for each tree the total height (measured after the tree is felled) is divided by this fixed number to decide the intervals at which diameter measurements will be taken. The result is that, the number of observations is the same for each tree while shorter trees will have shorter measurement intervals and vice versa. Regardless of which method is used, the DBH is always included since practical application of taper functions relies mainly on tree DBH and height to predict the diameter of the stem at any given height.

The heights at which the diameters are taken are expressed as proportions of the total tree height. Taper equations are then fitted by expressing the diameters as functions of the calculated proportions of heights at which diameter measurements are taken. As an example, the tree DBH will correspond to (1.3/H), where H is the total height of the tree in metres. Taper equations have developed in complexity over the years as their use became popular. An example of a taper equation is the Ormerod's function given as (Ormerod 1973):

$$d(h) = dbh\left(\frac{H-h}{H-1.3}\right)^{b} \tag{3}$$

5.3.6. Crown Parameters

Tree crown parameters are important metrics in forest measurements. A tree crown carries the foliage that is responsible for primary production and tree growth and hence is an excellent indicator of tree vigour. Trees with larger crowns typically have higher growth rates. Response to silvicultural treatments such as thinning is sometimes measured by the changes in tree crown parameters. In fact, tree crown size is used as an alternative to photosynthesis for predicting tree growth.

The most popular crown metrics are crown width, crown length and live crown ratio. In some advanced growth and yield modelling equations, crown volume and crown surface area may also be used (e.g., Mitchell 1969, 1975). Crown width is the measure of the average width of the area projected by the tree crown on a horizontal plane. The tree crown is assumed to be approximately circular, but this is hardly the case. Therefore, tree crown width is typically measured in two directions and averaged. It is always preferable to measure the minimum and the maximum widths and obtain the average.

Crown length is the vertical distance from the tip of the tree to the base of the live crown. The opposite measure is the height-to-live-crown described in Section 5.3.3. Technically, the sum of tree crown length and the height-to-live crown should equal the total tree height. For this reason, tree crown length is not usually measured in the field but calculated in the office by subtracting the height-to-live crown from the total tree height. The ratio of crown length to total tree height is the live crown ratio, which can also be used to measure tree vigour.

In a forest stand, tree crowns are classified into four major classes. Although the decision as to which crown class a tree belongs to is largely subjective, the classification is important in deciding if a particular tree is stressed, particular in thinning decisions. The four classes are the dominant, codominant, intermediate and suppressed. The dominant crown extends above the general crown level and receives light from every side. Codominants are usually in the general crown cover level; receive full light from the top and only partially from the side. The intermediates are shorter than the dominants and codominants, penetrate the crown cover but only receive direct light from above. The suppressed trees have crowns entirely below the general crown cover and do not receive direct light from above.

5.3.7. Standing Tree Volume

The most popular method of obtaining the volume of a tree is by felling the tree and measuring the bole diameter and length. However, if the desire is to obtain the volume without cutting the tree, this method is not suitable. In such a situation, tree volume may be obtained by using a volume table. The height and DBH measurements of the tree can be plugged into a standard or local volume table, and volume of the tree read directly. Volume equations that relate tree volume to the diameter and/or height are commonly used for tree species that already have these equations developed.

In recent times, a technology that uses camera images of the tree stem taken in several directions have been developed for measuring tree volume (e.g., Hapca et al. 2007). With a digital camera, the tree is photographed from at least two directions. A three-dimensional view of the stem is then produced by a specialised computer programme. The volume of the tree stem can then be determined, taking into account the fine detail of any irregularities along it (West 2009). A drawback of this technology is the difficulty in obtaining a clear, unobstructed image of the stem. It is also not possible to measure tree volume without the bark without using appropriate volume conversion formulae.

5.3.8. Estimating Log Volume

There are different ways of estimating log volume. The suitability of any of these methods will often depend on the desired level of accuracy and the field conditions. One method involves the complete immersion of the log in water of a known volume and measuring the volume of water displaced as the log volume. This method is known as xylometry. The method requires large immersion tanks, which may be difficult to transport for use in practical field conditions and may be suitable for laboratory conditions only for verifying the accuracy of other methods. There are various examples of the use of xylometry in research projects (e.g., Martin, 1984; Filho and Schaaf 1999).

In recent times, laser technology is increasingly being used in sawmills in advanced countries. The technology basically uses a laser to scan the log in several directions as the log enters the sawing line. With the scan, the diameter of the log is measured at very short intervals along its whole length. This information is then used by a specialised computer programme to calculate a precise, three-dimensional profile of the log, from which a very accurate volume of the log is determined. This technology is also used to detect and remove defects and also optimise the timber products that can be produced from the log. The technology is not practical for field applications.

The most common approach involves measuring a tree stem in short sections, determining the volume of each section and summing them to give the total volume (West 2009). For each log section, the diameters at the lower end (D_1), the middle (D_2) and the top end (D_3) are measured and the volume V of the section is calculated by any of the following formulae:

$$\text{The Smalian's formula; } V = \frac{\pi L(D_1^2 + D_3^2)}{8} \tag{4}$$

$$\text{The Huber's formula; } V = \frac{\pi L D_2^2}{4} \quad (5)$$

$$\text{The Newton's Formula; } V = \frac{\pi L(D_1^2 + 4D_2^2 + D_3^2)}{24} \quad (6)$$

The variable L in Equations (4) to (6) represents the length (m) of the section. Note that if the Smalian's formula is used, the middle diameter is not required while the Huber's formula requires only the middle diameter.

5.4. Measuring a Stand of Trees

Although most measurements are taken at the individual tree level, forest managers and owners are interested in stand-level variables such as volume and basal area per hectare to make decisions. However, it is not possible to measure every tree in the forest estate or even the stand. Sampling is often used to select some trees or groups (plots) of trees whose measurements are scaled up to obtain stand level variables. Some stand level variables are needed to facilitate the scale-up of tree measurements for stand-level measures.

A forest stand is defined as a relatively homogeneous patch of forest. Homogeneity may be related to such variables as species composition, age, average tree size (e.g., height) or average stem density. Depending on the management objective, some combinations of these variables may be used to define a stand. In plantation forestry, age and species composition are commonly used to define stands.

An important variable necessary for scaling up tree level information such as basal area and volume is the total area of the stand. If the stand is on sloping ground, the area is taken to be the equivalent horizontal ground area.

5.4.1. Stand Age

The age of the trees in a stand is usually relevant in determining the age of the stand when the stand is even-aged; that is, either all the trees in the stand regenerated naturally (in natural forests) or all the trees were planted (in a plantation) at or about the same time. In practical terms, however, a natural stand classified as even-aged, may take several months or even years for all the trees to regenerate and establish the stand. Sometimes even if the best of conditions necessary for regeneration exist, it may still take some time for all the available seeds to germinate. Thus, the concept of even-aged stands in natural forests is a relative term. Ages of such types of stands are usually pegged with the year the stand was treated (i.e., tilled and seeded). In cases where the treatment records are not available, the common practice is to age a sample of trees in the stand and obtain the average age as the stand age.

In the case of plantations, stand age is usually obtained from planting records. However, tree seedlings may grow for a few months in the nursery, sometimes up to a year or even more, before they are planted out. Aging plantations using the date of planting seems to provide adequate results for practical purposes, even though trees may be older than the age assigned to the plantation.

5.4.2. Species Composition

Species composition is a measure that is only relevant in a mixed species situation and is determined for each single species in the stand. Species composition may also be expressed in terms of species groups such as conifer species versus deciduous species. Although species composition is used to express the relative proportions of the individual species or groups of species in the stand, it is sometimes also used to measure the relative dominance of a species in the stand.

By definition, species composition is the relative proportion of a species or specific group of species in a mixed species stand. Both stem density and stand basal area are used to calculate species composition, but the most preferred variable is the basal area. Thus, species composition of a species A in a forest stand, denoted by SC_A is calculated from basal area on per hectare basis of the species A, denoted by BA_A and the total basal area of the stand, denoted by TBA as:

$$SC_A = \frac{BA_A}{TBA} \tag{7}$$

If the desire is to use stem density, the species basal area (BA_A) and the total stand basal area (TBA) in Equation (7) can be replaced by the corresponding measures of stem density, i.e., the stem density of species A and the total stem density of all species in the stand respectively.

5.4.3. Density and Stocking

Density and stocking are concepts that are sometimes mixed-up in the forestry literature. The two concepts, even though strongly correlated, are quite different. In simple terms, stem density measures the number of tree stems per unit area (usually hectare). Stem density is also sometimes referred to as stocking density. Stocking, on the other hand, is an expression of site occupancy of the trees in the stand. Stocking is commonly expressed as the percentage of the site occupied by trees. Thus, two stands of the same density may have totally different stocking percentages if the trees in one of the stands are evenly spaced while the trees in the other stand are clumped in only a fraction of the stand. The evenly spaced stand is expected to have higher percentage site occupancy (stocking). In a well-spaced plantation forest where planting survival is near 100%, stocking is expected to be very high. Mortality tends to reduce stocking in such situations especially if the mortality pattern is not random.

Stem density is estimated by counting the number of trees in a fixed area plot and dividing that number by the plot area in hectares. For example, if the total number of trees in a fixed area of 0.04 ha is 20, then the density of the forest stand from which the measurement was taken is 500 trees per hectare. The most common way of measuring stocking is to divide the stand into a fixed area grid (10 m^2 is common) cells. Stocking is then determined by expressing the total number of grid cells occupied by trees as a percentage of the total number of grid units. Stocking may also be assessed using high-resolution, large-scale aerial photographs. The interpreter visually delineates areas in the stand occupied by trees and expresses that area as a percentage of the total stand area. The grid system is suitable in very young stands while the photo interpretation methods are suitable for matured stands.

5.4.4. Regeneration Surveys

Regeneration surveys are special inventory surveys where the objective is to assess the presence, amount and the condition of a new crop of trees being established either in a recently harvested area or under a growing mature tree crop. Although these types of surveys are typically relevant in natural stand conditions where natural regeneration is the main method of stand establishment, in large plantation forestry, these surveys may be used to assess planting success and to identify areas of failures and the need for supplemental planting.

Regeneration surveys are very important for identifying areas of failures and taking corrective measures such as "beating ups" while the stands are still young to avoid stocking problems when the stand grows older. Depending on the objective, sampling or complete enumeration may be used. Even where sampling is used, and the results do indicate serious problems, the manager may order a 100% enumeration to identify problem areas that need attention. Stand establishment is an important step in forest management, and the need to get things right at this early stage is very important for the success of the plantation.

In addition to assessing seedling density, regeneration surveys also assess the stocking and the performance or competition the seedlings are experiencing. These variables are indicators of the chances of survival of the seedlings.

Typical data collected in regeneration surveys include seedling count per unit area, seedling stocking, the number of seedling per hectare taller than root-collar (30 cm) and competition from other vegetation. The most common approach is to use the grid system whereby the area is overlaid with a grid system of pre-determined spacing. The grid spacing is usually decided based on such factors as the amount of resources available, the area to be surveyed, the level of accuracy desired, and the management information needs. At each grid intersection, a regeneration survey plot (usually circular in nature) is established, and all the necessary information needed is collected. For a plantation forest, the grid system may be superimposed on the planting spacing grid.

5.4.5. Stand Basal Area

Stand basal area is an aggregate measure of individual tree basal areas, usually on per unit area basis. Stand basal area is an important stand parameter that usually reflects the level of crowding in the stand and hence can be used as an index of competition among the trees in the stand. Stand basal area, like tree diameter, also correlates very well with volume and is often used as a predictor of stand volume per hectare.

Stand basal area tends to increase as the trees age, grow and increase in size (DBH). This often results in increased competition among the trees. If not controlled, some of the trees end up dying as the stand basal area approaches a theoretical maximum. The forest manager typically monitors the stand basal area and will usually thin out some of the stems to keep the basal area below a management target value.

Stand basal area may be measured using fixed area sample plots or variable radius (point sampling) plots. These two methods are discussed in Chapter 6.

5.4.6. Stand Volume Estimation

According to Spurr (1952), few subjects in the entire field of forestry have received as much attention as the estimation of tree volume. The large number of approaches to the problem of volume estimation may be taken as an indication that no one approach has received more than partial recognition. To assess the volume of a complex and highly variable geometric solid like a tree in terms of very few measurements and by simple algebraic techniques is by its very nature difficult if not impossible (Spurr 1952).

There are three general approaches to the estimation of stand volume in use. In practically all inventory and management work, volume tables are used. To use a volume table for stand volume estimation, samples of trees are measured, but only from one to three measurements are taken on each tree (Spurr 1952). If only one measurement is taken, it is usually the DBH; if two, height is also measured; if three measurements; a measure of form is also included. From these few variables, the volume is estimated by reference to volume tables[1]. Volume tables give the mean regression of volume on DBH, height and form (in form-class tables only) for a series of carefully measured sample trees.

The second general approach is called the mean tree approach. The underlying theory of this method is that, if the tree of mean volume can be isolated, then the volume obtained by careful measurement of this tree can be multiplied by the number of trees in the stand to give the volume of the stand (Spurr 1952). One method of determining the tree of mean volume is to use the mean sample tree, which is based upon the assumption that the tree of mean basal area is also the tree of mean volume (Crow 1971). Although fairly good results have been obtained by this method, the fallacy of the basic assumption has long been recognised (Spurr 1952). Crow (1971) agreed that the mean tree method can be used for expedient estimates of tree biomass in plantations or uniform, natural stands or even-aged species. Especially promising are techniques based on individuals at one standard deviation from a mean stand dimension.

Another variation of the mean tree approach is first to estimate the volume of each sub-sample tree (mean tree) using standard formulae such as the Smalian's formula, which is popular because of its simplicity. Stand volume per ha can then be estimated from these volumes in three ways:

(a) A volume equation can be developed using the individual sub-sample tree DBH, height and calculated volume:

[1] It is assumed that volume tables for the species in that geographical area are already in existence otherwise this method cannot be applied.

$$V = \alpha + \beta D + \gamma D^2 + \lambda D^2 H \tag{8}$$

Where D is the DBH, H the total tree height, V is the calculated tree volume, α, β, γ and λ are regression coefficients. Equation (8) can be estimated by ordinary weighted least squares (OWLS) (Cunia 1964). The fitted volume equation is then used to predict the total volume of each tree in each sample plot. The individual tree volumes within each plot are then aggregated to estimate average stand level volume (in m^3/ha).

(b) When two-stage simple random sampling is used to obtain data from plantations, the stand volume can be estimated using the two-stage simple random sampling formula (Cochran 1977).

$$V = \frac{N}{n}\sum_{i-1}^{n}\frac{M_i}{m_i}\sum_{j=1}^{m_i} V_{ij} \tag{9}$$

where V is the total volume per ha, n the number of sample plots per plantation, N is the number of plots that could potentially be sampled in 1 ha obtained as $1/a$, where a is the individual plot area expressed in ha, M_i the number of trees per sample plot, m_i the number of sub-sample trees per plot, V_{ij} the volume of the jth tree in the ith plot.

(c) If however, a two-stage sampling with probability proportional to basal area at the second stage (PPG) of sampling is used then the formula for computing stand volume would be:

$$V = \frac{N}{n}\left(\sum_{i-1}^{n}\left(\sum_{j=1}^{M_i} BA_{ij}\frac{1}{m_i}\sum_{j=1}^{m_i}\frac{V_{ij}}{BA_{ij}}\right)\right) \tag{10}$$

where, V_{ij}, M_i, m_i, Y, N and n are as defined before and BA_{ij} is the basal area of the jth tree from the ith sample plot.

5.4.7. Fuelwood Volume Estimation

Fuelwood is wood that is used as fuel. Fuelwood may be in the form of firewood, charcoal, wood chips, pellets, or even sawdust. There are different sources of fuelwood. Community forestry programmes in most developing countries are designed at least in part to provide fuelwood for domestic and industrial needs. There are also individual woodlot owners and tree farmers who grow trees purposively for producing and selling fuelwood. Fuelwood may also be obtained from residue left after a timber harvest, in the form of branches, twigs and offcuts or even main stems that are unsuitable for use as timber or pulpwood.

Given the diverse sources of fuelwood, standardised methods of quantifying fuelwood have been widely researched, and various methods have been proposed. Some of these methods are based on the calorific value of the woody species. Other methods use dry weight or biomass while other methods use the volume to quantify the amount of fuelwood. For the purposes of this book, we will focus on the mensurational techniques of quantifying fuelwood in terms of volume.

In earlier sections of this Chapter, we discussed various methods of estimating the stem volume of a single tree and volumes for a stand of trees. These volume estimates did not include the volumes of the branches and twigs of the trees. However, for fuelwood, we need to extend the volume calculations to include as much of the volume of the whole tree as possible. The most challenging aspect of quantifying fuelwood volume is the irregular nature of tree branches – tree branches are often not straight and vary quite a lot in size and length, requiring several measurements before total fuelwood volume can be calculated. The labour intensive nature of this exercise requires that some form of sampling or indirect estimation techniques be applied to reduce cost.

Estimating Fuelwood Volume of Standing Trees

There are two common methods: one requires some form of sampling to estimate the volume of a few branches, which can then be extrapolated to obtain the branch wood volume of the entire tree and then the whole stand. A popular sampling method that is used extensively in forestry mostly for estimating foliage biomass but is also suitable for branch wood volume estimation is the randomised branch sampling (e.g., Valentine et al. 1984; Valentine and Hilton 1977; Jessen 1955; Marshall 1955). The second method involves measuring the total branch wood volume of some randomly sampled trees and then using tree variables such as DBH, total height, height-to-live crown to construct branch wood volume tables.

Regardless of whether all the branches are measured or a sample of branches is to be measured, accuracy is very important. The recommended tools include measuring tape, callipers of different sizes, permanent markers, tally card or digital data recording devices specially programmed for recording such data. A minimum of three people is needed to measure efficiently and tally individual branches or segments. Individual branches and segments must be tallied separately, and their individual volumes calculated using the Smalian's, the Huber's or the Newton's formula discussed in previous sections of this Chapter. For quality control purposes, it is recommended that each branch segment be labelled with the number corresponding to its entry on the tally card.

Estimating Fuelwood Volume of a Stack of Wood

In most applications, fuelwood volume is required for wood that has already been cut, hauled and stacked at home or for sale. While the dimensions of the stacks (or head loads) can be measured, and their volumes determined, determining the actual solid fuelwood can be challenging. A simple procedure is to use an appropriate measuring instrument such as a measuring tape to measure the dimensions of the stack and calculate the volume according to the geometrical shape formed by the stack. The volume measurement needs to account for voids that are included in the stack.. The values of the conversion factors of stacked wood in solid content depend mainly on the occurrence of form, curvatures, forking, knots, burls or swellings, care in stacking, and length of billets and their diameters.

As the value of fuel depends on the solid content, it is essential to know the methods of determining the solid contents of stacked wood. Adjustment factors are used to convert the volume of the stack to the actual solid wood volume in the stack. While universal conversion factors have been used, it is recommended that local conversion factors be developed. For example, in Kenya, the conversion factor for a headload of firewood was estimated based on xylometry (water displacement method) to be 0.38 (FAO 1983). The implication is that for a stack of firewood that is measured to be 2 m^3, the actual wood volume will be 0.76 m^3.

One way of estimating the conversion factor accurately for a particular geographic region is to measure the volume of individual billets in the stack. Secondly, pile the billets into a stack, each time, measuring the volume of the stack. The known volume of the wood is then expressed in each case as a fraction or proportion of the volume of the stack. The average ratio is then calculated and used as the conversion factor. Mathematically, the conversion factor is given as: $F_c = \frac{V_A}{V_S}$ where V_A is the actual volume measured by xylometry or other method and V_S is the volume of the stack of fuelwood.

5.5. Yield Estimation

5.5.1. Definition and Data Acquisition

Yield refers to the total amount of material available for harvest at a given time (Avery and Burkhart 1994). Mathematically, growth can be calculated as the first derivative of the yield equation while yield is calculated as the integral of the annual increments [sum of growth] (Clutter 1963). A yield table is usually a table showing the volume/ha at different ages for even-aged stands of trees growing on forest land of different productive capacities (Chapman and Meyer 1949). It is one of the oldest approaches to yield estimation, and modern yield tables often include additional information such as number of trees/ha, basal area/ha, stand height, diameter, and current and mean annual increments. A normal yield table shows the yields capable of being produced on forest sites when the "normal" capacity of the site is fully utilised by even-aged stands of forest-grown trees. Empirical [or variable-density] yield tables usually refer to yield tables that apply to "average" rather than full, or normal stocking (Avery and Burkhart 1994).

The best method of obtaining data for yield table construction is to measure permanent plots located in the stands of interest at intervals of 5 or 10 years over the entire period of growth of the stands (Chapman and Meyer 1949). The actual increases in the various yield variables and the number of trees that die from year to year and the volume of the surviving stand at any age are then a matter of record. However, this entire process would take a long time to complete, and the stand might be destroyed at any time by fire, disease or other agencies (Chapman and Meyer 1949). Yield tables constructed by this method are referred to as real growth series yield tables (Turnbull 1963).

The second type of yield tables, known as "abstract" growth series yield tables, are constructed by measuring a number of permanent sample plots (PSPs) located in stands of different ages and to re-measure these plots at 5- or 10-year intervals (Chapman and Meyer, 1949; Turnbull 1963). By the overlapping of the chosen ages, the trend of the development for one or two decades for plots of all ages is obtained after the second or third measurement. It is, however, essential that lands for PSPs be in stable ownership, preferably in a public forest, so that the owner's whim may not interfere with the experiment (Chapman and Meyer 1949).

Chapman and Meyer (1949) contend that the two methods described above are too time-consuming, and other methods are needed if yield tables are to be prepared for immediate use. To fill this need, a third standard method that has been extensively used is based on the principle of comparison of plots of different ages. The averages of stands for the same site but differing in age are combined into a curve assumed to show the trend of growth. If many plots of different ages on different sites are measured, each plot will show the yield obtained in a similarly stocked stand at the given age and for the site in which it is located.

5.5.2. Types of Growth and Yield Models

Growth and yield models are classified into three types: whole stand models, diameter class models, and individual tree models. Whole stand models usually include stand variables in the model such as age or basal area per ha, while diameter class models include average tree within each diameter class (Davis et al. 2001). Individual tree models use each tree in a sample or stand to estimate yield (Davis et al. 2001). Whole-stand models can further be divided into density-free and variable-density models. The basic Schumacher-type model given by $Q = f(A, S, D)$ is an example of a variable-density yield model, where Q is some measure of yield (height, DBH, basal area, volume, and fresh weight), A is stand age in years, S is some function of site index, D is some function of stand density. If however, stand density is fixed, then $Q = f(A, S)$ would be a density-free whole stand model.

Diameter class models often simulate the growth in each diameter class by calculating the characteristics, volume, and growth of the average tree in each class and multiplying this average tree by the inventoried number of trees in each diameter class (Davis et al. 2001).

The computed volumes in each diameter class are summed over all diameter classes in the stand to obtain the volume of the stand. Davis et al. (2001) draw a distinction between diameter distribution models and diameter class models. The former is considered a whole stand model because the number of trees in each diameter class is a function of the stand variables and all the growth functions are for stand variables. Diameter class models, on the other hand, have the number of trees in each class empirically determined and independently model the diameter classes subject to some aggregate stand influences (Davis et al. 2001).

The individual tree models are more complex than the whole stand or diameter class models. Most individual tree models calculate a crown competition index for each tree as a measure of how well the tree can compete for light and nutrients and growing space relative to other trees in the stand. This index is used to project whether the tree lives or dies, and if it survives, the index predicts its growth characteristics (Davis et al. 2001). The total yield of the stand is then the aggregate of the individual tree characteristics.

5.5.3. Volume Tables

One of the most important yield variables of interest to foresters is the amount of woody material available in a plantation, i.e., volume. For species that have volume tables, the volume is often read off the tables for a given site class using measurements of DBH and /or height of the plantation in question. Credit for the first modern volume table is generally given to Heinrich Cotta, who published one for beech in 1804 and in 1817, developed a set of standard volume tables. Three types of volume tables are recognised: local, standard and form-class volume tables (Husch et al. 1982). Local volume tables give tree volume in terms of DBH only. The term local is used because such tables are generally restricted to the local area for which the height-diameter relationship hidden in the table is relevant (Husch et al. 1982). Local volume tables are usually derived from standard volume tables though they can be prepared from raw data - that is from volume and diameter measurements for a sample of trees. Standard volume tables give volume in terms of DBH and merchantable or total height. Tables of this type may be prepared for individual species, or groups of species, and specific localities (Husch et al. 1982). Form-class volume tables give volumes in terms of DBH, merchantable or total height, and some measure of form, such as Girard form class or absolute form quotient (Husch et al. 1982).

5.5.4. Volume Table Construction

The problem of constructing volume tables is a statistical one. The construction of volume tables involves directly relating volume to diameter and height using graphs, alignment charts or equations (Spurr 1952). The possibility exist, however, for relating tree diameter and height to an indirect measure of volume such as a form factor or taper, and then constructing the volume table as a separate step from the form-factor table or taper curves (Spurr 1952). Historically, three methods that have been used for volume table construction: graphic, alignment-chart methods and mathematical modelling based on least-squares techniques.

Of the three general approaches to volume table construction, the graphic method is the oldest and requires less mathematical skill than the least-squares or alignment chart techniques (Spurr 1952). An advantage of the graphic technique is that the curves are fitted to the actual data rather than being forced to conform to any set pattern (Spurr 1952). Spurr (1952) however, notes that the method is not only subjective, but considerable experience is necessary especially for a small sample, and a large number of tree measurements are needed to provide good trends for each of the diameter and height classes.

Alignment charts provide an efficient means of expressing an equation (Spurr 1952). They were introduced as tools to correct for curvilinearity in multiple regression equations by Bruce and Reineke (1931).

Least-Squares Techniques

The graphic and alignment-chart techniques have been generally discarded in favour of mathematical functions and models. Almost all volume tables are now constructed using standard regression techniques based on the method of least squares. These techniques have gained wide acceptance for the construction of volume tables (Unnikrihnan and Singh 1984). This is because the method is free from the subjective bias of fitting curves by hand (Spurr 1952). A large number of different equations have been proposed for volume table construction, and considerable difficulty may arise in attempting to decide which equation is the most appropriate for a particular set of data (Furnival 1961). Considerable difference of opinion also persists regarding not only the function to be used but the proper criterion of comparison (Spurr 1952).

Classical least squares estimation is based on the assumptions of independently and normally distributed errors and the property of homoscedasticity, i.e., the variance of the dependent variable is constant for all values of the independent variables. The fourth assumption is that the sample is a simple random sample (Cunia 1964). In practice, however, most, if not all of these assumptions are not satisfied. Cunia (1964) found that tree volume for a given DBH is not normally distributed and highly skewed: large trees tend to deviate more on the average from the regression surface than do small trees (i.e., heteroscedasticity). It is also usual in forest inventory to replace the simple random sampling by stratified, cluster or systematic sampling (Bruce and Schumacher, 1950; Spurr, 1952; Cochran 1953).

Theoretically, weights should be employed that are inversely proportional to the variance of the residuals to achieve a homogeneous variance. But in practice, it may be difficult to determine the most appropriate way to weight a particular function (Furnival 1961). Another option is the use of logarithmic transformations of the equations. When logarithmic transformation is used to ensure that the assumptions of ordinary least squares are satisfied, it is often necessary to express estimated values of the variable of interest in original measured [i.e., untransformed] units (Baskerville 1972). However, the conversion of the logarithmic estimates back to measured units produces a bias that must be corrected. This results from the fact that if the distribution of the residuals about the log-transformed model is normal, then the distribution of the untransformed residuals is skewed (Furnival 1961; Baskerville 1972). It is argued that the transformation from the logarithmic form back to measured units by simply determining the antilogarithm has, by failing to account for skewness of the distribution in arithmetic units, yields the median rather than the mean value of the estimates (Finney 1941; Brownlee 1967; Baskerville 1972). This produces systematic underestimates of the dependent variable. Therefore, a correction must be made for this inherent bias that is proportional to the amount of variation associated with the regression (Schlaegel 1981).

The following correction for the skewness that results from this retransformation was proposed by Baskerville (1972).

$$\hat{Y} = e^{(\mu + \sigma^2/2)} \tag{11}$$

$$\sigma_A^2 = e^{(2\mu + 2\sigma^2/2)} - e^{(2\mu + \sigma^2/2)} \tag{12}$$

Where:

$\hat{Y}$ = estimated mean in measured units

σ_A^2 = estimated variance in measured units

μ = estimated mean in logarithmic units

σ = sample variance of the logarithmic equation

Beauchamp and Olson (1973) extended Baskerville's work and concluded that unless the variance of the residuals is quite large the correction provided by Baskerville (1972) gives close approximations to the unbiased value.

5.5.5. Model Selection Criteria

When modelling volume of forest stands, several options exist in terms of choice of functional forms and explanatory variables to be included in the regression equations. A natural question that arises is how to choose the most appropriate model among the potential models. Some of the criteria that have been suggested for selecting appropriate biomass and volume models are reviewed below. It is difficult however to find a single criterion or statistic to demonstrate that one model is definitely better than another (Schlaegel 1981).

Coefficient of Determination

Probably the most commonly used model selection criterion is the coefficient of determination, R^2 value. This statistic indicates the proportion of the total sum of squares of the dependent variable explained by the regression. One major disadvantage of this statistic is that it can be used to compare two or more models only if the units of the dependent variable are the same between models. Secondly, inclusion of additional independent variables never decreases the R^2 value, even though they may not be statistically significant (Schlaegel 1981). Almost all statistical packages today allow for the user to provide the R^2 as part of the output of the model estimation process.

Standard Error

According to Schlaegel (1981), the use of the standard error of the estimate ($S_{y.x}$) as a selection criterion for models is only second to the coefficient of determination. The standard error is calculated from the residual sum of squares of the regression of any equation by:

$$s_{y.x} = \sqrt{\frac{sum\, of\ squares\, residual}{n-p}} \tag{13}$$

Where:

n = the number of observations
p = the number of coefficients estimated in the model
$(n - p)$ = degrees of freedom of the residual term

The standard error is a measure of the variation in the observed dependent variable values not accounted for by the linear relationship with the independent variable(s) (Husch 1963). The standard error is a function of the number of coefficients estimated from the model since the denominator is dependent on sample size and number of regression coefficients in the model. Transforming the dependent variable changes the magnitude of the standard error for the same equation making it inappropriate for comparing equations in different units or with different dependent variables. As Schlaegel (1981) noted, the standard error is difficult to interpret without additional information such as the distribution of the data, the mean and range of the dependent variable.

Fit Index

An index of fit (Fit Index) comparable to the R^2 value can be obtained for every equation. In the case of untransformed linear regressions, the Fit Index is equal to the coefficient of determination (Schlaegel 1981). To calculate the FI, the predicted values are transformed back to the original units and corrected for bias if needed. The total corrected sum of squares is given by:

$$TSS = \sum (Y_i - \bar{Y})^2 \tag{14}$$

and residual sum of squares by:

$$RSS = \sum (Y_i - \hat{Y}_i)^2 \tag{15}$$

Where

Y_i = value of the ith observation in actual units

$\bar{Y}$ = arithmetic mean of Y, in actual units

$\hat{Y}_i$ = the ith predicted value of Y_i converted to actual units

The fit index is:

$$FI = 1 - \frac{RSS}{TSS} \tag{16}$$

Standard Error of Estimate in Actual Units

Using the residual sum of squares of the regression in the actual units of measure, a standard error of the estimate in actual units, SE may be calculated as follows:

$$s_e = \sqrt{\frac{\sum Y_i - \hat{Y}_i)^2}{n - p}} \tag{17}$$

Where Y_i and $\hat{Y}_i$ are observed and predicted values of Y in actual units respectively, n is the number of observations, and p is the number of coefficients in the model (Schlaegel 1981).

Coefficient of Variation

A useful statistic for making quick comparisons between models is the coefficient of variation (CV) expressed in actual units as percent:

$$CV = \frac{s_e}{\bar{Y}} \mathrm{x} 100 \quad (18)$$

The CV is an index of the variation among means of the predicted Y's after accounting for the variation due to the measured variables. This statistic should be much smaller than the coefficient of variation of the sample tree mean.

The Furnival Index

The usual index of fit, the root mean square residual (standard error), and the coefficient of determination can only be used to compare equations that have the same dependent variable, but are not suitable when transformations of the dependent variables are involved (Furnival, 1961; Crow 1971).

Furnivall (1961), proposed an index for comparing equations used in constructing volume tables with different dependent variables, based on the maximum likelihood principle. The index, known as the Furnival Index (I) is computed in three stages. First, the standard error of the residuals is obtained by fitting all equations to the data. Next, the geometric means of the derivatives of the several dependent variables with respect to the volume are computed with the aid of logarithms. Finally, the standard error is multiplied by the inverse of the appropriate geometric mean (Furnival 1961). The Index is given by:

$$I = [f'(V)^{-1}]s_e \quad (19)$$

Where I is the Furnival Index, $f'(V)^{-1}$ is the reciprocal of the derivative of the transformation applied to the dependent variable [volume] with respect to volume, and s is the standard error of the fitted regression. For the common transformations, the corresponding geometric means are presented in Table 1 (Furnival 1961; Alder 1980).

Table 1. Examples of geometric means of common transformations used in volume table construction

Transformation	Geometric mean
Log V	$anti\log = \frac{2.30\sum \log V}{n}$
lnV	$anti\log = \frac{\sum \log V}{n}$
V/D^2	$anti\log = \frac{\sum \log D^2}{n}$
V/D^2H	$anti\log = \frac{\sum \log D^2H}{n}$

Where D is the DBH, H is height and n is the number of observations. The equation with the smallest index is selected as the one that gives the best fit to the data. In the case of an untransformed equation (i.e., where the dependent variable is the volume), the Furnival Index reduces to the usual estimate of the standard error of the regression.

5.6. CONCLUSION

Successful plantation management relies on the availability of information at various levels: individual tree, stand and landscape levels. This information is usually obtained through different kinds of measurements in a process known as mensuration. Tree diameter and height are by far the two most important tree variables measured and used in forestry. It should be pointed out that tree and stand measurements should be carried out as carefully as possible to avoid measurement errors. The methods described in this Chapter are used in conjunction with the sampling methods discussed in the next Chapter. Using the best possible measurement without the correct sampling technique will still result in bias and inaccurate measurements. Conversely, using the correct sampling method and measuring variables incorrectly also results in biases as well.

Chapter 6

FOREST RESOURCES INVENTORY

This chapter provides descriptions of forest resource inventory methods and procedures with special emphasis on plantation forestry applications. Forest resource inventory is one of the most important tools of forest management. Unfortunately, inventory information is limited for the majority of forest plantations in Ghana today. The development of inventory tools and techniques has been slow, relative to the rate at which the natural forests decline and the need for such tools increase. The chapter also includes brief introduction to methods and practices that are applicable beyond plantation forestry and helps in the understanding of the concepts of forest resources inventory at various scales such as the landscape and the forest stand levels. Methods described include remote sensing and satellite imagery, photogrammetry, ground level inventory, including sample selection and basic statistical methods.

6.1. INTRODUCTION

Forest resource managers need information about the resource they have, how much they have, where those resources are, and how they are distributed across the landscape in order to prepare management plans for sustained production and maximum economic benefits. Information is required on such attributes as the productivity of the growing stock, the rate of increment of trees, the species composition, the size-class distribution and the quality of the growing stock. Forest resource inventories are designed to provide this information at varying levels of detail and scale (landscape level, stand level or sample plot level), in an efficient and cost-effective manner.

The concept of systematically collecting forest information to assist forest managers in planning and decision making began in Europe in the late 18th Century using what was referred to as ocular estimation techniques (a subjective way of visually estimating tree characteristics) (Loetsch and Haller 1973). The details of the forest information (i.e., volume and stocking) produced from ocular estimation was sufficient for producing maps to assist in planning and management. This was soon followed by the development of mensurational techniques throughout the 19th Century as the need for more detailed forest information became necessary due to increased demand for other uses. Even then, the developments in collecting information about forest resources were based on experience, early statistical knowledge and simple mensurational relationships. These developments were appropriate for collecting data on small units of forests only.

In contrast, developments in the 20th Century were dominated by advances in information technology and the desire for information on relatively larger forest units (Brack 1997). Today, information on forests is required at even much bigger scales such as the landscape or regional or even national levels. By combining efficient statistical sampling techniques and theories with advanced remote sensing techniques, forest inventory practitioners are able to provide information on forests at all levels (i.e., at the tree, plot, stand or polygon, landscape and regional levels).

Forest inventories may be repeated (recurrent) at regular intervals for purposes of monitoring and detecting possible changes, and providing data for forecasting growth and yield. Inventories may also be conducted at a point in time to provide the amount, condition and distribution of the resource. Information collected in such inventories is usually limited. As an example, in a timber cruise, the focus is usually on timber characteristics. In the distant past when the emphasis was on timber products only, forest resource inventories collected only a limited amount of information, primarily focused on timber products. The focus of recurrent inventory was on the regeneration and growth of timber trees; all other information obtained in recurrent inventory was designed to aid the determination of timber production. Recurrent inventory may be done using permanent sample plots, temporary sample plots or a combination of both. Present-day inventories are more complex and are designed to collect information on non-timber products as well.

It should be noted at this point that, forest inventory processes utilise various combinations of technology and tools to obtain data for forest management decisions. The technology and tools are applied at various stages of the inventory process including planning, the actual field data collection on individual trees or stands of trees, quality control of the data collected, data storage and retrieval, data analysis and presentation. At the planning stage, statistical sampling techniques are used to decide where inventory data should be collected and from how many locations the inventory data is to be collected.

6.2. Basic Statistical Concepts

The intent of this section is to introduce the reader to some basic statistical concepts frequently used in forest mensuration and inventories. For more detailed discussion of these concepts, the reader may refer to textbooks on sampling techniques with direct forestry applications, such as Shivers and Borders (1997), Schreuder et al. (1993), Avery and Burkhart (2002) or statistics books such as Cochran (1977).

We begin by briefly defining and describing some basic statistical concepts. These concepts are central to understanding the application of forest mensuration techniques in the conduct of forest inventory.

6.2.1. The Concept of Population

In everyday terminology, the term population is used to refer to the collection of individual items that have well-defined characteristics. In statistics, the concept of a population is viewed in a slightly different perspective; a population is viewed as the collection of individuals that are the focus of a particular study or data collection for statistical decision-making. A group becomes a statistical population if information is required about a certain characteristic of the group and a study has to be conducted or data collected from this group by taking measurements of this characteristic such that some inferences can be made about the specific characteristic of the group.

In forestry, for instance, a manager may be interested in deciding whether a stand of trees can be harvested and may wish to know what proportion of the trees is above a certain minimum size. The stand of trees becomes a statistical population from which data on tree diameters can be collected to support this decision. It is sometimes helpful to think of a population as a collection of unit values, where the "unit" (or element) is the thing on which the data (observations) are collected and the "value" is the property (data) observed on the unit (Freese 1962). In this example, the units are the trees and the values are the diameter measurements. In practice, however, population units in forestry are seldom individual trees but plots (fixed-area or variable area).

In statistics, populations are commonly classified as either *finite* or *infinite*. A finite population is one in which the number of elements or units is limited, such as a stand of trees. An infinite population is one in which there are an unlimited number of elements, such as the quantity of water in an ocean. An infinite population can also result where samples are drawn from a finite population and replaced before drawing the next sample, i.e., sampling with replacement. Theoretically, in a finite population it is possible to count all the units while it is not possible to count all the units in an infinite population.

Two other concepts related to population that are a source of confusion are population size and population total. Population size normally refers to the number of "final sampling" units in the population. For example, if we have a forest of area 20 ha that is divided into 2-ha plots for the purposes of a survey, then the population size is 10, given that there are 10 sampling units in this population. Population total, on the other hand, refers to the total of some variable over these units. If the total volume of wood in this 20-ha plantation is $100m^3$, then this is the population total in this example.

6.2.2. The Concepts of Sample and Sampling

A sample is a subset of a population that is selected either subjectively or most often, objectively using various sets of rules. The rules are crafted in ways that ensure that the sample possesses the same characteristics as the population. Sampling requires some form of randomization to ensure each element in the population has some chance of being included in the sample. In simple random sampling, for example, all elements in the population have an equal chance of being included in the sample. Other forms of randomization may result in modifications that will increase or decrease the chances of including some elements of the population into the sample. These set of rules are the basis of the different sampling methods in use in statistical literature, which will be discussed briefly in the sections that follow.

There are several good reasons why sampling is done instead of complete enumeration (tally) of a population. These include: reduced cost; greater speed; greater scope; greater accuracy; reduced respondent burden; and feasibility.

In the first place, complete measurement or enumeration may be impossible. For instance, a manager may be better informed on the total volume of all trees in a stand, but the destructive nature of cutting down all trees for measurements and the difficulty of accurately measuring the volume of all standing trees in the stand preclude the measurement of all trees.

In other instances, total measurement or count is not feasible. Consider the staggering task of measuring the diameters and heights of all the merchantable trees in a forest stand of say 1000 ha. The enormity of the task would demand some sampling of the population. Sampling can provide the needed information at a far lower cost than a complete enumeration, and this information may sometimes be more reliable than that obtained by a 100% count. With fewer observations to be made, more time will be available. Therefore, measurements of the sample units can be made with greater care, accuracy and with reduced data entry errors. In addition, as Cochran (1977) put it, taking a complete census of a population precludes an estimate of the amount of error and gives a false sense that no error was involved in the estimation (a zero error).

Finally, sometimes a manager or a national forestry regulator may need timely information to assess forest damage due to some natural events or even settle a dispute. Since sample data can be collected and processed in a fraction of the time required for a complete inventory, sampling may be a better alternative than a 100% enumeration which may take several months or years to complete.

6.2.3. Sampling and Surveys

In this age of electronic mass media, we are all exposed to the results of surveys or asked to participate in surveys of all kinds. Companies that want to sell us their products often want to know what we think about these products by conducting surveys on a sample of their clients. During elections, surveys are employed as the main tool to predict the likely outcome of the elections. When properly conducted, surveys provide very useful information to the conductor of the survey. The term *survey* is used to describe a method of collecting data from a sample of any population, be it individuals, plants or animals. *Sampling* is the process of selecting the units from the population on which to conduct the survey (or take measurements) to estimate characteristics of the whole population. In this case, sampling applied to collecting data on forest resources can also be considered surveys though the forestry literature generally uses the term sampling or survey sampling. In our view, a survey is a broader concept than sampling as the former includes measurements, sampling, and data analyses.

6.2.4. Differences between Sampling and Experimental Designs

Most students that are new to the application of statistical methods to forestry research are always faced with the question of what methods to use to either measure a forest attribute or understand factors that influence an observed forestry-related phenomenon. For example, if there is a desire to know the impact of pre-treatment of seeds on germination rates, should one use an experimental design or a sample survey? The answer to this question is simple, but first let us look at the differences between sampling and experimental designs. There are two key differences between the two:

- In sampling, one is interested in examining the population as it is (in its "natural" state), without changing or disturbing this population. In experiments, some parts of the population are deliberately disturbed or changed to examine the effect of the action.
- In experiments, the objective is to compare the mean response to changes in levels of the factors. In sampling, the objective is to describe the characteristics of the population (Scharwz 2011).

From these differences, it is clear that if one wants to know how much wood there is in a plantation; one will use a survey sampling approach. However, if one wants to examine the impact of pre-treating seeds with heat on germination rates, one will need to use an appropriate experimental design in which the investigator will vary the amount of heat and then analyse the impact of the pre-heating treatment on germination rates.

6.2.5. Population and Sample Statistics

A parameter is a unique characteristic of a population that is usually not known but can be estimated from a sample randomly drawn from that population. The mean height of trees in a population of trees is a parameter if obtained by averaging the heights of all the trees in the population. The proportion of living seedlings in a teak plantation is a parameter if obtained by observing all the seedlings in the plantation. The total number of units in the population is a parameter, and so is the variability among the unit values (Freese 1962).

Although population parameters are also known as the population statistics, some statistics literature may refer to population statistics as parameters and sample statistics as estimates. The rationale for referring to population statistics as parameters seems to reside in the fact that, population statistics are fixed; i.e., there can only be one population mean; while sample statistics do vary with the selected sample. By definition, a statistic is a measure that is descriptive of a population. In this context, a population parameter is also a statistic, thus making the distinction between sample and population statistics an important one. Population statistics are seldom known, and the process of estimating the population statistics using sampling is a major component of the subject of statistics.

Population and sample statistics are distinguished in statistics literature by different symbols. Typically, population statistics (or parameters) are represented by Greek alphabets, e.g., mean (μ), standard deviation (σ), while sample statistics are symbolised by the regular alphabets, e.g., mean ($\bar{x}$) and standard deviation (s).

6.2.6. Bias, Accuracy, and Precision

The concepts of *bias*, *accuracy* and *precision* are statistical measures that are used to assess the quality of a sample estimate. The objective of sampling is to estimate some characteristic of a population at a reasonable cost and the desire is to obtain an estimate that is as close to the true value as much as possible (accurate estimate). Without applying the proper sampling method, the estimate could include a systematic deviation called bias thus making the estimate inaccurate. Freese (1962) defined *bias* as a systematic distortion in the sample estimate. This distortion may be caused by measurement error, poor sample selection methods or by using the wrong parameter estimation technique.

A poor sample selection method could result in samples being consistently chosen from only a part of the population with the result that the estimated parameter say, the population mean being consistently either much higher or lower than the true value. For example, in a survey to assess reforestation success of a plantation project, a sampler who is unaware of the concept of sampling bias may consistently avoid placing sample plots in areas that are not well-stocked while placing plots exclusively in well-stocked areas. This practice could lead to a greater representation of the well-stocked areas of the plantation in the sample. The result will be a biased estimate of the average number of surviving seedlings per unit area, and the sampler could wrongly conclude that reforestation was successful, while, in fact, it was a failure. This wrong conclusion, resulting from bad sampling practice, can have major impacts on national policies that rely on the results of sampling for decision-making.

If we assume that the investigator in the above example is aware of the consequences of avoiding non-stocked areas on the sample estimates and took steps to ensure that all areas are well represented, the sample estimate could still be biased if the wrong method of calculation is used to obtain the average seedling survival per hectare. For instance, if the samples came from two separate blocks with fairly different stocking levels of the same plantation, the investigator may consider calculating a weighted average rather than a simple average particularly if the blocks are of different sizes. This procedure will reduce the bias in the estimate.

Accuracy is a measure of how close the sample estimate is from the true population value. Bias and accuracy are directly opposite; the larger the bias, the less accurate an estimate is. Accuracy can sometimes be confused with the concept of precision. As Freese (1962) puts it, "a badly biased estimate may be precise but can never be accurate." Precision is not synonymous with accuracy. *Precision* refers to the closeness of the individual sample values to their own mean.

In the first scenario of the stocking survey of the reforestation project, the sampler will likely come up with a very precise estimate of the average seedling survival. Since the entire sample plots were selected from well-stocked areas, plot to plot variation of seedling survival will likely be very low resulting in precise estimates that are badly biased and hence not accurate. Common measures of precision are sample standard deviation or sample coefficient of variation or any other statistic that measures how dispersed the sample values are from their mean. The ideal outcome of a sample survey is an accurate and precise estimate.

6.3. ESTIMATING SAMPLE SIZE

The amount of error associated with an estimate (sample statistic) is generally inversely related to the number of elements included in a sample (the sample size). As a general rule of thumb, the larger the sample size, the smaller the sampling error. This general relationship between sample size and sampling error is the basis for deciding the optimal number of samples to use in a data collection (inventory) process. One form of expressing this relationship is given in Equation (1):

$$n = \frac{t^2 CV^2}{AE^2} \tag{1}$$

Where:

AE = the allowable error expressed as a percentage,
CV = the Coefficient of variation, measured as the standard deviation of the population expressed as a percentage of the mean,
t = the probability factor usually from student t-distribution (t=2 is commonly used),
n = the target sample size.

The above formula assumes that simple random sampling is used. In some forest mensuration literature, this formula may be stated differently if the desire is to sample 5% or more of the population (i.e., if the intended sample fraction is greater than or equal to 5%). In that case, a finite population correction factor (FPC) is applied to Equation (1) to produce Equation (2):

$$n = \frac{t^2 CV^2}{AE^2}\left(\frac{N-n}{N}\right) \tag{2}$$

$$n = \frac{t^2 CV^2 N}{NAE^2 + t^2 CV^2}$$

Where:

N = the total number of sample units in the population (population size), $\left(\frac{N-n}{N}\right)$ = the finite population correction factor.

The process of sample size determination begins with the inventory manager deciding *a priori* what level of sampling error (the allowable error) is acceptable based on the level of risk he/she is willing to accept. This allowable error is normally stated as a percentage of the statistic to be measured. For example, an allowable error could be stated as ±10% of the mean. Typically, the standard error or a multiple of standard error (e.g., confidence interval) is used. Based on previous experience or knowledge of the population or data from a similar population, the coefficient of variation is estimated (a rough estimate is often enough). The information is then plugged into either Equation (1) or (2) to calculate the sample size n.

Example: suppose we are interested in sampling a 40-ha teak plantation to estimate the mean standing timber volume per ha, and we intend to use fixed area plots of 100 m^2 (0.01ha). Based on previous knowledge, we estimate that the coefficient of variation for standing timber volume is 75%, and we want a sampling accuracy of ±20% of the mean. Using Equation (1), the required number of sample plots will be:

$$n = \frac{(75^2 \times 2^2)}{20^2} \approx 57 \text{ plots}$$

Using Equation (2) will require an estimate of the population size, N. One way of calculating N is to divide the total area of the stand by the area of individual plots (both in the same units). This works out to $N = 4000$ plots and the estimated sample size is approximately 56 plots:

$$n = \frac{(2^2 \times 75^2 \times 4000)}{(4000 \times 20^2) + (2^2 \times 75^2)} \approx 56 \text{plots}$$

Compared to the population size (N=4000), the sample size is very small (only 1.5% of the population). Both methods give almost the same estimates, and so a finite population correction factor may not be necessary.

6.4. Sample Plot Selection

The focus of this section is on the various sampling techniques that are used in forest inventories. The objective is to introduce the reader to these concepts and briefly describe their advantages and disadvantages. Readers who are interested in exploring these topics further may consult forest mensuration textbooks e.g., Avery and Burkhart (2002), Shivers and Border (1997), Shreuder et al. (1993), Husch et al. (2003). Before getting into the details of the individual sampling techniques, a few introductory issues need to be explained to facilitate a better understanding of these techniques.

How is sampling done in practice? Specifically if we decide that we are going to select n sampling units from a population of N units, how do we go about doing that? While the answers to these questions are simple, they are sometimes not very apparent in many mensuration textbooks. In practice, sampling is done by drawing one unit at a time until all the n units to be included in the sample have been selected. There are two ways of drawing sample units from a population: with and without replacement. In sampling with replacement, once a unit is drawn to be in a sample, that unit is placed back in the population to be possibly sampled again. In sampling without replacement, once an individual unit is drawn, that unit is not placed back into the population for re-sampling. Both sampling methods are used in forest inventories to some extent, but sampling without replacement appears to be the most widely used. While the discussion of the merits of each method is beyond the scope of this book, it is sufficient to say that sampling without replacement tends to result in a smaller sampling error (standard error or variance). Further discussion in this section will focus mainly on the sampling without replacement.

For a given sample of size n drawn from a population of size N without replacement, there are usually a fixed number of all possible samples each of size n that can be drawn from the population. For simple random sampling without replacement, this number is given as:

$$\frac{N!}{n!(N-n)!} \tag{3}$$

Assuming we have a population of size N = 10 and we wish to draw samples of size n=3. The total number of all possible samples of n =3 are:

$$\frac{10!}{3!(10-3)!} = \frac{10\times9\times8\times7\times6\times5\times4\times3\times2\times1}{3\times2\times1\times7\times6\times5\times4\times3\times2\times1} = 120$$

Suppose we can draw all the 120 possible samples and calculate the sample mean of each sample, we will have a collection (population) of sample means. We can then compute the population statistics (e.g., mean and standard deviation) of this "new population" (of sample means). We could also collect the sample standard deviations of the 120 samples into a population of standard deviations. In general, it could be any other statistic. The distribution of this new population is termed the sampling distribution of a statistic. For the purpose of this section, however, we will focus on the sampling distribution of means.

A unique property of the sampling distribution of means is that the mean of the sampling distribution of means is the mean of the original population, and the standard deviation of means is the standard error of the mean. In practice, since it is neither possible nor wise to collect all possible samples from a population, standard error of the mean is normally estimated from the sample standard deviation and the sample size:

$$s_{\bar{x}} = \frac{s}{\sqrt{n}} \tag{4}$$

Where:

$s_{\bar{x}}$ = the standard error of the mean,

s = the sample standard deviation given by;

$$s = \sqrt{\frac{\sum_{i=1}^{i=n}(x_i - \bar{x})^2}{n-1}} \tag{5}$$

Readers may notice the inverse relationship between sample size (n) and the standard error $s_{\bar{x}}$ .Without getting into further details, this inverse relationship is the basis for the sample size determination described in Section 6.3.

6.4.1. Simple Random Sampling

A *random sample* is one in which the selection of a particular sample does not depend on whether another sample has been chosen. *Simple random sampling* (SRS) is the process of selecting individuals from a population where each individual in the population has an equal chance of being included in the sample. Simple random sampling is the easiest to understand but the most difficult to implement in practice, particularly in forestry. How do you really implement a random selection process on a large tract of forested land to ensure that every tree on the land or every part of the land has an equal probability of being included in the sample? The honest truth is that it cannot be done without some modifications and assumptions.

One way of implementing SRS at the stand level is to use a Geographic Information System (GIS) tool to locate randomly a point in the stand. With the knowledge of the location information (co-ordinates), the point is located using either Global Positioning System (GPS) or compass and a sample plot is established at that location. The process is continued until all the *n* sample plots have been selected for measurements. To avoid the overlap of plots, a restriction is often built into the GIS program such that subsequent points do not fall within a certain minimum distance from an already selected point., Restrictions are made to avoid selecting points from within a certain distance from the edge of the stand. This restriction is made to avoid an issue known in forest mensuration as the "edge effect" a situation whereby trees growing at the edge of a stand are not necessarily representative of the whole stand due to differences in environmental factors affecting tree growth at the edge (such as light). For more discussion on this issue, readers are encouraged to refer to Schreuder et al. (1993).

Simple Random Sampling is often used in forestry to evaluate the variance of a forest unit when we do not know either what the stand boundaries are or what underlying environmental variables may influence the variability within the forest. The data from the SRS are used as preliminary data to plan an appropriate sampling design and estimate the sample size and shapes.

The implicit assumption required for simple random sampling to be efficient in estimating the population characteristic is that the values being observed on the individual sample units do not vary extremely across the entire population and that any variation in the population is fairly random and does not follow a specific pattern. In forestry, this assumption could be violated in a situation where a forest stand is made of a mosaic of different soil or soil moisture conditions resulting in high variation in tree growth. The assumption is also violated if the forest stand to be sampled is located on a slope where growth conditions and the consequent tree sizes vary systematically along the slope.

The procedure for calculating sample statistics from a simple random sample is pretty straightforward:

$$\text{Mean: } \bar{x} = \frac{x_1 + x_2 + x_3 + \ldots\ldots + x_n}{n} = \frac{\sum_{i=1}^{i=n} x_i}{n} \tag{6}$$

$$\text{Sample standard deviation: } s = \sqrt{\frac{\sum_{i=1}^{i=n}(x_i - \bar{x})^2}{n-1}} \tag{7}$$

$$\text{Standard error of the mean: } s_{\bar{x}} = \sqrt{\frac{s^2}{n}} \tag{7a}$$

Where:

$\bar{x}$ = Sample mean,
$s_{\bar{x}}$ = Standard error of the mean,
s = Sample standard deviation.

If sampling is done without replacement, a finite population correction factor must be applied, and Equation 7a becomes;

$$s_{\bar{x}} = \sqrt{\frac{s^2}{n}\left(\frac{N-n}{N}\right)} \tag{8}$$

Example: Suppose we want to use SRS to survey and estimate the mean volume of wood per ha on a 50-ha teak plantation on a private woodlot. In addition to the mean volume, we also want to estimate the other stand parameters associated with the plantation. How would we go about it?

To begin with, we have to define or decide the number of sample plots to measure, the sizes and shapes of the sample plots and the unit values to measure within each sample plot. For this example, assume that we use 15 sample plots, each of square size 50 m by 50 m and measure plot volumes as the unit values. Once these decisions are made, then SRS is applied to select the plots for measurement. The 50 m by 50 m units are plotted on a map of the plantation and assigned numbers. Note that because there are 50 ha in the population that is being divided into one-quarter ha of sample plots, the total number of units in the population (N) will be 200, with n = 15. Fifteen random numbers are drawn from the total of 200 using a random number table or any other random number generator. The units drawn constitute the 15 sample plots needed for this sampling task as shown in Figure 1.

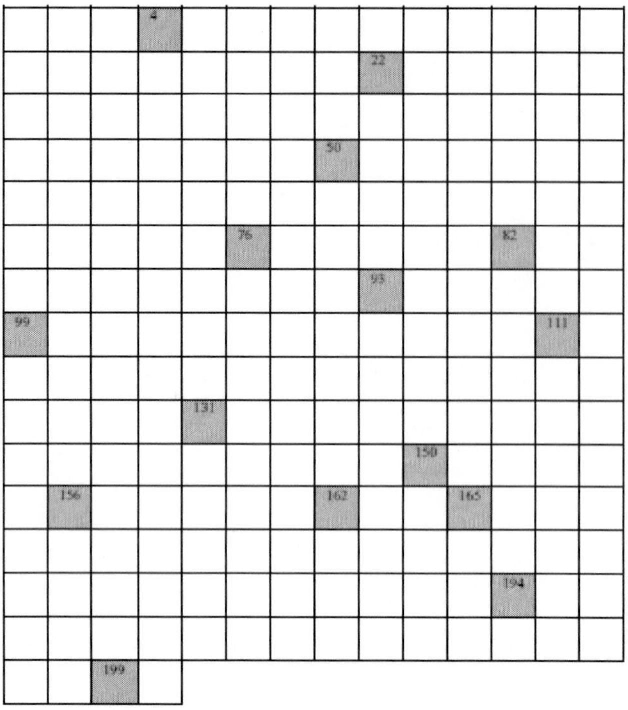

Figure 1. Sample selection using simple random sampling.

In Table 1, the plot number in the first column represents random numbers drawn between 1 and 200 and do not impact in any way the calculations. With sampling without replacement, no unit (quarter-ha) will be drawn more than once. In each of the sample plots, we will measure parameters such as diameters and heights of each tree and use these to compute the sample plot volumes using the methods described in Chapter 5. The results are recorded in a table, similar to Table 1.

Estimates

Mean: Using Equation (6), the mean volume is given as the sum of all sample plot volumes divided by the number of samples. That is, 180/15 = 12m^3. But our objective is to estimate the mean volume of wood per ha, not per plot, so we need to scale up (or convert) this mean volume of the samples to volume per ha. Recall that the sizes of the sample plots used to derive the values in Table 1 were one-quarter of a ha, and hence the estimated volume per ha will be multiplied by 4 and that gives us 48 m^3/ha as our answer. The total volume in the plantation (population volume) will then be 50 (ha) times 48 m^3, which equals 2400m^3.

Standard errors: The first step in estimating the standard error of the mean (Equation (7a)) is to estimate the sample standard deviation of the individual values of x. This is achieved using Equation (7).

$$\text{Variance} = s^2 = \frac{\sum_{i=1}^{i=n}(x_i - \bar{x})^2}{n-1} = 180/14 = 12.857 \text{ m}^3$$

In our example, sampling was done without replacement. As a result, a finite population correction factor must be added, implying we have to use Equation (8) to calculate the standard error of the mean as:

$$\text{Standard error} = s_{\bar{x}} = \sqrt{\frac{s^2}{n}\left(\frac{N-n}{N}\right)} = \sqrt{\frac{12.857}{15}\left(\frac{200-15}{200}\right)} = 0.89 \text{ m}^3$$

This (0.89) is the standard error of the mean per sampling plot (quarter-ha). By the rules for the expansion of variances and standard errors, the standard error for the mean volume per ha will be (4) (0.89) = 3.562 m^3. Similarly, the standard error for the estimated total volume will be $Ns_{\bar{x}}$ = (200) (0.890) = 178.00 m^3.

Table 1. Example of the data generated from simple random sampling

Plot number	Plot volume (m^3) x_i	$(x_i - \bar{x})^2$
194	8	16
50	17	25
99	12	0
111	18	36
165	9	9
76	11	1
131	7	25
93	17	25
162	9	9
156	12	0
4	10	4
150	9	9
22	14	4
199	16	16
82	11	1
	$\sum_{1}^{n} x_i = 180;\ \bar{x} = 12$	$\sum_{i=1}^{i=n}(x_i - \bar{x})^2 = 180$

Confidence intervals: The mean of 48 m^3/ha for this plantation is a single number that does not tell us much about the potential variability within the population. To make this mean estimate more meaningful, foresters usually calculate the confidence interval, which defines the limits that indicate the range within which we might expect (with some specified degree of confidence, usually 95%) to find the mean. The confidence interval is often calculated as *Estimate* $\pm$ *2 (standard error of estimate).*

Using this definition, we can determine the confidence intervals for the volume per ha and the total volume of the plantation as:

Mean volume per ha: 48 $\pm$ 2 (3.562); the confidence interval is from 40.88 to 55.12 m^3
Total volume in plantation: 2400 $\pm$ 2 (178.08); the confidence interval is from 2043.84 to 2756.16 m^3

The interpretation of these confidence intervals is that unless a 5% chance has occurred in sampling, the population mean volume per ha is somewhere between 40.88 and 55.12 m^3, and the true total volume is between 2043.84 and 2756.16 m^3. Because of sampling variation, the 95% confidence limits will, on the average, fail to include the mean (or any other parameter of interest) in 1 case out of 20 (i.e., 5% of the time). It must be emphasised, however, that these limits and the confidence statement take account of sampling variation only. They assume that the plot values are without measurement error and that the sampling and estimating procedures are unbiased and free of computational errors. If these basic assumptions are not valid, the estimates and confidence statements may be nothing more than a statistical hoax (Freese 1962).

6.4.2. Systematic Sampling

In some cases, it is not logistically convenient to randomly select sample units from the population. Systematic sampling is a non-probability sampling technique that attempts to avoid some of the practical problems associated with simple random sampling by drawing samples in a predetermined pattern. The implementation of systematic sampling in forestry avoids the shortcoming of locating simple random samples in practical field situations by randomly selecting the location of the first sample plot and then choosing the location of subsequent plots at predetermined intervals (distance). Theoretically, systematic sampling is suitable in situations where there is a systematic variation in the population such that, there is a risk of a bias that might be associated with simple random sampling picking samples within a narrow portion of the population thereby generating a sample that may have very small variance and yet very different from the population mean (i.e., a sample that is not representative of the population).

The application of systematic sampling in forest inventory often employs a grid system with the sample units in equally spaced rows with a constant distance between the sample units (Freese 1962). The grid covers the entire forest estate to be inventoried forming *N* potential sampling points. The number of grid points is divided by the sample size to estimate sampling interval (K).

Systematic sample selection begins with the random selection of the first grid point from the first K consecutive grid points. This grid point then constitutes the first sample unit. Continuing from the selected grid point, the subsequent sample units are selected by picking every K^{th} grid point following a predetermined order. This process is known as 1-in-K systematic sampling with a random start. In Figure 2, we choose the sample units by starting randomly from unit number 5 and selecting the 10th unit from that point forward to produce the 15 sample plots. We could have also divided the number of grid plots (N=200) by the sample size (n) to determine the sample interval to be approximately every 13th grid. Curious readers can check where the 15 sample units would have been on the grid in Figure 2. The positions of the sample units explain why we did not use the interval of 13 in this example.

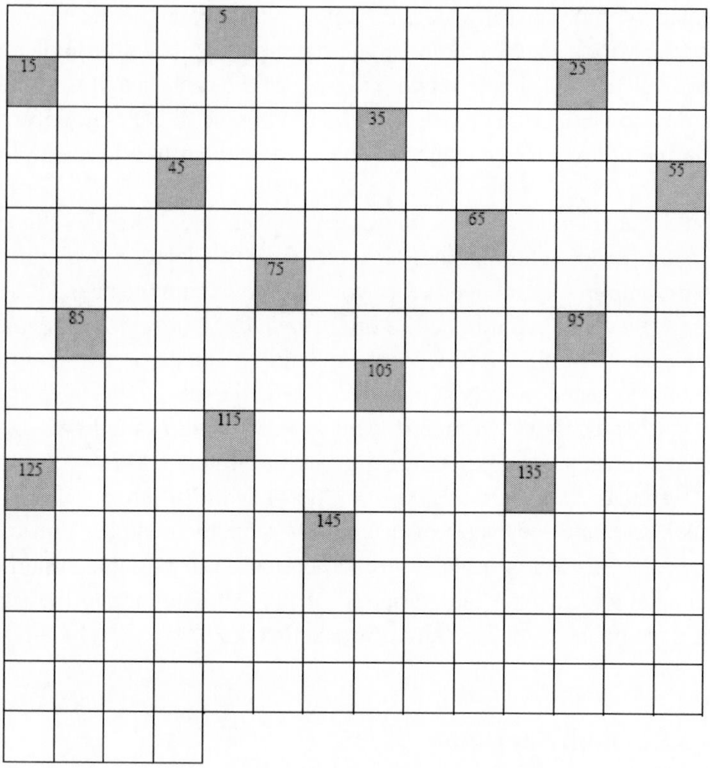

Figure 2. Sample allocation in a systematic sampling with a random start at the 5th unit.

Systematic sampling is used for various reasons. The most common reason is that sample units are easier and cheaper to locate and measure. There is also the general feeling that systematic sampling by design deliberately spreads the sample over the entire population thus ensuring more representation than say, simple random sampling. Apart from operational convenience and the desire for representation, in a population that falls into a definite pattern, systematic sampling is seen as the only sampling method that can provide unbiased and precise estimates. However, if the pattern in the population is cyclical in nature and the design of the systematic sample pattern happens to coincide with this cyclical pattern, the estimate may still be biased.

Randomisation is restricted only to the selection of the first unit point. The inclusion of any of the remaining points in the sample is determined by picking every Kth unit. The calculations of the sample statistics are based on the assumption that systematic sampling is the same as simple random sampling, and the basis of this assumption is the random start of the systematic sampling process. Therefore, Equations (6) to (8) can be used to estimate mean and standard error similar to simple random sampling.

The main reason for analysing the data from systematic sampling as if they were from a random sample stems from the random selection of the first sample. However, the true precision of an estimator from a systematic sample can be either worse or better than a simple random sample of the same size, depending on whether units within the systematic sample are positively or negatively correlated among themselves (Schwarz 2011). If the units are positively correlated within the sample, the sample variance will underestimate the true variation in the population. It must be pointed out that while logistically simpler, a systematic sample is only 'equivalent' to a simple random sample of the same size if the population units are 'in random order' to begin with (Krebs 1989). Even worse, there is no information in the systematic sample that allows the manager to check for hidden trends and cycles (Schwarz 2011).

So how should one handle a situation where you are not sure that the data from a systematic sample are random and that these may be biased? There are two options. One is to consult an expert in sampling for advice before you start collecting the data. However, if there is no one to consult, then one would need to apply *repeated systematic sampling*. Repeated systematic sampling is an empirical method of obtaining a standard error from a systematic sample. In this method, instead of choosing one systematic subsample of every K^{th} unit as discussed above, you would select *m* independent systematic sub-sample of size *n/m*. As an example, instead of choosing a single systematic sample of size 200 from a population, you can choose four systematic samples of size 50 (200/4), with different starting points each time. You will then estimate the mean of each sub-systematic sample. You can now treat these means as a simple random sample from the population of possible systematic samples and use the usual sampling theory. The variation of the estimate among the sub-systematic samples provides an estimate of the standard error (Schwarz 2011).

6.4.3. Stratified Random Sampling

Stratified random sampling is a method of sampling that involves the division of a population into smaller relatively more homogeneous groups known as strata. In stratified random sampling, the strata are formed based on similarity of attributes or characteristics of the individual units, resulting in relatively more homogeneous strata compared to the entire population. A random sub-sample from each stratum is taken in a number assigned to each stratum based on a specified allocation rule. These sub-samples from the strata are then pooled to form a random sample. Common characteristics used for stratifying forests for inventory purposes are: stand age, soil type, average tree height, percentage composition of target species type, average stem density, etc., all of which tend to affect the standing timber volume. Therefore, stratified random sampling is used when we know that the population is heterogeneous for any of these reasons.

Once the strata have been defined, samples can be drawn from each stratum using simple random sampling or systematic sampling. The key distinguishing feature of stratified random sampling is the stratification and the restriction of randomisation within each stratum; randomization within each stratum is independent of the randomisation in the other strata.

Stratification is done mainly to improve the precision of the estimates of the whole population. The amount of gain depends on how stratification is done. If the population is heterogeneous and it can be divided into strata, using prior information about the population, each of which is internally homogeneous, i.e., the characteristic under consideration varies little from one unit to another, a precise estimate (an estimate with smaller variance) of any stratum parameter can be obtained from a small sample in that stratum. These estimates can then be combined to obtain a precise estimate for the whole population. If the variation within strata is less than the variation among strata, the population estimate will be more precise than if sampling had been done at random over the entire population (i.e., if simple random sampling had been used).

Despite this advantage, it is important to note that in implementing stratified random sampling, each sampling unit in the population must be assigned to one and only one stratum. Also, the size of each stratum must be known, and that sampling must occur in each stratum. In cases where there is insufficient knowledge about the sizes of the strata and the amount of variability within the population to define clearly the strata, it would be difficult to derive the full benefits of this sampling method.

The following notations are commonly used in stratified random sampling:

The suffix h (h=1, 2, 3, . . .,L) denotes the stratum, and i denotes the unit within the stratum.

N = total number of units in the whole population

N_h = Total number of population units in stratum h.

n_h = Total number of sample units in stratum h

$W_h = N_h / N$ = the weight of stratum h or the proportion of population units in stratum h.

$\overline{X}_h = \dfrac{\sum_{i=1}^{N_h} X_{hi}}{N_h}$ = The population mean for units belonging to the stratum h.

$\overline{x}_h = \dfrac{\sum_{i=1}^{n_h} x_{hi}}{n_h}$ = The sample mean for units belonging to the stratum h.

$s_h^2 = \dfrac{\sum_{i=1}^{i=n_h} (x_{hi} - \overline{x}_h)^2}{n_h - 1}$ = The sample variance for units belonging to the stratum h.

$\overline{x} = \dfrac{\sum_{h=1}^{L} N_h \overline{x}_h}{N}$ = The sample mean of all units.

$s_{\overline{x}}^2 = \dfrac{1}{N^2} \sum_{h=1}^{L} \left[\dfrac{N_h^2 S_h^2}{n_h} \left(\dfrac{N_h - n_h}{N_h} \right) \right]$ = The variance of the sample estimate of the population mean.

There are three common methods of allocating the overall sample size n to all the L strata in a population. The method used to allocate samples among the various strata will depend on the main objectives of the inventory and the amount of information we have about the forest (population).

1. Equal allocation – where the n sample units are allocated equally among all strata irrespective of the sizes of the strata such that, each stratum sample

$$n_h = {}^{n}\!/_{L}$$

2. Proportional allocation – where the n sample units are allocated to each stratum in proportion to the size of the stratum. For example, if one of the strata contains half of the units in the population, then under this allocation rule, half of the sample observations would be made in that stratum. i.e.:

$$n_h = n\left(\frac{N_h}{N}\right)$$

3. Optimum allocation based on variability within each stratum - n is allocated to each stratum based on some measure of variability within each stratum, e.g., the standard deviation of the stratum (σ_h).The main goal of optimum allocation is to achieve the smallest standard error possible with a total of n observations i.e.:

$$n_h = \frac{w_h \sigma_h}{\sum_{i=1}^{L} w_h \sigma_h} \tag{9}$$

If our objective is to get the most precise estimate of the population mean for a given cost, then we will allocate sampling units in each stratum with a cost c_h as:

$$n_h = \frac{(w_h \sigma_h)/\sqrt{c_h}}{\sum_{i=1}^{L}\left[w_h \sigma_h)/\sqrt{c_h}\right]} \tag{10}$$

Example

Continuing with the example under simple random sampling (SRS), suppose that we are aware of the heterogeneity in the 50-ha teak plantation, and we want to take this into account in our sampling framework. A quick look at Table 1 shows that there is a lot of variation in sample volumes ranging from 7-18 m^3. Let's assume that these differences in volume are due to differences in ages of the stands. We can stratify this plantation into three age groups based on the previous knowledge that the trees were planted in different years. How should we estimate the parameters and how different would the results be from the SRS method?

Based on our knowledge of the boundaries of the plantation and when each age group was planted, the plantation can easily be divided into three strata using age as the defining characteristic. Within each stratum, we will now apply simple random sampling as before to select 5 sample plots for measurements. We will keep the same quarter-ha square plots of sizes 50 m by 50 m. A total of 15 sample plots will be chosen as before (five per stratum using the equal allocation rule). We will assign the following areas to each stratum: Stratum I has 10 ha; Stratum II has 15 ha, and Stratum III has 25 ha. In this case N_1 to N_3 are defined in terms of the units by multiplying the area of each stratum by 4 (since each ha contains 4 sampling units). In this case we have $N_1 = 40$; $N_2 = 60$; $N_3 = 100$ and $N = 200$. The result of this selection process is shown in Figure 3, where simple random sampling is used in each stratum to select the plots for measurement.

Although the equal allocation rule is used here for illustrative purposes, note that it would have been more appropriate to use the proportional allocation rule in that case, Stratum III would have had half of all sample units rather than just five. One important point to note is that within the strata, samples are treated exactly like simple random sampling. The only computational complication introduced by stratified random sampling is how to aggregate the estimates from each stratum to the total sample and population levels.

Estimates

Mean: Again, in this example, our objective is to estimate the mean and other parameters of the population (50-ha plantation). An estimate of the population mean begins with an estimate of the sample means within each stratum. The procedure is exactly the same as in SRS. The results are shown in Table 2. It is clear from Table 2 that the means vary widely among the three strata (8.4, 11.2, and 16.4 m^3).

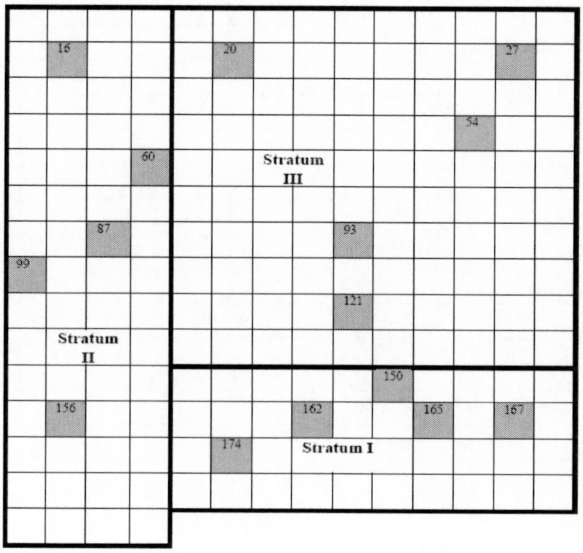

Figure 3. Sample selection using a stratified random sampling technique.

Table 2. Example of the data generated from stratified random sampling

Plot number	Plot volume (m^3) x_{h_i}	$(x_{h_i} - \bar{x}_h)^2$
Stratum I		
174	7	1.96
167	8	0.16
165	9	0.36
162	9	0.36
150	9	0.36
Total	**42**	**3.2**
Mean = 8.4		
$s^2 = 0.8$		
N_1 =40		
Estimated stratum volume/ha = 33.6		
Estimated total stratum volume = 336.0		
Stratum II		
16	10	1.44
60	11	0.04
87	11	0.04
99	12	0.64
156	12	0.64
Total	**56**	**2.8**
Mean = 11.2		
$s^2 = 0.7$		
N_2 =60		
Estimated stratum volume/ha = 44.8		
Estimated total stratum volume = 692.0		
Stratum III		
20	16	0.16
27	14	5.76
54	17	0.36
93	17	0.36
121	18	2.56
Total	**82**	**9.2**
Mean = 16.4		
$s^2 = 2.3$		
N_3 =100		
Estimated stratum volume/ha = 65.6		
Estimated total stratum volume = 1640.0		

The mean of the stratified sample is now calculated using:

$$\text{Mean: } \bar{x} = \frac{\sum_{i=1}^{i=h} N_h \bar{x}_h}{N} = \left(\frac{(40\times 8.4)+(60\times 11.2)+(100\times 16.4)}{200} \right) = 13.24\text{m}^3$$

Where N is defined as $N = \sum_{h=1}^{L} N_h$ =200. Careful readers will observe that this mean is different from the mean derived from simple random sampling. Volume per ha is now 4(13.24), which is 52.96 m^3/ha and estimated volume for the plantation is given as 52.96 m^3/ha times 50 ha = 2648 m^3. The main reason for the difference between the mean here and that of the SRS is that Stratum III which has a higher mean also has the largest area of the whole plantation. As a result, the mean from that stratum increases the overall mean. Think of the areas covered by each stratum as a weight and the estimated mean from the stratified sample as a weighted mean rather than a simple mean (as in SRS).

Standard error: The first step in determining standard errors is to obtain the estimated variance among individuals within each stratum (s_h^2). These variances are computed *within a stratum* in the same manner as the variance of a simple random sample as shown in the last column of Table 2, Using the sample variance equation, which is identical to Equation (7). The sample variance for units belonging to the stratum h is calculated as:

$$s_h^2 = \frac{\sum_{i=1}^{i=n_h} (x_{hi} - \bar{x}_h)^2}{(n_h - 1)}$$

After obtaining the variances of all three strata, we find the standard error of the mean of a stratified random sample by the formula:

$s_{\bar{x}} = \sqrt{\frac{1}{N^2} \sum_{h=1}^{L} \left[\frac{N_h^2 s_h^2}{n_h} \left(\frac{N_h - n_h}{N_h} \right) \right]}$, using the values in Table 2, this equals:

$$\sqrt{\frac{1}{(200)^2} \left[\frac{(40)^2 \times 0.8}{5} \left(\frac{40-5}{40} \right) \right] + \left[\frac{(60)^2 \times 0.7}{5} \left(\frac{60-5}{60} \right) \right] + \left[\frac{(100)^2 \times 2.3}{5} \left(\frac{100-5}{100} \right) \right]} = 0.356$$

The standard error of the stratified random sample is 0.356 compared to 0.890 for the simple random sample. This comparison clearly demonstrates the advantages of the stratified random sample over the simple random sampling. Apart from providing separate estimates of the mean and variance of each stratum, for a given sampling intensity, stratified random sampling often gives more precise estimates of the population parameters than would a simple random sample of the same size (Freese 1962). From this standard error estimate, the appropriate confidence intervals can now be determined using the same procedure as described in Section 6.4.1. Note that in the above example, simple random sampling was applied in each stratum for simplification. However, different sampling methods may be used in each stratum for cost or convenience reasons.

A practical example of stratified random sampling will be if we are interested in estimating the mean volume of neem plantations in northern Ghana. By stratifying northern Ghana into the three regions (Northern, Upper East and Upper West), we not only reduce the overall standard error of the estimated mean, but we can also estimate the mean volume in each region, which is information that will be useful for regional planning.

6.4.4. Cluster Sampling

In some cases, units in a population occur naturally in groups or clusters. Cluster sampling, as the name implies, is a sampling method in which sampling units are drawn in clusters rather than as individual units. In cluster sampling the population is divided into groups (or clusters) and clusters are chosen at random, and every unit within the chosen cluster is measured. Although seldom the case, if, for some specific reason, individual trees in a forest stand are considered as the basic sampling units, then sample plots become cluster samples of trees. The benefit of selecting plots of trees instead of single trees to estimate say, average tree diameter or height may be that of operational efficiency; i.e., it may be more cost-effective to measure a group of trees close to each other than to measure individual trees located several metres apart.

In forestry, however, the sample unit is seldom the individual tree but rather plots of trees even in cases where average tree parameters may be part of the end product of forest inventory. In such a case, cluster sampling may be used to select clusters of 2 or more sample plots in situations where this practice is deemed cost efficient. In certain larger scale forest inventory situations where access to various parts of the forest estate may be challenging, the inventory manager may randomly select a relatively fewer points in various parts of the forest estate and establish groups of plots in order to limit costly movement of equipment. In natural regeneration surveys, it may be too onerous to tally seedlings on larger plots while smaller plots may be too small to capture the overall variability of the regenerating stand. The surveyor may choose clusters of smaller plots per location such that individual plots are small enough for easy tallying while capturing enough variability per location.

Sample statistics for cluster sampling are calculated as follows;

Sample mean per element in a cluster: $\bar{x}_i = \dfrac{\sum_{j=1}^{m_i} x_{ij}}{m_i}$ (if different cluster sizes)

Sample mean per element in a cluster: $\bar{x}_i = \dfrac{\sum_{j=1}^{m_i} x_{ij}}{m}$ (if cluster sizes are the same)

Sample mean per cluster: $\bar{x} = \dfrac{\sum_{i=1}^{n} \bar{x}_i}{n}$

Sample standard error of the mean: $s_{\bar{x}} = \sqrt{\left(\dfrac{N-n}{N}\right)\dfrac{\sum_{i=1}^{n}(\bar{x}_i - \bar{x})^2}{n(n-1)}}$ (11)

Where:

x_{ij} = the jth observation in the ith cluster,
N = the total number of clusters in the population
n = the total number of clusters sampled
m_i = the number of sample units in cluster i

The assumptions used above are that there is no sampling within each cluster; i.e., all units in a selected cluster are measured. Therefore only the variation between clusters is included in the determination of sample variance. Also, note that the calculation of the sample mean from the cluster means as presented above presupposes that the number of units in each cluster is the same (i.e., cluster sizes are the same). In forestry, however, the above method of estimating sampling mean may be biased as cluster sizes are seldom the same. The bias may occur if cluster size affects average cluster values. A more realistic sample mean in such situations is given by:

$$\bar{x} = \frac{\sum_{i=1}^{n}\sum_{j=1}^{m_i} x_{ij}}{\sum_{i=1}^{n} m_i} \quad (12)$$

Example

To illustrate the concept of cluster sampling, suppose we wish to survey the same 50-ha teak plantation that is made up of distinct 1-ha square blocks (50 blocks total) using 0.25-ha sample plots. The conditions existing in the plantation are such that access is relatively difficult, and we wish to minimise the movement of personnel and equipment across the stand as much as possible. We then decide that we will select and survey 5 blocks, each block consisting of 4 plots, each of area 0.25-ha. In this case, each of the 5 blocks is a cluster. We proceed by first selecting 5 blocks at random and within each block, we demarcate and measure all the four plots and obtain the following individual plot volumes (Table 3). In Figure 4, each of the shaded clusters is a one-ha plot with four quarter-ha sample plots within them.

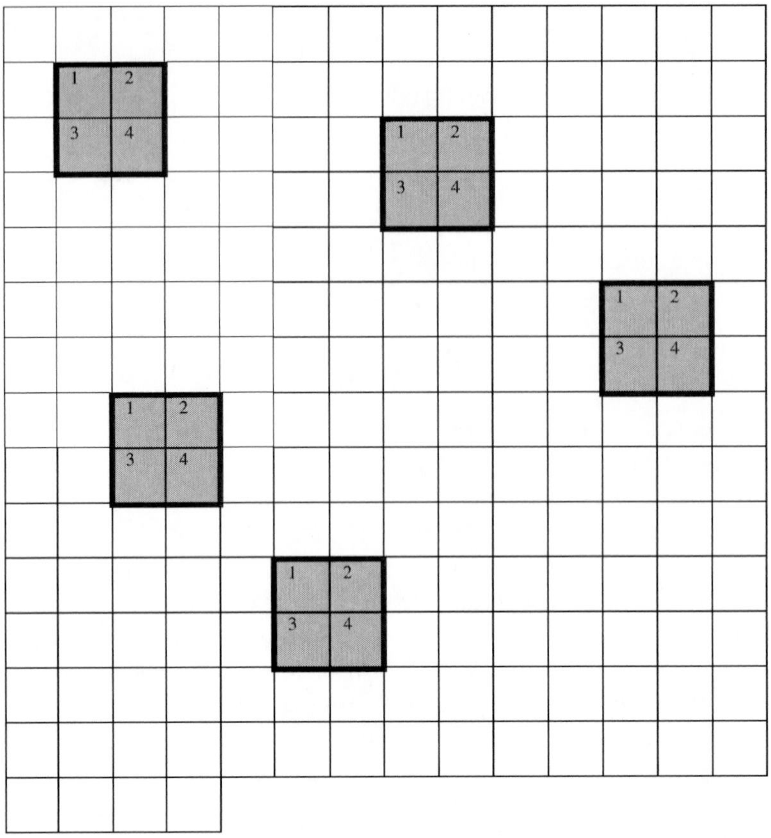

Figure 4. Sample plot selection in cluster sampling.

Table 3. Example of the data generated from cluster sampling

Sample Block (i)	Sample Plot (j)	Plot volume (m^3) x_{ij}	$(\bar{x}_i - \bar{x})^2$
1	1	8	
1	2	17	
1	3	12	
1	4	18	
$\bar{x}_1$		**13.75**	**5.06**
2	1	9	
2	2	11	
2	3	7	
2	4	17	
$\bar{x}_2$		**11.00**	**0.25**
3	1	9	
3	2	12	
3	3	10	
3	4	9	

Sample Block (i)	Sample Plot (j)	Plot volume (m^3) x_{ij}	$(\bar{x}_i - \bar{x})^2$
$\bar{x}_3$		**10.00**	**2.25**
4	1	14	
4	2	16	
4	3	11	
4	4	12	
$\bar{x}_4$		**13.25**	**3.06**
5	1	9	
5	2	11	
5	3	10	
5	4	8	
$\bar{x}_5$		**9.5**	**4.00**
$\bar{x}$		**11.5**	

Sample statistics:

We estimate sample statistics by first identifying the required variables:
The total number of clusters in the population is N=50,
The number of clusters sample n=5;
Number of sample units (plots) per cluster, m=4

$$\text{The sample mean: } \bar{x} = \frac{\sum_{i=1}^{5}\sum_{j=1}^{4} x_{ij}}{\sum_{i=1}^{5} m_i} = \frac{230}{20} = 11.50 \text{ m}^3.$$

$$\text{Standard error of the mean: } s_{\bar{x}} = \sqrt{\left(\frac{N-n}{N}\right)\frac{\sum_{i=1}^{n}(\bar{x}_i - \bar{x})^2}{n(n-1)}}$$

$$s_{\bar{x}} = \sqrt{\left(\frac{50-5}{50}\right)\frac{(5.063+0.25+2.25+3.063+4.0)}{5(5-1)}} = 0.746 \text{ m}^3$$

On a per hectare basis, the sample mean and standard error are obtained by multiplying these estimates by 4 to get 46 m^3/ha and 2.984 m^3/ha respectively.

6.4.5. Multi-Stage Sampling

In multi-stage sampling, the selection of the final sampling units takes place in stages. A multi-stage sampling is a form of cluster sampling in which only a sub-sample of the units of the chosen clusters are measured. Instead of measuring all the units contained in the selected clusters, the researcher randomly selects some units from each chosen cluster for measurements. This describes a two-stage sampling design. Selecting the n clusters is the first-stage sample, and the clusters are the first-stage sample units or the primary units. The second-stage sample units (or the secondary units) are the individual units selected at random from each sampled cluster. By extension, in a three-stage sampling design, primary units are selected first, then secondary units and finally the tertiary units. Two-stage sampling is the most commonly used multi-stage sampling design in forest inventory and has been chosen for further discussion.

Appropriate primary and secondary sample sizes must be decided on prior to implementing two-stage sampling. Ideally, the number of primary sample units (psu) and secondary sample units (ssu) should be chosen in an optimum combination that minimises overall sample variance. However, this approach is seldom used due to the computational complexity and because there is no guarantee that the chosen psu/ssu combination will be operationally feasible to implement. The two sample sizes are usually determined by considering the amount of variability at each stage. In cases where the primary sample units vary quite significantly while internally each unit is relatively homogeneous, it is recommended that the number of secondary units be kept at a very minimum size while the number of primary units increased to effectively estimate the between primary units variability.

There are some reasons why two-stage (or multi-stage sampling in general) may be used in forest inventory in preference to other sampling designs. One of the most obvious is where it is either impossible or too costly to measure all the primary sample units thus necessitating some form of sub-sample of secondary units from within the primary units. A common example in tropical forestry is the use of randomised branch sampling for estimating branch-wood volume left in the forests after logging. The branches may be too many to completely measure and usually require the surveyor to randomly select a few main branches (primary units) and then sample some of the smaller branches (secondary units) from the main branches.

In sampling a very large forest, locating and getting to sampling plots can be quite expensive while measurement of the plots may be relatively cheaper. It may seem logical in that circumstance to make measurements on two or three plots at or near each location. In national forest inventories, forest stands may be located in different regions requiring the surveyor to have to move equipment and personnel over long distances to visit and measure sample plots in every stand. The travel cost may be too high to bear. The surveyor may be compelled by the high travel cost to select a random sample of stands (primary units) to visit and then establish and measure sample plots (secondary units) in each selected stand.

For example, if one is interested in estimating the standing volume of timber in teak plantations in the Tamale Forest District, it would be difficult if not impossible to measure every tree in every plantation in the District. The most logical process would be to obtain a list of all teak plantations in the District, and then select a sample of these plantations (primary units). Within the plantations selected (primary units), sample plots will now be laid (secondary units), and the necessary measurements taken to determine the standing volume.

Finally, if properly designed, two-stage sampling may yield estimates of a given precision at a cost lower than simple random sampling.

The following notations are commonly used in two-stage sampling:

N = Total number of primary sample units (psu) in the population,
n = the number of primary units sampled from the population
M_i = Total number of secondary sample units (ssu) in the ith primary unit,
m_i = the number of secondary sample units (ssu) sampled from the ith psu

The sample mean and variance are given by (see Cochran 1977):

$$\bar{x} = \frac{1}{nm}\left(\sum_{i=1}^{n}\left(\sum_{j=1}^{m_i} x_{ij}\right)\right) \tag{13}$$

(If equal number of ssu, *m,* are selected in each psu)

$$\bar{x} = \frac{\left(\sum_{i=1}^{n}\left(\sum_{j=1}^{m_i} x_{ij}\right)\right)}{\sum_{i=1}^{n} m_i} \tag{14}$$

(If varying numbers of ssu, m_i, are selected at each psu)

$$s_{\bar{x}}^2 = \left(\frac{N-n}{N}\right)\frac{s_1^2}{n} + \frac{n}{N}\left(\frac{M-m}{M}\right)\frac{s_2^2}{nm} \tag{15a}$$

$$s_1^2 = \frac{\sum_{i=1}^{n}(\bar{x}_i - \bar{x})^2}{(n-1)} \tag{15b}$$

$$s_2^2 = \frac{\sum_{i=1}^{n}\sum_{j=1}^{m_i}(x_{ij} - \bar{x}_i)^2}{n(m_i - 1)} \tag{15c}$$

Where:

$\bar{x}$ = The sample mean,
$\bar{x}_i$ = the mean of the ssu in the ith psu

x_{ij} = the measurement or observation on the jth ssu in the ith psu
$s_{\bar{x}}{}^2$ = the variance of the sample mean
$s_1{}^2$ = the variance of the psu,
$s_2{}^2$ = the variance of the ssu.

The variance formula in Equations (15a) to (15c) above assumes that the size (M) of the psu is the same requiring an equal number of ssu within each psu. In real forestry applications, however, this is hardly the case. Therefore, the variance of the estimated sample mean for two-stage sample where the number of ssu (Mi) varies amongst psu, is stated a bit differently as follows:

$$s_{\bar{x}}^2 = \frac{N^2}{M_0^2}\left(\frac{N-n}{N}\right)\frac{s_1^2}{n} + \frac{N}{nM_0^2}\sum_{i=1}^{n}\left[M_i\left(\frac{M_i - m_i}{M_i}\right)\frac{s_{2i}^2}{m_i}\right] \tag{15d}$$

$$s_1^2 = \frac{\sum_{i=1}^{n}(\bar{x}_i - \bar{x})^2}{(n-1)} \tag{15e}$$

$$s_{2i}^2 = \frac{\sum_{i=1}^{n}(\bar{x}_i - \bar{x})^2}{(m_i - 1)} \tag{15f}$$

Where M_0 is the total number of ssu in the entire population, given by $s_1^2 = \sum_{i=1}^{N} M_i$.

Notice that Equations (15b) and (15e) are the same. The variation between psu is calculated the same way as in the case where psu sizes are assumed to be the same. However the within psu variation (variation between ssu in psu) is calculated a little differently.

Even if Equations (15d) to (15f) were used to calculate the variance where psu vary in size, Freese (1962) warns that there could still be a bias in the estimate of the sample variance as the variation between psu are weighted equally in both Equations (15a) and (15d). Freese (1962) recommended some approaches for dealing with this problem, two of which are briefly discussed below. One approach is to use stratified two-stage sampling, where primary units are grouped into size classes. The standard two-stage methods and calculations are then applied to each stratum. Population estimates can then be obtained by combining the individual stratum estimates according to the stratified sampling formulae already discussed in the previous section. This option requires that the sizes of the psu be known.

The second approach, which also requires knowledge of the sizes of the psu involves selecting the psu with probability proportional to the size of the psu. The ssu within each psu can then be selected with equal probability. This procedure ensures that the representation of the psu in the sample is weighted by size such that the larger psu have a higher chance of being included in the sample. The practical procedure for selecting psu with probability proportional to size and the corresponding formulae for estimating mean and variance can be found in regular sampling textbooks such as Cochran (1977), Freese (1962), Shreuder et al.

(1993) and Shiver and Borders (1997). The basic requirement is that the size of each psu must be known.

Example

Suppose that a 50-ha teak plantation is subdivided into 50-square blocks of 1-ha with each block uniquely identified and marked using say, four corner posts. The individual blocks could have been created by different planting time (years), planting stock, or some tending operation. Just as in the other examples, our objective is to estimate the average volume of wood per hectare. Using a two-stage sampling design, our desire will be to sample some of the 50 blocks at random and then establish and measure 0.25 ha sample pots in each selected block. The total number of plots that can be sampled from each 1 ha block is 4 plots (200 plots for the entire estate). Figure 5 shows the three secondary sample plot selection in each block (shaded) after the 5 blocks have been randomly selected at the primary stage.

Suppose we settle on sampling 5 blocks (primary units) and then measuring 3 plots (secondary units) per sample block each with an area of 0.25-ha. If we assume that access to any part of the forest estate is not an issue, we can assign each block a unique number, place all numbers in a container and randomly draw 5 numbers. The blocks with their numbers drawn constitute our primary sample. Three sample plots can then be selected at random from each of the 5 blocks using, for example, the process described in the simple random sampling example. Suppose the survey yields the results in Table 4 below.

Table 4. Example of the data generated from a two-stage random sampling

Sample Block (i)	Sample Plot (j)	Plot volume in m³ (x_{ij})	$(x_{ij} - \bar{x}_i)^2$	$(\bar{x}_i - \bar{x})^2$
1	1	6	25	
1	2	12	1	
1	3	15	16	
Total		**33**	**42**	
Mean ($\bar{x}_i$)		**11**		**1**
s_{2i}^2			**21**	
2	1	13	16	
2	2	24	49	
2	3	14	9	
Total		**51**	**74**	
Mean ($\bar{x}_i$)		**17**		**25**
s_{2i}^2			**37**	
3	1	13	25	
3	2	5	9	
3	3	6	4	
Total		**24**	**38**	
Mean ($\bar{x}_i$)		**8**		**16**
s_{2i}^2			**19**	

Table 4. (Continued)

Sample Block (i)	Sample Plot (j)	Plot volume in m^3 (x_{ij})	$(x_{ij} - \bar{x}_i)^2$	$(\bar{x}_i - \bar{x})^2$
4	1	16	36	
4	2	8	4	
4	3	6	16	
Total		**30**	**54**	
Mean ($\bar{x}_i$)		**10**		**4**
s_{2i}^2			**28**	
5	1	14	0	
5	2	18	16	
5	3	10	16	
Sample Block (i)	**Sample Plot (j)**	**Plot volume in m^3 (x_{ij})**	$(x_{ij} - \bar{x}_i)^2$	$(\bar{x}_i - \bar{x})^2$
Total		**42**	**32**	
Mean ($\bar{x}_i$)		**14**		**4**
s_{2i}^2			**16**	

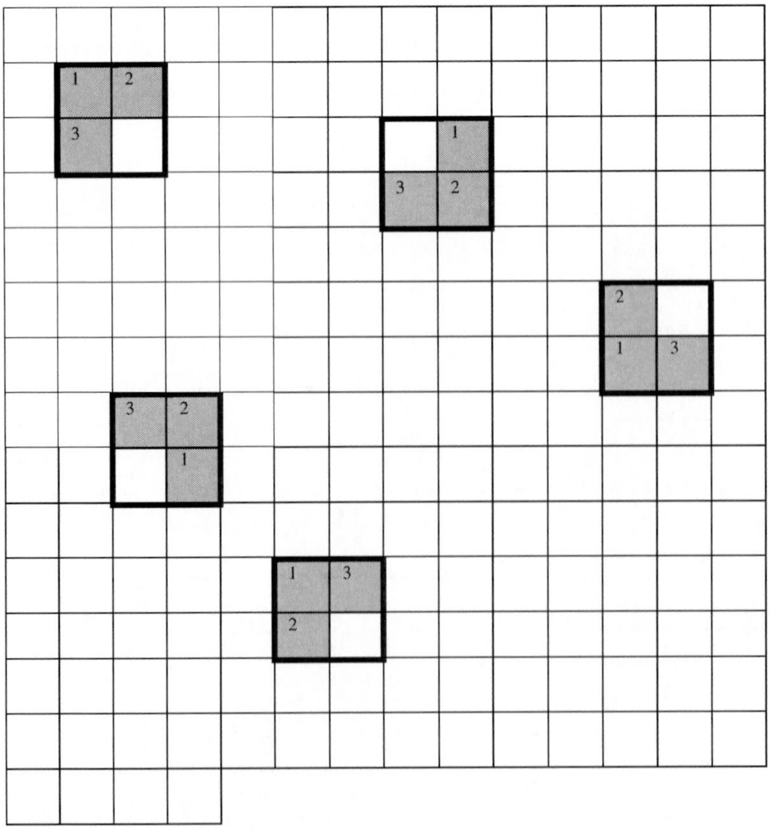

Figure 5. Sample allocation in a two-stage sampling design.

Estimates

From Table 4 above, an estimate of the population mean can be obtained by summing up the 15 individual plot volumes and dividing the total by the number of plots. However, from Table 4, we already have the total plot volume for the individual primary units (blocks). So we obtain the total by summing up the plot total per block and then divide by 15:

$$\bar{x} = \left(\frac{33 + 51 + 24 + 30 + 42}{15} \right) = 12\,\text{m}^3$$

Volume per hectare can be determined by multiplying the average plot volume in m^3 by 4 (as before), which gives us 48 m^3/ha.

We then estimate the standard error of the mean by first estimating the variance of the primary and the secondary units;

The variance of the primary units is:

$$s_1^2 = \frac{\sum_{i=1}^{n} (\bar{x}_i - \bar{x})^2}{(n-1)} = \left(\frac{1 + 25 + 16 + 4 + 4}{5 - 1} \right) = 12.5$$

The variances for the secondary units are given in Table 4 for each primary unit. However Equation (15c) can be re-written as;

$$s_2^2 = \frac{\sum_{i=1}^{n}}{n} \left(\frac{\sum_{j=1}^{m_i} (x_{ij} - \bar{x}_i)}{(m_i - 1)} \right) \tag{16}$$

This way it is easy to recognize that the variance (s_2^2) of the secondary units is the average of the variances for the secondary units in each primary units (s_{2i}^2), i.e.,

$$s_{2i}^2 = \frac{\sum_{j=1}^{m_i} (x_{ij} - \bar{x})}{(m_i - 1)} \tag{17}$$

We, therefore, substitute the term in the larger bracket of Equation (16) with s_{2i}^2 for the primary units:

$$s_2^2 = \frac{\sum_{i=1}^{5}\left(s_{2i}^2\right)}{5} = \frac{1}{5}(21+37+19+28+16) = 24.2.$$

To complete the estimation of the sample variance, first determine the total number of primary units N and the total number of secondary units per primary unit M. these are N=50 and M = 4 (each primary unit contains 4 secondary units) and m =3 (as only 3 units are sampled in each primary unit). Plugging all the values into Equation (15a) and taking the square root, we obtain the standard error as:

$$s_{\bar{x}} = \sqrt{\left(\frac{50-5}{50}\right)\frac{12.5}{5} + \frac{5}{50}\left(\frac{4-3}{4}\right)\frac{24.2}{5\times 3}} = 1.513 \text{ m}^3$$

The standard error for the average volume per hectare is 1.549 m^3 times 4 = 6.054 m^3/ha.

A common error in the use of multi-stage random sampling is that individuals not familiar with the correct procedures of data analyses often analyse the data as if they arose from a simple random sample. Notice that in multi-stage sampling, if a particular block was not selected at the primary stage then none of the plots within that block can be chosen in the secondary stage for measurement. The consequences of this error are that the estimated standard errors are too small and do not fully reflect the actual imprecision in the estimates. For this example, equal numbers of ssu were selected in each psu. However using the variance formula for the case where an unequal number of ssu is selected for each psu will yield the same results. The reader may verify this as a practice exercise.

6.4.6. Multi-Phase Sampling

A closely related sampling method to multi-stage sampling is multi-phase sampling. In some surveys, multiple measurements of the same survey units are performed. Multi-phase sampling, like multi-stage sampling, involves several stages of sample selection. Unlike multi-stage sampling, sample selection at every stage involves the same basic sample unit. In a multi-phase sample, we first start by taking a larger sample of units strictly to obtain auxiliary information. The next step is to take a second sample, usually a subsample of the relatively larger sample, where other variables that may be more expensive to measure for the entire sample are observed. In forestry, two-phase sampling (also known as double sampling) is commonly used, although the use of three-phase sampling is becoming increasingly popular (e.g., Von Luke and Saborowski 2012). Only double sampling will be discussed further.There are two common situations in which multi-phase sampling are used in forestry: to provide basic information for stratification, and to reduce cost.

The first situation arises where it is difficult to stratify a population in advance due to lack of information on the stratification variables. In this case, the first phase of sampling is used to measure the stratification variable on a random sample of units. The selected units are then stratified, and further samples are taken from each stratum as needed to measure a second variable. For example, if we are interested in measuring the amount of standing volume following fire damage to a plantation, we may start by using simple random sampling to quantify the damage. We will then stratify the plantation, according to the amount of damage. In the second phase, we will concentrate the allocation of units on plots with low fire damage to measure total standing volume of wood. By doing this, we save time and effort by concentrating the second phase measurement only in areas of less damage, given that we are only interested in areas that contain a usable volume of wood. In this case, though two phases of sampling took place, each phase measured a different variable on the same plots.

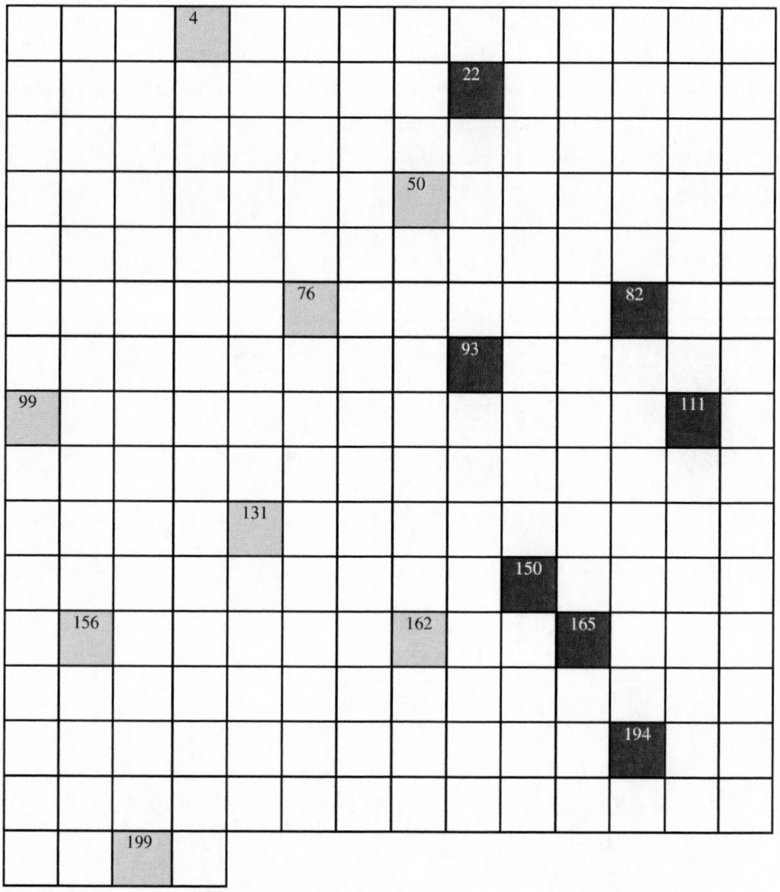

Figure 6. Sample selection using multi-phase sampling.

The second motivation for using double sampling is to reduce cost. If the variable of interest is expensive to measure, but a related variable is much cheaper, we might first sample many of the sampling units and measure the cheaper (auxiliary) variable, and only measure the response variable on a subsample (or smaller sample) (University of Montana, undated). A common example of the use of double sampling in forestry is in aerial photo interpretation. Relatively imprecise estimates of stand level variables such as species composition and the basal area can be obtained from air photographs at a relatively cheaper cost. Using double sampling, the surveyor may select some of the delineated polygons for detailed field measurements to obtain more accurate measurements of these same variables. These more accurate estimates can be used to adjust the estimates from the aerial photographs to improve the estimates of sample statistics of these variables.

Figure 6 shows the same 50-ha plantation that we used to illustrate simple random sampling. Here, we assume that the simple random sampling was done to gather information needed to stratify the plantation for further measurement in the second phase. In the second phase, only seven of the 15 initial simple random samples are now measured, not all of them (shaded darker). The remaining eight are not measured at all in the second phase. The data is analysed using the information collected in the second phase.

There are two common methods by which these adjustments can be done; the ratio and regression methods (or estimators). In the example above suppose there are *n* polygons from the aerial photograph and the surveyor selects *m* polygons for field measurement for total stem density per hectare. If we let the photo interpreted values for the variable of interest be X and the corresponding ground-based values to be Y, we can use *ratio estimates* to estimate sample mean as follows:

1) First, compute the sample ratio (r) using the matching values in the second (smaller) sample:

$$r = \frac{\sum_{i=1}^{m} Y_i}{\sum_{i=1}^{m} X_i} \tag{18}$$

2) Estimate the mean of the variable using the larger sample and photo-interpreted values (X_i):

$$\overline{X} = \frac{1}{n}\sum_{i=1}^{n} X_i \tag{19}$$

3) Obtain a more accurate estimate of the sample mean by using the ratio r obtained in Step (1) above to adjust the mean obtained in Step (2) above:

$$\overline{Y} = r\overline{X} \tag{20}$$

The sample variance can be estimated as:

$$s_{\overline{Y}}^2 = \frac{(N-n)S^2}{n} + \left(\frac{n-m}{n}\right)\frac{S_r^2}{m} \tag{21}$$

Where:

$$s_r^2 = \frac{1}{(m-1)} \sum_{i=1}^{m} (Y_i - rX_i)^2$$

$$s^2 = \frac{1}{(m-1)} \sum_{i=1}^{m} (Y_i - \overline{Y})^2$$

Example

Assuming that aerial photography was used to produce a map of the forest in the 50-ha teak plantation example. Assume there is a total of n =200 polygons on the aerial photograph and the surveyor chooses m =15 of these for ground measurement. Let the estimated volume of standing timber from the aerial photograph be represented by X_i and the estimated volume from the ground plots to be Y_i. If $\sum_{i=1}^{m} Y_i = 500$ and $\sum_{i=1}^{m} X_i = 2000$, then $r = 0.25$. Given that the mean of X ($\overline{X}$) is 2000/200 = 10, therefore, $\overline{Y} = 0.25(10) = 2.5$

However, we could also decide to apply a *regression estimator* based on the relationship between the auxiliary variable X and Y to get an estimate of the principal variable Y. The relationship between X and Y could take many different forms (linear, polynomial, etc.). However, for purposes of this discussion, let us assume that the relationship between them is linear. The relationship can be characterised as:

$$\overline{y}_{reg} = \overline{y}_s + b(\overline{x}_l - \overline{x}_s)$$

Where:

$\overline{y}_{reg}$ = regression estimate of mean of Y from the double sampling

$\overline{y}_s$ =estimate of mean of Y from second, small sample

$\overline{x}_l$ =estimate of mean of X (auxiliary variable from first, large sample)

$\overline{x}_s$ =estimate of mean of X from second, small sample

b =linear regression coefficient

n = number of sampling units in first, large sample

m = number of sampling units in second, small sample

In our example above on estimating standing volume using aerial photography, instead of using a ratio of Y to X to estimate the principal variable (Y), we could have used the estimated volumes from the first phase and second phase to develop a regression, and based on this regression, estimate the volume of Y.

6.5. Sample Plot Layout

6.5.1. Fixed Area Plots

Fixed area plot sampling is a method of using sample plots with a fixed size (area) for selecting the trees to be measured. The plots are normally circular or square, although rectangular plots or strips are also in use. Fixed area plot sampling is a type of sampling without replacement since trees are not included in more than one sample plot, i.e., fixed area sample plots do not overlap. In fixed area plot sampling, all trees (regardless of size) weigh equally when tree data is used to estimate stand level variables such as basal area or volume per ha. The weight of a tree in determining stand level volume or basal area is called tree factor or plot expansion factor, usually estimated as the inverse of the plot area in ha. For example, in a 400 m^2 sample plot, each tree factor (TF) is calculated as;

$$TF = \frac{1}{400/10000} = \frac{10000}{400} = 25$$

Tree factor plays a critical role in deciding the size of fixed area plots to use in sample surveys. The rule of thumb is to pick a plot size with the smallest TF possible while mindful of the fact that larger plots may require that more trees be measured. Since TF is the factor by which individual tree estimates are magnified, the use of smaller plots with larger TF can result in high variability and hence larger sampling errors especially in stands where tree sizes vary significantly. In natural stands where tree density may be low, and tree sizes may vary quite substantially, larger plots are typically used to measure large mature tree volume or basal area.

The process of estimating average stand volume or basal area from a fixed area plot is quite straight forward:

1) Calculate the volume or basal area of each tree measured in the plot using the tree diameter and height and a volume equation or table;
2) Expand to obtain the per hectare estimate of volume or basal area for each plot by multiplying each tree volume or basal area by the TF and then obtain the total sum for the entire plot; and
3) Use the appropriate equation based on the sampling method employed, to obtain the average volume or basal area per ha for the target sample population.

6.5.2. Variable Area Plots

Variable-radius plot (VRP) sampling is a method of selecting trees to be tallied based on their size and not the frequency or density of the trees in the stand. The main advantage of using the VRP sampling instead of the fixed area method is that the probability of tree selection is proportional to the size (the basal area at breast height) of the tree. VRP are more efficient to measure than fixed area plots because a plot perimeter is not required. If only basal area per ha is required, the cruiser does not need tree DBH measurements to determine stand basal area. This method of sampling was invented in 1948 by an Austrian forester, Walter Bitterlich. While it is referred to in this book as variable-radius plot sampling, it is also called angle count sampling, angle gauge sampling, plotless sampling, Bitterlich sampling or point sampling.

The principle behind VRP sampling is quite simple; a point is selected at random in the stand and marked as the cruise point. Each tree within the vicinity of the cruise point is imagined to have a given plot (circular) size in relation to its size (DBH or height). The sample trees are those whose imaginary plots include the selected cruise point. That is those trees whose imaginary plots the location of the cruise point falls into. Larger trees with proportionately larger imaginary plots tend to have higher probabilities of being selected at any cruise point.

The cruiser determines whether a tree is included in the plot or not by using an angle gauge device, a prism wedge or a relascope. The surveyor views through the angle gauge device from the cruise point and selects trees that are larger than the width of the angle gauge. These trees are then counted and/or measured for DBH, height and other tree parameters, depending on the survey objective. Most angle gauges used for VRP sampling have calibrated basal area factors (BAFs) that are used to expand tree level estimates to per ha estimates. By definition, BAF is the number of units of basal area per ha represented by each tree tallied in point sampling. Common basal area factors used in North America are 5, 10, 20 and 40, and some angle gauges may two or three of these factors as options. As a rule of thumb, smaller basal area factors will sample more trees. Wedge prisms are typically calibrated to one basal area factor. If not calibrated, the surveyor may do so by following the simple steps below:

1. Select a target tree;
2. Measure "D," the diameter of the subject tree to the nearest tenth of a centimeter or as accurately as possible;
3. Sight through the prism and adjust your distance from the target until a borderline image is obtained, i.e., until the edge of the offset image at breast height viewed through the prism and the edge of the actual image of the tree have some coincide;
4. Measure the distance A from the prism to the target tree's center (in metres). Repeat this measurement 5 times and calculate the average distance; and
5. Calculate basal area as $BAF = (d/2A)^2$.

With a prism, the PRISM is the centre of the point and must be held over the cruise point marker. The above steps may also be used to calibrate the cruiser's thumb or any other device to be used for the inventory, if no angle gauge or wedge prism is available to the cruiser.

6.5.3. Line Transects

The goal of this section is to introduce the concept of line transect sampling. As the name implies, the survey is conducted along one or more transect lines. The observer moves along the transect line(s) and records sightings of sample units and their distances (perpendicular distance) to the survey line. Depending on the scale and the ground conditions, the surveyor may move on foot, in a car or a boat or even in an aircraft. Whatever the method of movement is, the surveyor must adhere to two strict requirements:

1) No sample units must be recorded more than once,
2) Observations must be mutually independent; that is, the observation of a particular element must not impact in any way the observation of another element.

Line transect sampling is popularly used in wildlife surveys or some other ecological surveys. The requirements stated above are difficult to meet in wildlife surveys as some animals and birds in general can be quite elusive such that a particular animal sighted at one point may be sighted and recorded again at some other point in error.

The probability of observing a sample unit in line transect sampling is not constant but varies for the following reasons:

1) Visual obstacles along the transect line and the limited field of vision may impact the ability of the surveyor to observe a sample unit; and
2) In a game survey, the elusiveness of the animals may result in double counting and thereby modify the probability of observing such elements.

In general, it is assumed that all things being equal, the probability of observing a sample unit is inversely proportional to the perpendicular distance from the transect line. That is, the detection could be perfect at the transect line up to a critical distance after which it declines monotonically with distance (Ramsey and Harrison 2004).

A closely related sampling method worth mentioning in this section is the line intercept method used in many ecological studies. The distinction between the two methods is warranted here as the methods have been constantly mixed up. In the Line intercept method, a horizontal line or transect is established, and measurements are taken of sample units at intercepts along the course of the line. Any sampling unit that is not intersecting with the horizontal line is not measured. Line intercept sampling is implemented by extending a measuring tape to create a transect across the site. The observer then proceeds along the line-transect, identifying any target element (say plants) intercepted by the tape and recording intercept distances (the portion of the tape covered by the element). The cover is calculated by adding all intercept distances and expressing this total as a proportion of tape length. Each transect is regarded as one sample unit, so multiple transects must be measured to estimate sample variance and conduct statistical analyses of cover data. Common forestry application is for estimating the amount of woody debris on the forest floor.

6.6. Types of Ground Inventory Surveys

Ground surveys have traditionally been the main ways by which foresters carried out inventory of forest resources. These surveys involve going out physically into the field and collecting data from one or more of the sample plots described above.

6.6.1. Static or Cross-Sectional Surveys

The aim of static or "point-in-time" forest inventory is to provide information about the growing stock of a forest at the time the inventory is taken. The information collected is usually in the form of volume estimates for management decisions regarding the use of the resource. It is also useful for the estimation of relationships that are not time-dependent (allometric relations), but are useful tools for the assessment of the resource (Alder 1980). However, when growth rings are present, it is possible to estimate time-dependent relationships such as growth rates by employing stem analysis (Alder 1980).

In static forest inventories, sample plots are established across a representative portion of the forest estate and measured for individual tree information such as diameter at breast height (DBH), total height (H), and records of other none quantitative tree characteristics such as stem form, tree damage codes. Also measured are stand level variables such as percent slope, aspect, elevation, the average age of the stand. In plantation forestry, these measurements can provide information for building yield curves and tables, which can be used as inputs into forest harvest planning and scheduling.

The use of stem analysis is not very popular in moist tropical regions where most species do not show annual or seasonal growth rings as a result of the absence of seasonal periods of growth alternating with rest periods (Loetsch and Haller 1973). However, those parts of the tropics where marked seasonal droughts occur (e.g., the savannah regions), annual rings may be quite clear and can be used for stem analysis. Tropical deciduous plantation species such as teak exhibits what is termed "false" growth rings which may be relied on for estimating individual tree growth from which stand volume growth can be derived in static forest inventories.

In uniform plantation forests, growth can also be estimated using the temporal sample plot data from stands of different age classes and the same species. The basic requirement is that, there should be sufficient number of observations for each age class, so that, the estimates from plots of the same site quality but different age classes can be combined into a yield curve. Growth can be obtained from this curve by the fact that, growth is a differential of yield (Avery and Burkhart 2002).

6.6.2. Recurrent or Continuous Surveys

Recurrent or continuous forest inventory is defined to comprise all inventory designs in which sampling is used on successive occasions, with the primary objective of estimating the present characteristics of the forests both at the first and the second occasions, and to estimate the change in the forests during the period between the inventories. This change could be a change in volume, recruitment of new trees through regeneration, or the exclusion of trees through mortality. Recurrent inventories are carried out at fixed time steps, usually between 5 and 20 years in length. The length of the interval is chosen based on the dynamics of the forest estate being inventoried; shorter intervals are used for relatively young and fast-changing forests so as to not miss some important changes. In relatively older and slow growing forests, longer intervals are typically chosen to observe any noticeable changes, be it growth or mortality. Another important consideration is cost; since ground inventories are relatively cost intensive, the fewer the number of times it is done the better. However, ultimately, a balance has to be sought between capturing the most important dynamics of the forests and costs.

In recent times, recurrent forest inventory programs are large scale in nature with the intent of generating information on the overall dynamics of the forest estate at a much larger scale. Such large-scale inventory programmes use a combination aerial and ground surveys; the aerial survey helps provide the general picture of the overall landscape and help stratify the forest into relatively homogeneous units typically called polygons. Ground level surveys are then used to collect detailed information on trees, seedling, shrubs and herbs from selected polygons. This detailed information can then be complemented with the aerial survey data to provide the manager with the necessary information needed to plan for management activities over the entire landscape.

In the context of this book, only the ground survey component will be discussed in some detail. The ground survey methods utilise three well-published sampling plot schemes for collecting data: permanent sample plots (PSP); temporary sample plots (TSP); and semi-permanent sample plots (SPSP). The permanent sample plots (PSP) scheme is the oldest and most popular method used in recurrent forest inventory.

As the name implies, permanent sample plots (PSP) are usually fixed area plots and can be circular, rectangular or square in shape (square and circular shapes are the most common shapes) that are established, permanently marked and re-measured periodically. PSPs are established by first deciding on where to locate the plots using an appropriate sampling technique and then the plot size and shape to be used. The plots are then located in the field and marked with permanent markers in the form of posts at the plots corners or at the plot centres or both. These posts can be metal or any other material that does not easily degrade and must be clearly labelled with permanent marks for easy identification. The objective is to facilitate easy location in the future for re-measurements. Also, the location of each PSP or group of PSPs in relation to a landmark feature (natural or man-made), called *Tie Point*, is recorded. The tie point is usually in the form of bearings and distance from the tie point to the centre or a corner of the PSP. In natural forest PSP system, individual trees above a certain minimum size (usually diameter at breast height) that are considered mature trees are usually numbered and tagged with permanent tags bearing the numbers assigned to each tree. In plantation forestry, all trees are tagged and numbered. The idea of tagging and assigning permanent numbers to individual trees is to be able to track the growth and mortality (death) of each tree.

Temporary sample plots (TSP) are the direct opposite; they are usually established and measured only once and no effort is made to re-measure the same plots at subsequent times. The use of TSP in recurrent inventory is less common than the use of PSP. Basically, at each cycle of the recurrent inventory, TSPs are established in numbers that are enough to provide the most accurate estimate of the resource at that point. No permanent markers are left for the purpose of future identification and measurements. In comparison with PSP, TSPs are typically smaller in size but larger in numbers.

The semi-permanent sample plots are in concept what most PSP programs are; a set of plots is established at the start of a recurrent inventory as PSP. After a few re-measurements, some or all of the PSPs are replaced with new plots. The rationale for replacing some or all of the PSPs are typically the following:

1) The stands in which PSPs were originally located grow into stand types that are well represented in the sample. The manager may be required to assess the data at each inventory time to determine if certain stand types are under-sampled since the objective is to obtain an accurate estimate of the forest resource; and
2) The harvest schedule in the forest management plan may dictate that stands in which the PSPs are located be harvested, necessitating that plot distribution among stand types be re-evaluated, and new plots are established in new stands.

Another version of the semi-permanent sampling system is what is termed sampling with partial replacement originated by Bickford (1956) and popularised by Ware and Cunia (1962). In this system, a fixed number of PSPs are established at the start of the recurrent inventory such that subsequent re-measurement, some of the PSPs, termed the “matched plots” are retained while the rest are dropped in favour of new plots, termed “unmatched plots.” The idea is to benefit from the advantages of both PSP and TSP and that is, the ability to estimate growth while at the same time obtain efficient estimates of the current growing conditions.

6.7. Introduction to Remote Sensing

This section introduces some of the common and most recent remote sensing technologies being used in natural resources management. The intent is to identify what these technologies are, briefly describe their general operational principles and how they are applied to collect forestry data. Remote sensing refers to all technologies used for acquiring data from an object without getting into physical contact with the object. By this definition, remote sensing includes technologies such as photogrammetry, satellite imagery, LiDAR and all technologies that involve sensing and recording reflected or emitted energy and processing, analysing, and applying that information. More formally, however, remote sensing is defined as the science and the art of acquiring information about the earth's surface without actually being in contact with it. Remote sensing is a very broad field with very wide applications covering several different fields and professions including agriculture, forestry, geology, hydrology and air pollution monitoring.. Although details may vary, the general principles are essentially the same; the process involves an interaction between incident radiation and the target of interest and the collection of the energy that has been scattered by, or emitted from the target objects using sensors. The energy recorded by the sensor is transmitted, to a receiving and processing station where the data are processed into an image or non-image records. These records are then interpreted to extract information about the target. The information could be stored in the form of digital numeric or geographic information systems (GIS) data that can subsequently be used for further processing.

6.7.1. Aerial Photography

One of the oldest and perhaps the cheapest of all the remote sensing technologies is photogrammetry – the science or art of obtaining reliable measurements through photography. Although there are several variations and applications of photogrammetry, this section will focus on the aerial photography, which has wide applications in forest inventory in many jurisdictions. In broad terms, aerial photographs are taken from the air, normally, taken vertically from an aircraft using cameras capable of recording highly accurate images of features on the ground from higher altitudes. The quality of aerial or air photographs depends on factors such as the type of film, the scale, and overlap and the quality of the photograph determine what type of information can be obtained from the photograph.

The use of aerial photography in forestry involves taking multiple overlapping photos of the ground from an aircraft as the aircraft flies along a pre-planned route called the flight path. These photos are then processed by stereo-plotter into stereo pairs, allowing the interpreter to see two photos at once in a stereo view. The stereo pairs must be geometrically corrected such that the scales on the photos are uniform. This correction allows the interpreter to be able to measure true distance on the photograph. The process is known as orthorectification, and the adjusted photos are call orthophotographs or simply orthophotos. Orthorectification is used to adjust the photographs for lens distortion and camera tilt. There are two ways of collecting inventory data using aerial photographs: directly interpreting the hard copy orthophotograph using a pair of stereoscopes or interpreting the digital orthophotographs on a specialised computer screen using special computer software. The second approach is often called the Softcopy. If hard copies are used, a process call digitising is used to convert the interpreted information into Geographic Information System (GIS) database for further processing. The conversion process for the Softcopy is quite straightforward as the softcopy software easily translates the interpreted data into GIS format.

In forest inventory done using aerial photographs, information such as species composition, stem density, average tree height, percent canopy cover and sometimes basal area per hectare are obtained on relatively homogeneous units called polygons. Polygons are often delineated on an orthophotograph as objectively as possible; following specific guidelines designed to ensure accuracy and repeatability of the interpretation process. The guidelines usually specify among other things, the minimum polygon size, the specific characteristics such as species composition, density and average height to use for delineating the polygons. The interpreted values are calibrated using ground-based measurements from a very few sample plots purposively selected at some target locations deemed to be representative of the various stand types.

6.7.2. Satellite Imagery

The traditional use of the term remote sensing usually refers to this technology. However, remote sensing includes more than the use of satellite imagery to collect data on the earth's surface. In the context of remote sensing, a satellite is a man-made machine that is launched into space and orbits the earth. Satellite imagery is an effective means of observing and quantifying the complexities of the earth's surface. This technology has evolved quite rapidly and has become a cost-effective means of acquiring data on land cover, land use, air pollution, infrastructure, etc. Some of the advantages of satellite imagery include:

1) Transparency and auditability - meaning the process of data acquisition is repeatable and can produce similar results if the same process is followed by different people.
2) Cost effective –due to the altitudes at which the images are taken, the coverage can be quite broad while at the same time producing data at a very high spatial resolution.

3) The process of feeding the data acquired through satellite imagery into existing GIS-based processing systems can be fully automated thus avoiding manual processing and saving time and money.

6.7.3. Satellite LiDAR

Light Detection and Ranging (LiDAR), is a remote sensing technology that uses light in the form of a pulsed laser to measure variable distances to the earth. These light pulses, combined with other data recorded by the airborne system, generates precise, three-dimensional information about the shape of the earth and the characteristics of the earth's surface.

A LiDAR instrument is made up of a laser, a scanner, and a specialised Geographic Positioning System (GPS) receiver. In the field of forestry and other natural resource applications, airplanes and helicopters are the most common platforms used for collecting LiDAR data over broad areas. The process of acquiring LiDAR data begins with an airborne laser pointed at a targeted area on the ground, and a beam of light is reflected by the surface the laser encounters. A sensor then records the reflected light to measure a range. Other data recorded by the airborne system include the position and orientation data, the scan angles, and some calibration data. When the laser ranges are combined with these data, a dense, group of elevation points, called a "point cloud" Is generated. Each point in the point cloud has three-dimensional spatial coordinates (latitude, longitude, and height) that correspond to a particular point on the Earth's surface from which a laser pulse was reflected. The density of the point cloud (measured by the number of points per unit area) determines the type and accuracy of data that can be generated.

In forestry applications, LiDAR is commonly used to measure tree heights from which tree diameter can be generated, stem density, and most recently, with further process and the use of ancillary data from sources such as aerial photographs, species composition can also be determined.

6.8. Main Steps in Conducting Survey Sampling

A successful survey is a result of careful planning and attention to detail at the beginning of the process. Careful planning also avoids problems with data analyses after the completion of the sampling task. It is easy to jump right into data collection without careful planning, with the hope that once the data are collected, one would figure out how to analyse them. This is a mistake, and to avoid this pitfall, we have outlined below the principal steps to be used in designing and implementing a sampling survey. This process is not unique to forestry, and can be applied to any survey (modified from Scharwz 2011).

1. The first step in any survey process is to formulate the objectives of the survey. A concise statement of why the survey is being conducted is essential. An example of a survey objective would be "to estimate the amount of standing volume of wood on a 20-ha plantation."

2. Secondly, one needs to define the population to be sampled. For example, what trees are to be measured and where are they located? Be clear what the population size is, recalling the definition provided in Section 6.2.1. The population could be the 20-ha plantation.
3. Decide on what data is to be collected. It is critical that this decision be made with the research objectives in mind. It is not unusual for researchers to collect data and later on during data analysis to regret that a particular measurement was not taken. The data could be diameter, height, the weight of sample trees, etc. It is more important to collect a few measurements well and accurately than to collect many poorly.
4. Establish the degree of precision required for the survey. For example, in estimating the standing volume of the 20 ha plantation, one could decide that the target mean volume should be $\pm 10\%$ from the population mean. This will be the basis for deciding the sampling intensity (size).
5. Establish the sampling frame, which is a list of sampling units that is exhaustive and exclusive. For example, in cluster sampling, one needs to decide what the clusters will be and what will be measured within each cluster. Alternatively, in the case of two-stage sampling, decisions regarding what will constitute primary and secondary sampling units will have to be made at this stage.
6. Decide on what sampling design to use. Carefully review your sampling objectives, the degree of precision needed, the nature of the population, etc. to help you in choosing among the various designs discussed in this Chapter. Note that in most cases, designs are improved considerably by combining different designs rather than applying only one design-type. For example, most designs can be improved by stratification and this should always be considered at the design stage.
7. Always do a pre-test of your survey. It is very important to try out field methods and questionnaires to help detect any defects in the equipment or misinterpretation of questions to be administered.
8. Organize the field work. Training of field personnel if they have not done this kind of work before will be essential to ensure that measurements are taken consistently and correctly. Training also ensures that safety procedures are followed, and equipment pre-tested to ensure they are in good working conditions.
9. Summary and data analysis. Once all data are collected, they are summarised and prepared for analyses. Keep in mind that the data must be analysed according to the design that was applied in collecting them. No switching of designs at this stage of the game.
10. A post-mortem is the final step in looking back at the design and implementation of the survey to assess what went well, what did not go well and what was tricky. The essence of this is to plan better for next time.

Sampling is so widely used and its results so important that the application of wrong sampling processes and techniques can have far-reaching consequences. Everyone who uses sampling as a way to obtain information for decision-making has to take this task seriously to safeguard the scientific credibility of the results.

6.9. Case Study 1: Modelling Growth and Yield of Neem Plantations

In this section, an application of stand dynamics modelling techniques is presented. The data used in the analyses were collected from neem plantations planted in the Guinea savannah zone of Ghana.

6.9.1. Sampling Procedure

The data were collected from 120 temporary sample plots selected from 30 plantations within the Tamale Forest District area using a stratified two-stage sampling design. Table 5 shows the distribution of sample plantations, plots and the number of trees which were measured.

Table 5. Distribution of sample plantations, plots and trees measured by town/village

Town/Village	No. of Plantations Observed	No. of sample plots measured	Total No. of trees measured
Tamale	8	32	580
Nyankpala	6	24	563
Kumbungu	5	20	500
Kumbungyili	3	12	300
Katariga	2	8	185
Vitin	1	4	100
Nyashie	1	4	100
Tarikpaa	1	4	100
Jangyili	1	4	100
Dalogyili	1	4	100
Choggo	1	4	65
Total	30	120	2, 693

All plantations of neem in the study area were stratified by age into five groups (1-, 2-, 3, 4-, and 5-year age groups). Five plantations were selected randomly for measurement in each stratum. Also, three and two plantations from 6- and 9-year age groups respectively were also selected randomly for measurement. In all plantations selected for study, the following procedures were carried out:

(a) four square plots with a 10 m x 10 m sides (1/100 ha) were set up in each stand. Plot boundaries were laid exactly half-way between the lines of trees to avoid the effect of errors in locating them. Since the usual initial spacing for neem and other plantation species in the area is 2 m x 2 m, in the absence of mortality, each sample plot contained 25 trees. It was realised after measuring a few plantations that selective felling had been done in some of the plantations above three years. It was therefore decided that instead of setting up random plots within the plantations, selective sampling should be done in areas as fully stocked as possible. It was therefore not possible to estimate mortality as this was confounded with human effects.

(b) all trees in each plot were measured for diameter at 50 cm above ground and diameter at 130 cm (DBH) with either digital callipers or a diameter tape (for the larger trees) to the nearest mm, and for total height with height poles to the nearest cm. For trees that were less than or equal to 130 cm in height, only diameter at 50 cm and height were measured. Each stem of a tree forking below breast height was measured at 130 cm and recorded separately and the height of only the tallest stem was measured. A single diameter corresponding to the diameter of the tree of the same basal area as the total basal area of all the forked stems was then calculated using the formula:

$$D_m = 2x\sqrt{\sum_{i=1}^{n} D_i^2} \qquad (22)$$

where

D_m is the mean diameter, D_1, D_2,.. D_n are the diameters at breast height of the first, second, etc. to the *n*th forked stem.

(c) the basal areas of all trees measured on each plot were calculated using the diameters at 50 cm. One tree from each plot with dimensions as close to the tree of mean basal area (at 50 cm) and height as possible was selected and felled close to the ground with a cutlass. Just after felling the sampled tree, the following variables were recorded:
   i) diameter at 50 cm above ground level,
   ii) diameters at 0, 25, 50 and 75% of the total height,
   iii) diameter at breast height,
   iv) total height,
   v) fresh weight of the whole tree,
   vi) fresh weight of a 10 to 30 cm stem section in the centre of the stem for determination of dry to fresh weight ratio.

The stem sections were air dried and their air-dry weights taken.

6.9.2. Data Analysis

Individual Tree Volume Computation

Smalian's formula was used in conjunction with the DBH, height and diameters at 0, 25, 50 and 75% of total height to compute the volumes of the 120 sample trees. As the sample trees were measured at equal intervals along the stems, the formula was reduced to:

Figure 7. Dr. Thompson Nunifu (left) and Mr. James Amaligo taking measurements of a 5-year-old neem plantation.

$$V = \frac{H}{8}(B_1 + 2B_2 + 2B_3 + 2B_4) \quad (23)$$

Where:
V = total stem volume
H = total height
B_1, B_2, B_3, and B_4 = basal areas at 0, 25, 50 and 75% of the total height

Individual Tree Volume Equations

The individual tree volumes were used to develop local and standard volume table equations using regression analysis. Local volume equations estimate volume using only diameter, while standard volume equations use both diameter and height as explanatory variables. Prior to the analysis, 12 of the sample trees were discarded because they were shorter than 130 cm and therefore had no diameters at breast height. The remaining 108 trees were used for the development of the volume tables. Fifteen of the common regression models used in volume estimation (Unnikrihnan and Singh 1984) were fitted to the computed volumes to determine the most appropriate model. Six of these models have single independent variables and the remaining nine have two independent variables. The independent variables used were either DBH or height or combinations of both. The criterion for selecting the best regression models was the Furnival Index (Furnival 1961).

Plot Volume Computation

The best standard volume equation among the 15 models compared on the basis of the Furnival Index was:

$$V = \frac{H}{8}(B_1 + 2B_2 + 2B_3 + 2B_4) \tag{24}$$

where V is the total tree volume in dm^3, D is the diameter at breast height and H is the total height. This equation was used to estimate the volumes of each tree on each measured plot. Plot volumes were then calculated as the sum of the volumes of all individual trees on that plot.

The computed Chi-square value between the actual sample tree volumes and the volumes predicted by Equation (24) was 28.78, which was not statistically significant at the 0.05 probability level [$\chi^2_{(0.95,\ 104)}$ = 124.34]. The average error in using Equation (24) to predict volumes was - 6.07%. This implies that on average, Equation (24) underestimated individual tree volumes by about 6%.

Allocation of Stands to Site Classes

All plantations were designated Site Class I, II or III in relation to their mean dominant height/plantation age. There are two main ways to measure the dominant height of forest trees: top height and predominant height. Top height is defined as the average height of a specified number per unit area of the trees with the largest diameters at breast height. Predominant height, on the other hand, is the average height of a specified number of trees per unit area of the tallest trees in the stand (West 2003). Practically, it is easier to determine top height than predominant height, because it is easier to identify the largest diameter trees in a stand than the tallest ones (West 2003). Therefore, in this study, top height was used.

The method used the minimum - maximum procedure described by Alder (1980). In each age class, the minimum, mean and maximum top heights were calculated. Three separate linear regressions were then fitted to the minimum, mean and maximum sets of observations using the logarithmic transformation of Schumacher's (1939) equation:

$$H_0 = H_{max} \exp(\beta A^{-k}) \tag{25}$$

where H_o is the mean top height, H_{max} represents the maximum height the species could reach on the site, A is the age of the stand, α , β and k are coefficients. Taking natural logarithms of both sides gives:

$$\ln H_0 = \ln H_{max} + \beta A^{-k} \tag{26}$$

For each site class, this equation was fitted using regression techniques. Pecked lines were then drawn half-way between the site classes to show the boundaries between them. In this study, k was estimated to be 1 using the procedure suggested by Alder (1980). This procedure requires that the residual sum of squares for Equation (26) for various trial values of k be computed. The value of k at which the minimum sum of squares is observed provides the best estimate of k. The regressions are then recalculated using this value of k to give the corresponding best estimates of the α and β parameters. If $\alpha = \ln H_{max}$, the equation becomes

$$\ln H_0 = \alpha + \beta A^{-1} \tag{27}$$

6.9.3. Yield Table Construction

The basic form of the Schumacher (1939) equation was used for the development of yield tables for Site Classes I, II and III:

$$\ln Q = \alpha + \beta_1 A^{-k} + \beta_2 S + \beta_3 D_s \tag{28}$$

where Q is some measure of yield (height, DBH, basal area, volume, or fresh weight), A is stand age in years, S is some function of site index, D_s is some function of stand density and α, β_1, β_2, and β_3 are regression coefficients.

The site and stand density were excluded as explanatory variables from Equation (28) since individual equations were developed for each site class and a uniform stand density of 2500 stems/ha was used in all site classes. The reciprocal of stand age had a high correlation with lnQ. For these reasons Equation (28) was reduced to:

Table 6. Diameter and height statistics of measured neem plantations

Age (years)	DBH (cm)			Height (m)		
	Mean	SE	Range	Mean	SE	Range
1	0.82	0.02	0.16 - 2.65	1.42	0.02	0.53 - 3.11
2	2.26	0.03	0.35 - 4.60	2.14	0.03	0.60 - 5.21
3	4.22	0.05	1.14 - 7.05	4.67	0.04	2.11 - 7.76
4	5.43	0.08	2.00 - 8.72	4.74	0.04	3.11 - 9.34
5	6.07	0.09	2.50 - 11.61	5.41	0.05	3.63 - 10.61
6	8.18	0.14	4.71 - 12.26	5.85	0.10	3.71 - 11.73
9	11.40	0.27	9.05 - 16.81	9.21	0.17	7.64 - 12.57

$$\ln Q = \alpha + \beta_1 A^{-1} \tag{29}$$

According to Alder (1980), the β□ parameter is always negative and α is usually between 2 and 7.

6.9.4. Results

Diameter and Height Statistics

The diameter and height statistics of the sample plantations are summarised in Table 6.

The annual diameter increment ranged from 0.64 to 2.11 cm and averaged 1.41 cm. The standard error of the mean DBH ranged from 0.02 to 0.27 cm and increased with age. The mean annual height increment was 0.93 m and ranged from 0.07 to 2.53 m. The standard error of the mean height increased with age from 0.02 m in the first year to 0.17 m nine years after planting (Table 6). The mean diameter at 4 years was 5.43 cm. This is slightly higher than the mean diameter at the same age of 5.14 cm reported for neem in north-eastern Nigeria by Verinumbe (1991). In this study, the mean heights at 2 and 5 years were 2.14 and 5.41 m respectively. These were lower than mean heights reported for the same ages as 3.6 and 7.5 m respectively by Streets (1962). It is difficult to compare these statistics with those of Streets (1962) because it is not clear which part of northern Ghana nor the type of neem plantations from which the measurements were taken.

A simple linear relationship between height and DBH was developed to predict height for any given diameter. This equation is given as:

$$H = 1.167 + 0.675D \qquad (30)$$
$$R^2 = 0.97$$

where H is the total height and D is the diameter at breast height.

Table 7. Summary statistics of form factors

Age (years)	Cylindrical Form Factors			Conical Form Factors		
	Mean	SE	Range	Mean	SE	Range
1	0.25	0.01	0.20 - 0.33	0.76	0.05	0.59 - 0.99
2	0.29	0.01	0.26 - 0.32	0.87	0.03	0.78 - 0.96
3	0.31	0.01	0.21 - 0.48	0.94	0.02	0.69 - 1.44
4	0.30	0.01	0.25 - 0.34	0.90	0.02	0.74 - 1.04
5	0.28	0.01	0.24 - 0.32	0.93	0.03	0.73 - 0.95
6	0.38	0.04	0.28 - 0.42	1.15	0.11	0.84 - 1.35
9	0.41	0.05	0.35 - 0.48	1.25	0.15	0.99 - 1.42

Stem Form Factors

The ideal tree trunk from a timber production point of view is a cylinder. The comparison of tree bole form with cylinders may be expressed as form factors. Form factors are usually measured as the ratio of an upper-stem diameter with the DBH. Higher form factors indicate lower rates of stem taper and vice versa. In this study, both cylindrical and conical form factors (when the stem is compared to a cone) were computed for each sample tree and the means for each year calculated. The summary is presented in Table 7.

The mean cylindrical form factors varied from 0.25 in the first year of growth to 0.41 in the ninth year. The conical form factors ranged from 0.76 to 1.25 in the ninth year. The form factors tend to increase with age though they are uniform within the age groups as shown by their standard errors (Table 7). The current initial spacing of 2 m x 2 m for neem appears to be wider than optimum because the form factors show that the trees are more conical than cylindrical. Closer initial spacing will ensure straighter and more cylindrical boles that are both desirable qualities for poles and rafters.

Relationship between Stem Dry Weight and Fresh Weight

From this study, a simple linear relationship was established between air-dry weight and fresh weight of the stems of the sample trees. A linear regression model that can be used to predict stem dry weight from fresh weight is given by:

$$W_D = 0.716 W_F \qquad (31)$$

$$R^2 = 0.96$$

where W_D is the dry weight of the stem and W_F is its fresh weight.

Local and Standard Volume Table Equations

The Furnival Index was used as the criterion to select the best equations from 15 equations compared for volume table construction. The following two equations were judged the most appropriate for the construction of local and standard volume tables, respectively:

$$\ln V = -0.780 + 1.711 \ln D \qquad (32)$$

[n = 108, R^2 = 0.96, Correction Factor = 1.038]

$$\ln V = -1.689 + 1.165 \ln D + 1.124 \ln H \qquad (33)$$

[n = 108, R^2 = 0.97, Correction Factor = 1.027]

where:

V = overbark stem volume in dm^3,
D = DBH in cm, and
H = total height in m.

To correct for log-normal bias in Equations (32) and (33), correction factors were calculated as $e^{s^2/2}$ where s^2 is the variance of the respective logarithmic equations (Baskerville 1972).

Equation (32) was selected as the best local volume table equation to provide quick volume estimates in places where height measurement is difficult or not possible to undertake. Separate local volume table equations were developed for each site class and compared with Equation (28) using the test of coincidence. The test showed that there was no significant difference between the coefficients of the overall equation (Equation (32)) and the separate equations. Equation (32) underestimates volumes compared to the standard volume table (Equation 33) but is satisfactory for field volume estimations. It is however recommended that the standard volume equation be used whenever precise volume estimates are required.

Table 8. Local volume table for neem in northern Ghana

Diameter Class (cm)	Volume (dm^3)
1	0.48
2	1.56
3	3.12
4	5.10
5	7.47
6	10.21
7	13.29
8	16.70
9	20.42
10	24.46
11	28.79
12	33.41

Equation (33) was selected for the standard volume table (Table 9) because it had the lowest Furnival Index. This equation had the following characteristics: DBH and height explained 97% of the variation in stem volume. Test of coincidence showed that there was a significant difference between Equation (33) and separate equations for the different site classes. Though these differences were statistically significant, they were so small that there is no practical justification for the use of separate equations for each site class.

Without the necessary correction for log-normal bias, Equations (32) and (33) underestimate volumes by only 3.8 and 2.7% respectively and therefore the correction factors can be neglected when precise estimates are not required. To correct for bias, the correction factor is multiplied by the geometric mean to obtain the unbiased arithmetic mean. These correction factors were applied in the construction of both volume tables in Tables 8 and 9. The addition of form factors to the independent variables in both Equations (32) and (33) showed that a change in form had an insignificant effect on volume estimates.

Top Height/Age by Site Class Curves

Equation (20) was used to develop the top height/age curves in Figure 8. This figure shows means and ranges of top height in Site Classes I, II, and III. The top height over age by site class curves shown in Figure 8 are to be used to classify stands into site classes for yield estimation. To determine the site class of a particular stand, it is necessary to determine the top height and age of the plantation. For example, a plantation that is 7 years old with a top height of 11m will be classified as Site Class I, while it will be classified as Site Class III if the same plantation had a top height of 8m, using Figure 8. It may be difficult to determine plantation age on the field since there are no clearly visible annual growth rings. The ages of plantations can usually be obtained from records at the Forestry Office in Tamale or by the Forestry Technical Officer in charge of the area where the plantation is located. Individual farmers and community leaders will also be able to help determine the ages of their plantations.

Table 9. Standard volume table for neem in Northern Ghana (dm^3)

Diameter Class (cm)	Height (m)									
	1	2	3	4	5	6	7	8	9	10
1	0.19	0.42	0.65	0.90						
2	0.43	0.93	1.46	2.02	2.59	3.19				
3		1.49	2.35	3.24	4.17	5.11	6.08			
4			3.28	4.53	5.82	7.15	8.50	9.88		
5			4.25	5.88	7.55	9.27	11.02	12.81		
6				7.27	9.34	11.46	13.63	15.84		
7					11.18	13.72	16.31	18.96		
8					13.06	16.03	19.06	22.15		
9						18.39	21.72	25.40		
10						20.79	24.72	28.77	32.79	36.90
11							27.62	32.10	36.64	41.24
12							30.57	35.52	40.55	45.64

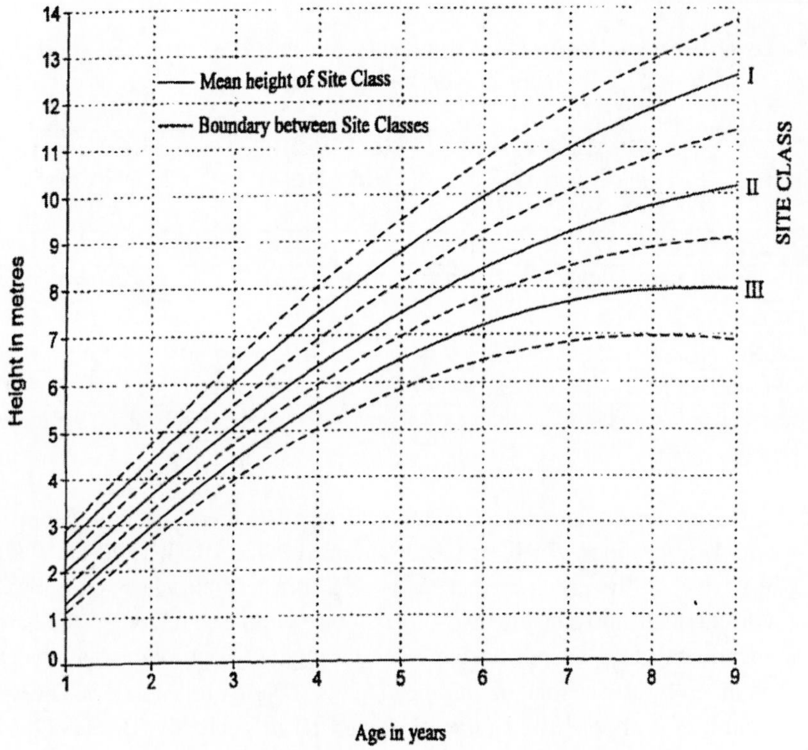

Figure 8. Top height/age curves site class curves for neem in Northern Ghana.

The accuracy of the method used for the site index curves depends critically on the assumption that all sites have an equal likelihood of being represented in each age class (Alder 1980; Clutter et al. 1983). Since the plantations were randomly selected with no conscious effort to include any particular site type, the above assumption is likely satisfied for the 1-5 year age groups. However, it cannot be guaranteed for the 6- and 9-year age groups where only 3 and 2 plantations were measured respectively. For example, if older-age plantations are found on poorer sites and the younger ones on good sites, it can lead to bias in the site index curves. Several factors operate to prevent an equal representation of all sites in the various age classes. Good sites produce bigger and taller trees more rapidly than poorer sites and, therefore, both selective felling and harvests take place earlier on good sites. Secondly, most of these lands on which neem is planted have been removed from agricultural production and since the poorer lands are removed first, older age plantations are more likely to be found on poorer sites. For these reasons, the site index curves produced by this method should be regarded as provisional. Permanent sample plot data will be needed to validate these curves.

Yield Tables

The coefficients of Equation (29) given in Table 10 can be used to develop empirical yield tables for neem in Northern Ghana. The R^2 values for the various yield parameters were high, ranging from 0.78 to 0.94. This means that between 78 - 94% of the variation in yield was explained by variation in age.

Table 10. The coefficients of Equation (29) used to develop the empirical yield tables for neem in Northern Ghana

Variable	Site Class I [Yield Class 12]			Site Class II [Yield Class 8]			Site Class III [Yield Class 4]		
	α	β_1	R^2	α	β_1	R^2	α	β_1	R^2
Diameter	2.75	-2.78	0.94	2.56	-3.17	0.92	2.34	-3.18	0.89
Height	2.50	-1.89	0.93	2.37	-2.14	0.87	1.99	-2.10	0.94
Basal Area	3.90	-5.45	0.89	3.45	-6.33	0.84	3.07	-6.23	0.82
Volume	5.20	-5.17	0.90	4.97	-6.53	0.78	4.80	-10.52	0.98
Fresh Weight	5.29	-4.83	0.92	4.84	-5.26	0.94	4.74	-7.42	0.91

Note: When these equations are used to estimate basal area, volume and fresh weight, the results are on a per ha basis.

The yield table is divided into Site Classes I, II and III; these are equivalent to Yield Classes 12, 8 and 4 m^3/ha respectively at the optimum rotation. Both yield classes and site classes are used to refer to the productive capacity of a forest. Yield classes are measured in m^3/ha/year and are divided into steps of two m^3/ha. For example, Yield Class 12 has a mean annual timber increment of 12 m^3 per hectare. The biologically optimum rotation length for each site class is given by the estimated volume β□□ coefficients in Table 10. These rotation lengths are 5, 7 and 11 years for Site Classes I, II and III respectively. Because the MAI at optimum rotation for Site Class III occurs outside the range of the data (at 11 years), it was estimated as 4.23 m^3 /ha from Equation (29) using the coefficients in Table 10. Yields can be estimated by use of Equation (29) (coefficients in Table 10) for each stand attribute. The first step in estimating yields is to determine the site class to which the stand belongs. To do this, it is necessary to determine the mean height of the dominant and co-dominant trees and the stand age. Once the mean height and age are determined, Figure 8 is used to estimate the site class to which the stand belongs. The yield equation or table applicable to that site class is finally used to estimate the required yield variable.

The Yield Table probably underestimates the growth and yield potential of neem plantations in Northern Ghana as some of the plantations measured had been selectively harvested. Selective harvesting usually removes the largest trees for poles and rafters. Secondly, most of these plantations are established on either abandoned or infertile agricultural lands. The mean diameters and heights given for the plantations on Site Classes I and II indicate that trees on these good and medium sites can be used as small rafters at age three. Site Classes I and II can probably be thinned in the third year, Site Class III in the fourth. The time for thinning suggested here coincides with that being practised by the farmers.

In using either the equations or tables, caution should be exercised in predicting yields beyond the range of the data used in their derivation. This is important because the Schumacher model used here assumes asymptotic convergence and hence predictions of yield beyond age 9 years show that the yields flatten out after about 10 years of plantation age. It should also be appreciated that the yield equations represents averages to be expected for all stands in a given site and age class. Finally, it should be noted that these yield table/equations are applicable to pure neem stands with an initial spacing of 2 m x 2 m. Neem plantations with different initial spacing or those intercropped with other tree species or food crops will be expected to differ in their growth and yield characteristics. For these reasons, it is possible for observed yields for a given stand on the field to differ from those predicted by this equation.

6.10. Case Study 2: Growth and Yield of Teak Plantations in Ghana

This section presents the application of forest inventory and analysis techniques for data collection and analysis to develop growth and yield models. The case study was conducted on teak plantations in the Guinea Savannah Zone of Ghana.

6.10.1. Sampling Procedure

This empirical field study was conducted in the savannah zone in 1996. Sample plantations were taken from four Forest Districts: Tamale, Savelugu, Yendi and Damongo districts, all located in the Guinea Savannah vegetation zone. Data were collected from a total of 100 temporary sample plots from 25 plantations, ranging in ages from 3 to 40 years. Plantations were stratified into one-year age classes and sampled with an effort to equal allocation of three sample plantations to each age group. For each age group, an effort was also made to include the full range of site conditions (from the poorest to the best). A pre-sample inspection was done to assess the conditions of each sampled plantation. Plantations that were found to be badly understocked due to mortality or selective harvesting were replaced.

For each plantation selected, four circular plots each of radius 7 m were sampled at random. For each plot, all teak trees were measured for diameter at breast height (DBH) and total height. Three trees were selected at random from each plot as sub-sample trees. These were felled for detailed measurements of total height and diameters at stump level, breast height and half the height above breast height. In all 12 trees were felled and measured for each plantation. In addition, plantation records from the Rural Forestry Division in Tamale of the Ghana Forestry Department were examined.

Growth and yield survey data from about 42 plantations in the high forest zone (HFZ) of Ghana as well as records of planting and stand tending practices, were obtained from the Forest Services Division Planning Branch, Kumasi Ghana. The ages of these plantations ranged from 13 to 26 years and were mostly from the Offinso Forest District of the Ashanti Region. The data were mainly stand-level summaries including top height, average DBH, basal area per hectare, stem density per hectare, age and gross volume. No individual tree data could be obtained from this zone due to time and resource limitations.

6.10.2. Data Analyses

Volume Estimates for Teak Plantations in the Savannah Zone

Using the diameter and height information obtained from the field, volume estimates of each sampled plantation were obtained. The total volume of each sub-sample tree was computed using the Smalian's formula. The Smalian's formula was used because of its simplicity and for the fact that it was found to be more suitable for the types of measurements taken on the sub-sample trees. Stand volume per hectare was estimated from these volumes in three ways as presented in Nunifu (1997):

1) Two-stage simple random sampling (SRS) (Cochran 1977):

$$V = \frac{N}{n}\sum_{i-1}^{n}\frac{M_i}{m_i}\sum_{j=1}^{m_i}V_{ij} \tag{34}$$

Where V is the total volume per hectare, n the number of sample plots per plantation, N is the number of plots that could potentially be sampled in 1 hectare obtained as $1/a$ where a is the individual plot area expressed in hectares, M_i the number of trees per sample plot, m_i the number of sub-sample trees per plot, V_{ij} the volume of the jth tree in the ith plot.

2) Two-stage sampling with probability proportional to basal area at the second stage (PPG):

$$V = \frac{N}{n}\left(\sum_{i-1}^{n}\left(\sum_{j=1}^{M_i}BA_{ij}\frac{1}{m_i}\sum_{j=1}^{m_i}\frac{V_{ij}}{BA_{ij}}\right)\right) \tag{35}$$

Where, V_{ij}, M_i, m_i, Y, N and n are as defined before and BA_{ij} is the basal area of the jth tree from the ith sample plot.

3) A volume equation was developed using the individual sub-sample tree DBH, height and volume measurements and used to estimate stand volume per hectare (see Equation 6.36);

$$V_t = \alpha + \beta D + \gamma D^2 + \lambda D^2 H \tag{36}$$

Where D is the diameter at breast height (DBH), H the total tree height, V_t is tree volume in cubic decimetres, α, β, γ and λ are regression coefficients. Equation [6.36] was estimated by ordinary weighted least squares (OWLS) (Cunia 1964).

Detailed descriptions of the variance estimators of methods 1) and 2) are given in Nunifu (1997) and Nunifu and Murchison (1999).

6.10.3. Yield Table Construction

Yield estimates for the plantations were used to fit empirical yield curves for teak with site index (SI) included as a predictor variable. The functional form by Schumacher (1939) given in Equation (37) was used for the development of the yield models:

$$Y = \alpha \mathrm{SI} \exp\left((\beta + \gamma/\mathrm{SI})A^{-k}\right) \tag{37}$$

The measure of yield Y could be the mean DBH, mean height, basal area per hectare or volume per hectare; A is the age of the plantation; SI is the site index and α, β, γ and k are constants. Parameter estimates of the yield curves were obtained using the model procedure in SAS/ETS (SAS Institute Inc., 2004). Site index models presented in Nunifu and Murchison (1999) were used to estimate the site index of each plantation. The value for k was determined iteratively and was found to be closer to 0.5 and was fixed at that value for subsequent analysis. Problems of obtaining a unique solution with the data necessitated the grouping the data into three site classes for yield curve fitting, thereby eliminating the need for using SI as a continuous predictor variable. The three site classes I, II and III in order of decreasing productivity, were defined using the site index estimates of individual plantations, as follows: I = SI 16m +, II = SI 12m to 16m and III = SI 12m or less. Two dummy variables d_1 and d_2 were defined such that, d_1=1 site class II and 0 for I and III; d_2=1 site class III and 0 for I and II. The addition of these dummy variables into the linearized form of Equation (37) and eliminating SI as a variable resulted in Equation (38).

$$\ln(Y) = (\theta_0 + \theta_1 d_1 + \theta_2 d_2) + (\beta_0 + \beta_1 d_1 + \beta_2 d_2)A^{-0.5} \tag{38}$$

Where θ_0, θ_1, θ_2, β_0, β_1 and β_2 are parameters to be estimated

Mean annual volume increments (MAIs) were determined as the ratio of volume yield models to age. The current annual increments (CAIs) models were obtained as partial derivatives of volume yield models with respect to age. From these, the ages of maximum MAI and CAI were determined as indicators of potential biologically optimum rotation ages. The Schumacher model was considered for yield modelling because of its simplicity, its wide application in most forestry literature and it's suitability for the data type used (temporary sample plots).More flexible models such as the Chapman-Richards model (Richards 1959) are available but are more suitable for re-measured permanent sample plot data.

6.10.4. Results

The individual tree volume equation developed for teak in this study is given by Equation (39). Individual tree volume estimates from this equation are given in cubic decimetres for convenience and predicted volumes using this equation must be converted into the standard cubic metres (m^3) for further use. Table 12 presents the summary of stand level volume estimates using simple random sampling (SRS), sampling with probability proportional to basal area (PPG), and the volume equation (VE) for the SZ. The stand volume estimates derived using the volume equation was considered the best as it was derived from predicted volume of all trees in the sample plots. They were therefore used in constructing the volume yield model for the SZ.

$$V = -0.36 + 0.96D - 0.13D^2 + 0.05D^2H \quad R^2 = 0.99 \quad SE = 0.11 \tag{39}$$

Summaries of stand level characteristics of all the teak plantations used in this study are presented in Table 11. The stand characteristics presented are average DBH, top height, basal area, stem density and gross volume per hectare. The results show a sharp contrast between the SZ and the HFZ in terms of stem density and average tree size. Plantations in the SZ are relatively much denser (by up to 5 times) than those in HFZ and while average tree sizes (DBH) in the SZ are relatively smaller, the net effects of the higher densities are higher basal areas and gross volumes in the SZ teak plantations of comparable ages.

Field observations and records show that while the older (17 years and older) teak plantations in the SZ were government owned the younger ones (9 years or younger) were private-owned plantations established mostly during the Rural Afforestation Programme. Teak plantations in both cases were not thinned as the government owned plantations were established more for protective than productive reasons and the private individuals were not equipped with the requisite technical know-how to thin stands. All plantations included in the HFZ data were government-owned. The relatively low densities in these plantations were a combination of relatively wider planting spacing to avoid the need for pre-commercial thinning and mortality due to inadequate protection and weeding (Anon. 1992).

Yield Models

The regression coefficients of the yield models are given in Table 11. Models for estimating MAI and CAI are Equations (40) and (41) respectively.

Table 11. Summary of stand parameters of teak plantations in Ghana

Stand age	Average DBH (cm)	Top Height (m)	Density (trees/ha)	Basal area (m^2/ha)	Gross volume (m^3/ha)	Trees / m^3
3	2.76	4.10	1510	0.99	4.13	366
4	3.25	4.82	1613	1.50	5.49	294
6	5.95	7.36	2100	6.87	29.49	71
7	7.67	9.45	1765	8.34	44.33	40
8	7.69	9.62	1705	7.94	43.12	40
9	7.82	9.33	1591	8.22	38.09	42
17	10.78	10.29	1488	14.21	73.74	20
26	18.21	16.82	1413	37.89	339.42	4
31	21.06	17.19	1396	50.14	472.28	3
38	23.58	24.53	1299	58.83	776.72	2
40	19.53	18.37	1429	43.83	220.33	6

Table 12. Stand volume estimates (m^3/ha) by; Two-stage simple random sampling (SRS), probability proportional to basal area (PPG) and volume equation (VE)

Plantation Age (years)	Estimate of gross volume (m^3/ha)		
	SRS	PPG	VE
3	4.50	3.79	4.13
4	7.48	6.07	5.49
6	27.38	26.01	29.49
7	42.47	37.43	44.33
8	35.56	37.10	43.12
9	33.04	32.03	38.09
17	57.06	64.32	73.74
26	303.56	293.75	339.42
31	428.39	437.74	472.28
38	650.31	675.60	776.72
40	160.93	185.49	220.33

Table 13. Regression coefficients for component yield models for Plantation teak from Equation (38)

Yield	Site Class I		Site Class II		Site Class III	
	$\alpha = \alpha_0$	$\beta=\beta_0$	$\alpha=\alpha_0+\alpha_1$	$\beta=\beta_0+\beta_1$	$\alpha=\alpha_0+\alpha_1+\alpha_2$	$\beta=\beta_0+\beta_1+\beta_2$
Average DBH(cm)	4.2011	-4.8790	3.9412	-5.5012	3.5811	-6.2020
Average Height (m)	3.6721	-4.3133	3.6030	-4.9899	3.1015	-5.2314
Basal Area (m^2/ha)	6.1010	-9.1985	5.7272	-10.7312	4.9544	-11.7521
Volume (m^3/ha)	8.1023	-11.1292	7.8691	-12.3323	7.2029	-14.1232

$$MAI = \frac{1}{A}\exp\left(\theta + \beta A^{-0.5}\right) \tag{40}$$

$$CAI = -\frac{1}{2}A^{-1.5}\beta\exp\left(\theta + \beta A^{-0.5}\right) \tag{41}$$

CAI and MAI are estimated to peak at ages; 14 and 31, 17 and 38, and 21 and 48 respectively for Site classes I, II and III. The peak MAI ages are considered to be the biologically optimum rotation ages under the management regimes that characterized the data. As the more productive sites are expected to have relatively shorter biologically optimum rotation ages, it is not surprising that site class III has relative shorter MAI culmination age.

6.11. Conclusion

Forest inventories are important tools used by forest managers for planning and sustainable management of forests. This Chapter not only described some important statistical concepts used in forest inventory but also provided examples of how modelling is applied to actual plantation data from Ghana. The results of these examples have applications to the two plantation species discussed (teak and neem). In most countries, the availability of growth and yield models for their major tree species is considered the basic information for forest management. In Ghana, this basic information is still lacking, and efforts must be made by forestry researchers to develop these tools to support forest plantation development efforts.

Part III. Plantation Silviculture and Management

Chapter 7

CRITICAL SILVICS OF SELECTED PLANTATION SPECIES

7.1. INTRODUCTION

Critical silvics is the vital information on the biological behaviour of a tree species that is needed to grow it as a managed crop (Day 1996). This information usually includes the genetic behaviour of the species when grown as a crop, the most suitable and dependable reproduction methods, the types of stand management required and any difficulties in growing and managing the species (Day 1996). The information provided by the critical silvics of potential tree species for plantation forestry is indispensable for successful plantation establishment and management. While this information is available for some plantation species in Ghana, it does not exist for other species. Secondly, there is no single repository of such information for landowners, plantation developers and forestry practitioners. Through a review of the literature, this chapter provides the critical silvics of selected tree species that are currently used or promoted for use in national plantation development efforts. Based on the critical silvics, the suitability of each of the species for the major vegetation zones in Ghana is suggested.

7.2. CRITICAL SILVICS OF TEAK (*TECTONA GRANDIS*)

7.2.1. Teak in Ghana

Presently, teak is the most important plantation species in Ghana in terms of the areas planted and the value of its wood products. In 2013, about 43,000 m^3 of teak lumber was exported, earning €19 million (Ghana Forestry Commission 2014). Teak was introduced into Ghana between 1900 and 1910 (FAO and UNEP 1981). In fact, trials of teak in Ghana date back to 1905 under the German administration in the Volta Region (Kadambi 1972). Teak has since acclimatised well and has been widely grown in both industrial plantations and small community woodlots. But large-scale plantations of teak in Ghana only started in the late 1960's, under a plantation programme that was initiated with the help of the Food and Agricultural Organisation (FAO) of the United Nations to supplement the supply of wood products from the indigenous natural forests (Prah 1994). By 1987, these plantations were estimated to cover over 45,000 ha (Drechsel and Zech 1994; FAO 2002a) with over 30,000 ha planted with teak, of which an estimated 10,000 ha survived (FAO 2002a).

A further increase in teak plantations occurred following a five-year Rural Afforestation Programme in 1989 under the erstwhile Ghana Forestry Department, which saw a boost in teak planting through the establishment of new plantations and small-scale community woodlots in Northern Ghana. Apart from electric and telephone transmission poles, teak is also valued by small-scale farmers and local communities particularly in Northern Ghana as poles for construction, fencing, rafters, fuelwood, stakes and windbreaks. It has also become an important source of income for private organisations and individuals especially farmers who plant the species on their farms.

Several private companies, firms and individuals have established teak plantations at various scales and for various purposes using various partnerships, programmes and strategies in Ghana in the past decades (FAO 2002a). Notable among these are the Pioneer Tobacco Company Ltd. (PTC) which owns about 5,000 ha of teak plantations mainly in the Brong Ahafo Region, the Ashanti Goldfield Company Ltd., with over 1,400 ha of teak and the Bonsu Vonberg Farms Ltd., which owns about 500 ha of teak (FAO 2002a). Of the more than 50,000 ha of newly planted forests that were established between 2000 and 2004, 60% was teak (ITTO 2005a).

7.2.2. General Description

Teak, also known commercially as *teek* or *teca* (Spanish) belongs to the family *Verbenaceae*. Teak varies in size according to locality and conditions of growth. On favourable sites, it may reach a height of about 40 to 45 m, with a clear bole of up to 25 or 27 m, and a diameter of between 1.8 and 2.4 m (Farmer 1972). Teak is generally resistant to drought and heat and, can survive and grow in a wide range of climatic and edaphic conditions (Hedegart 1976). According to Kadambi (1972), records from Thailand reported a teak tree, claimed to be the world's largest tree (in 1965), with approximately 6.6 m diameter at breast height and 45 m total height. In drier regions, trees are generally smaller. The boles are generally straight, cylindrical and clear when young, but tend to be fluted and buttressed at the base when mature (Kadambi 1972). They tend to fork when grown in isolation but are generally shade intolerant.

7.2.3. Natural Distribution

Teak grows naturally in Southern Asia, from the Indian Subcontinent through Myanmar (Burma) and Thailand to Laos, approximately 9° and 25° latitude and 73° to 103° E longitude (Troup 1921). As an exotic species, teak grows in several parts of the world. First naturalised outside its natural range in Java some 400 to 600 years ago (Kadambi 1972), the earliest plantation of teak has been traced back to 1680 by Perera (1962), when it was successfully introduced to Sri Lanka. Since then, teak has been introduced and acclimatised well in other parts of Asia. Horne (1966) stated that Nigeria was the first place outside of Asia where teak was introduced in 1902, and it quickly spread to other parts of tropical Africa, reaching Ghana in 1905.

7.2.4. Site Requirements

Teak grows on a variety of geological formations and soils (Kadambi 1972; Seth and Yadav 1959), but the quality of growth depends on the depth, structure, porosity, drainage and moisture holding capacity of the soil (Kadambi 1972). Teak grows best on deep, well-drained and fertile soils with a neutral or acid pH (Kadambi 1972; Watterson 1971), generally at elevations between 200 and 700 m, but exceptionally at elevations of up to 1300 m above sea level (Troup 1921). Warm tropical, moderately moist climate is best for teak growth. Optimum annual rainfall for teak is 1200 to 1600 mm, but it endures rainfall as low as 500 mm and as high as 5000 mm (FAO 1983; Hedegart 1976; Kadambi 1972; Troup 1921).

7.2.5. Propagation

Teak is relatively easy to establish in plantations and because of the enduring global demand for products from teak it has good prospects as a plantation species (Krishnapillay 2000). Plantation grown teak is established using stump plants rather than direct sowing of teak seeds which does not always give satisfactory results (Borota 1991). Depending on desired product (fuelwood, poles, lumber or a mixture of products) and the site quality, the initial planting spacing generally range from 1.8 x 1.8 m to about 3 x 3 m (Kadambi 1972). When planted in a taungya system, spacing could be as wide as 4.5 m between rows. Generally, on good soils, wider spacing is used. This results in better diameter and height growth, and also reduces nursery, planting and early thinning costs (Kadambi 1972). On sloping terrain, wider spacing has been suggested to encourage ground cover and to avoid erosion (Weaver 1993).

7.2.6. Management

Teak is shade intolerant but needs training for improved form. Accordingly, plantations must be thinned regularly and heavily, particularly in the first half of the rotation. Initial planting density is usually between 1 200 and 1 600 plants per hectare (Krishnapillay 2000). Closer planting spacing is sometimes adapted to ensure quick canopy closure, thereby achieving training and reducing weeding cost (Adegbehin 1982; Kadambi 1972). However, this practice necessitates early thinning. The time of the first thinning is determined by site quality. Lowe (1976) noted that although thinning may be delayed for 10 to 15 years after planting without unduly affecting the growth potential of the final crop, very heavy thinning becomes necessary if the growth of the final tree crop is to be maintained at satisfactory levels. Teak also coppices quite vigorously, making post-harvest re-establishment much easier than from seed.

7.2.7. Growth and Yield

Teak is fast growing when young, but its overall growth rates on rotation basis are not outstanding (FAO 1956). It is considered moderate to fast growing (Briscoe and Ybarra-Conorodo 1971). A study of the standing biomass of teak in India showed height growth to be most rapid between 10 and 50 years after which it declined (Weaver 1993).

The rotation of teak in India is a function of forest type and management systems (Ghosh and Singh 1981). Plantation crops have rotations between 50 and 80 years, whereas, in areas where teak occurs in mixed stands, the rotation age is about 70 to 80 years. Coppice systems or coppice with standards have rotations of between 40 and 60 years (Weaver 1993). FAO (1985) quotes the peak ages for the mean annual volume increment at 50 and 75 years, respectively, for site classes I and II in Kerala, India, based on stem wood volume. In Indian yield tables for teak (Laurie and Ram 1940), the maximum total volume growth occurs at ages between 5 and 15 years depending on site class. Similar estimates in Trinidad (Miller 1969) are between 7 and 12 years. At Mtibwa, Tanzania, Malende and Temu (1990) estimated the peak ages of mean and current annual increments for teak to be at 42 and 55 years respectively. At base age 20, the site index for teak was estimated by Malende and Temu (1990) to be between 16 and 25 m. In Miller (1969), the estimate is between 15 and 23 m. Akindele's (1991) estimate for North-western Nigeria was between 10 and 29 m. At the same base age, data from Laurie and Ram (1940) ranged from 28 m for site class I to 12 m for site class V. Similar results have been reported by Keogh (1982), Friday (1987) and Drechsel and Zeck (1994). In Ghana, a similar study for teak in the high forest zone reported indices ranging from 17 to 26 m (Anon 1992).

Logu et al. (1988) estimated the aboveground biomass production for teak to be between 2.1 and 273 t/ha for ages 5 and 97 years, respectively. The mean annual biomass increment was estimated to peak at between 10 and 40 years depending on site conditions. Detailed information on the growth and yield of teak in Ghana are provided in Chapters 6 and 9 of this book.

7.2.8. Uses

Teak has gained importance worldwide as a top quality timber species with attractive physical properties such as high natural durability, ease of seasoning without splits and cracks, and attractive grain and colour (Pandey and Brown 2000). These physical properties make teak suitable for various purposes including; electric transmission poles, furniture, shipbuilding and decorative panelling. The growth characteristics of the species make it particularly attractive as a plantation species.

7.2.9. Pests and Diseases

Many kinds of pests and diseases have been identified for teak in other regions of the world. For example, the teak defoliator and skeletoniser (*Hyblaea puera* and *Eutectona machaeralis*) cause extensive damage to young plantations, while root rot due to *Polyporous zonalis* is also common in plantation grown teak (Anon 2009). The pink disease fungus causes cankers and bark flaking, and a powdery mildew caused by *Olivea tectonae* and *Uncinula tectonae* leads to premature defoliation of teak trees (Anon 2009). Fresh leaf extracts of *Calotropis procera, Datura metal* and *Azadirachta indica* have been found to be most effective against teak skeletoniser. This method is of immense importance in insect pest control considering its harmless and pollution free implications on the environment further avoiding the operational and residual hazards that are involved in the use of organic and inorganic insecticides (Anon 2009).

7.3. Critical Silvics of Neem (Azadirachta indica)

7.3.1. Neem in Ghana

According to Streets (1962), neem was introduced into Ghana *circa.* 1915. It was first planted in small plots scattered throughout the country along roadsides and in amenity belts in towns and villages. Neem has acclimatised well throughout Ghana and is popular as a source of fuelwood and as poles and rafters for building construction. Neem extracts are also valued for their medicinal properties in treating malaria and as an insecticide for storing grain. As neem is not browsed by animals and grows rapidly, it is easy to grow and maintain even within villages. Local processing of the many products from neem are essential to meeting basic needs at the rural household and community level and can give rise to the establishment of thriving village industries and thus generate employment and additional income (Radwanski and Wickens 1981). Based on their research of various aspects of neem in Nigeria, Radwanski and Wickens (1981) concluded that there was enough scientific evidence on the subject to warrant the launching of a well-coordinated multidisciplinary research and development programme leading to agricultural, industrial and commercial exploitation of neem. A survey of community woodlots in some districts of Northern Ghana in 1997 showed that over 60% of all tree species planted under the Rural Afforestation Programme (RAP) were neem (Ayamga 1997).

7.3.2. General Description

Neem or nim (*Azadirachta indica* A. Juss.), which is synonymous with *Melia indica* (A. Juss.) Brand., and *Melia azadirachta* L. belongs to the family *Miliaceae*. It is a deep-rooted, small- to medium-sized tree, broad-leaved and evergreen, except in periods of extreme drought (National Academy of Sciences 1980). Mature trees attain heights of 7 – 20 m with a spread of 5 – 10 m and may live for more than 200 years (Ketkar 1976). With its widely extended branches, the tree forms an ovate to round crown on a straight stem (Maydell 1990). The bark is brown-grey, of medium thickness and longitudinally and obliquely fissured; the slash is reddish brown (Maydell 1990). The heartwood is hard and durable with a specific gravity varying from 0.56-0.85 with an average of 0.68. The wood is resistant to termites and other wood-destroying insects even in exposed areas (Maydell 1990).

The leaves are imparipinnate, alternate, 20-40 cm long and 1 – 3 cm wide on slim petioles; 6- 10 cm long. The flowers are white, yellowish or cream-coloured, small, numerous and honey-scented (Maydell 1990). The fruit is an ellipsoidal drupe, with one, rarely two, seeds, 1.2 -1.8 cm long, green-yellow when ripe, with a thin cuticle and juicy fruit pulp (National Academy of Sciences 1980).

7.3.3. Distribution

In its native Indo-Pakistan subcontinent, neem is found in a large belt extending southwards from Delhi and Lahore to Cape Comorin. In South Asia, it is also found in Bangladesh, Upper Burma, and in the drier parts of Sri Lanka (Ahmed and Grainge 1986). In Southeast Asia, it occurs scattered in Thailand, southern Malaysia, and in the drier Indonesian islands east of Java (Ahmed and Grainge 1986). In Africa, neem is particularly widespread in Nigeria and Sudan; it is also found along the East African coastal plains stretching from Ethiopia across Somalia, Kenya and Tanzania to Mozambique and in West Africa in the sub-Sahelian region of Mauritania, Togo, Ivory Coast and Ghana (National Academy of Sciences 1980).

7.3.4. Site Requirements

Neem is very drought resistant and grows with as little as 150 mm of annual rainfall. The optimum annual rainfall, however, is 450 – 750 mm (Maydell 1990). The tree has been successfully grown in regions with up to 2,000 mm annual rainfall and temperatures of between 0°C and 44°C though the tree is frost tender in the seedling and sapling stages (National Academy of Sciences, 1980; Radwanski and Wickens 1981). Neem also grows better on dry, stony, shallow and nutrient-deficient soils than other species. Neem is salt-tolerant and can be grown on marginal soils with low fertility (Ahmed and Grainge 1986).The tree requires neutral to alkaline soils and will not grow well where the pH is less than 6 (Laurie 1974). It tolerates altitudes of 50 – 1,500 m above sea level. Generally, lateral roots may extend radially to 15 m. The tree, however, is intolerant to frequent inundation and lateritic outcrops (Maydell 1990). Poor soil drainage may retard the growth of neem.

Figure 1. A nine-year-old neem plantation near the Business Secondary School, Tamale (in the picture is Dubik Moisob –picture taken in 1996).

7.3.5. Propagation

Propagation is generally by seeds which should be sown immediately after maturity, i.e., December to the end of February (Maydell 1990). This is because the seeds are short-lived and do not retain their viability for long periods (Troup 1921), with a two-week upper limit. The seeds begin to germinate as soon as they fall from the trees (Evans 1992). Loss of viability appears to be due to and is accompanied by the fermentation of the unopened cotyledons inside the inner seed case. If the cotyledons are green, the seeds are good and will germinate, but if the cotyledons have turned brown or yellowish, they are not likely to germinate (Smith 1939). Nagaveni et al. (1987) recommend the collection of neem fruits when they are greenish-yellow and still on the tree, as opposed to the usual practice of collecting fallen fruit. If this is followed by depulping or drying, immediate germination will be delayed and, therefore, permit longer storage (Evans 1992). It is advisable to use only swollen seeds and to transplant when seedlings are 30 – 50 cm high. Neem trees start bearing fruits from the fifth year onwards, and a mature tree produces more than 20 kg of fruit, corresponding to 10 – 15 kg of seeds per year. There are about 4,000 – 6,500 seeds/kg (Evans 1992). The bare-root method has been the traditional way of raising planting stock in the Sahel-Sudan zone of Africa though in dry years survival of the seedlings is often poor (Evans 1992).

In a study by Oboho et al. (1985) in northern Nigeria, germination time did not vary with seed weight, but seedling height at time of appearance of leaflets did, being tallest for the high weight and lowest for the low weight class. The number of leaflets, biomass production and growth rate varied with seed weight. It was realised in this study that the biomass production of the high weight class was three times that of the medium weight class, due to the appearance of twin seedlings from the high weight class.

7.3.6. Management

Neem plantations can be grown successfully in all parts of Ghana. In Northern Ghana, it can be grown on biologically optimum rotations for fuelwood and poles that range from 5 to 11 years. It is probable that neem may be grown to a larger size on longer, less productive rotations (Nanang 1996). Weeds do not affect growth as neem is very resistant to competition and may become a noxious weed under favourable site conditions since the seeds are widely distributed by birds (Maydell 1990). However, in Nigeria a study revealed that freeing young neem plants of grass had a striking effect on planted seedlings. The study showed that strips that had been hoed to rid them of weeds, with little breaking of the soil, showed a spectacular difference in health and survival. The results appeared to be due entirely to the eradication of grass and not to disturbance of soil as cultivation around plants had no better effect than light surface hoeing (Anon., 1952). The tree coppices freely and early growth from coppice is faster than growth from seedlings (Maydell 1990).

Given the invasive potential of neem, it is important that the management of neem includes some practical advice on how to reduce its population when so desired. This advice will ensure that the benefits of neem are maximised while the negative impacts are minimised. Neem is a prolific seeder, characteristic of invasive species. The fruit ripens at the onset of the rainy season. They then germinate and establish themselves while there is available soil moisture (CAB International 2005). This means that neem has the potential to occupy the most fertile lands and displace native vegetation. Neem has a large root system, and when the tree is felled, it re-sprouts easily and difficult to kill. The best way to kill neem seedlings is to uproot them while they are still in their early stages. If the understory of the mature tree is cultivated or cleared, neem seedling survival is much lower than when it is left uncultivated. This means that neem can compete with the other native grasses. It also indicates that with timely weeding, neem can be reduced in the field (Judd 2004). The threat of neem to become an invasive species and spread uncontrollably has to be balanced with the potential benefits that can be derived from its sustainable and controlled management on degraded, less fertile, soils.

7.3.7. Growth and Yield

Neem is a fast growing tree: two-thirds of the height may be reached after 3 to 5 years (Maydell 1990). The rate of development of young neem plants after the first season is fairly rapid. As a rule, the trees put on an average annual girth increment of 2.3 – 3.0 cm (0.73 – 0.96 cm in diameter), though more rapid growth is easily attained (National Academy of Sciences 1980). In four different test plots in West Africa, the height of neem trees varied from 4 to 7 m after the first three years and from 5 to 11 m in 8-year-old stands. According to the National Academy of Sciences (1980), in West Africa, cropping is usually done on an 8-year rotation, with original spacing between the plantation trees most commonly 2.4 m x 2.4 m. In Ghana, first rotation yield (at 8 years) was 108 – 137 m^3 of fuelwood/ha and in Samaru (northern Nigeria), the yield at the same rotation was 19-169 m^3/ha (National Academy of Sciences, 1980; Maydell 1990). Streets (1962) reported mean heights for Northern Ghana as 3.6 and 7.5 m after 2 and 5 years respectively. In Cuba, a small stand of neem trees planted on a fertile soil reached a mean height of 14.2 m and a mean DBH of 27 cm eight years after planting (Betancourt 1972). Mean tree diameter after four years of growth in Nigeria was reported as 5.14 cm and survival was more than 75% (Verinumbe 1991). In the semi-arid Sahel of Africa, neem typically achieves growth rates of 5m^3/ha/yr (Evans 1992). Under favourable Sahel conditions, the annual leaf biomass production may reach more than 10 tonnes/ha (Maydell 1990).

7.3.8. Uses

Due to the multipurpose use of neem, it has been hailed as a wonder tree. Its uses span from improving soil fertility to curing various forms of ailments, giving it the name, the "pharmacy tree." Some of the major uses of neem are outlined below.

Energy and Fuelwood

Neem seeds contain up to 40% oil, which is used as fuel in lamps and as a lubricant for machinery (National Academy of Sciences, 1980; Maydell 1990). Neem has long been used as fuel in India and Africa. It has become the most important plantation species in northern Nigeria and is planted for fuelwood and poles around the large towns (National Academy of

Sciences 1980). The high wood production capacity of neem, coupled with its high calorific value of 51.1 MJ/tree and fuelwood value index of 3.9 at 56 months after planting, make the species an ideal choice for this purpose (Lamers et al. 1994).

Construction, Shade and Windbreaks

The natural durability of neem makes it a good choice for use in building construction and for making furniture in rural areas. The wood is tougher than teak and very similar in characteristics to mahogany (National Academy of Sciences 1980). In rural areas, supply of poles for building is frequently the most pressing need and for many purposes, pole diameters at breast height of 5-10 cm are required along with a smooth surface, freedom from snags and resins and good natural durability (Evans 1992). Neem possesses most of these qualities. Neem has been used successfully as a windbreak and as a source of shade for humans and cattle. It is a splendid street tree for the arid tropics. In Niger, shelter belts of neem interplanted with other species reduce wind speeds by up to 65% (National Academy of Sciences 1980).

Pest Control

A pool of biologically active constituents, including the triterpenoids azadirachtin, margosan-o, salanin, and meliantroil, are found in neem leaf, fruit, bark, and seed (Schmutterer, 1982; Warthen 1979; Evans 1992). These compounds reportedly control more than 100 species of insects, mites, and nematodes – including such economically important pests such as the desert and migratory locusts, rice and maize borers, pulse beetle and rice weevil, root knot and reniform nematodes and citrus red mice (Grainge et al. 1985; Warthen 1979; Jacobson 1958, 1975). Modes of control include antifeedant, growth regulatory, repellent, hormonal, or pesticidal action in larval and/or adult stages of these pests.

Traditionally, Indo-Pakistani farmers simply mixed 2 – 5 kg of dried neem leaves/100 kg of grain to control stored-grain pests (Ahmed and Koppel 1985). Alternatively, empty sacks were soaked overnight in water containing 2 – 10 kg of neem leaves/100 litres water and then dried these sacks before filling them with the grain (Ahmed 1984). Evans (1992) indicated that mulches of neem foliage inhibit termite attack of newly planted trees. It is reported that the insecticidal effect of azadirachtin is as good as DDT and is not toxic to man (Maydell 1990).

Soil Improvement

In a case study in northwest Nigeria by Radwanski and Wickens (1981), soil analysis showed that the average pH value of red sands without neem was 5.4 in the first and second layer, and the average pH in the corresponding layers under neem was 6.8. A substantial increase in the soil pH values under neem was due to the accumulation and decomposition of leaf litter and is a surface phenomenon. The content of organic carbon had risen from 0.12% under fallow to 0.57% under neem, and the total nitrogen content increased from 0.013% in the top layer of the fallow soil to 0.047% in the corresponding layer of the neem phase. Neem is a nonleguminous tree, and there is no symbiotic nitrogen fixation in the soil. However, free-living, nitrogen-fixing bacteria are known to be C heterotrophic and thus dependent on the supply of organic carbon for their energy (Mengel and Kirkby 1978). It is possible therefore that a substantial increase of the organic carbon content in the neem soil may create conditions favourable for the proliferation of those bacteria and a consequent increase in the supply of nitrogen (Radwanski and Wickens 1981).

Studies in the Gambia report a steady rise in the mean values of phosphorous content as neem matures at the site. Young neem trees do not drop much litter when establishing at a site, and no other plants vegetate the site and hence the organic matter is less for those locations with young stands. It is not surprising that neem increases phosphorus. As the neem grows older, it changes the site, producing more and more organic matter, adding litter to the soil, thereby increasing the phosphorous and organic matter (Judd 2004).

To evaluate the effects neem plantations have on the yield of food crops, surface soil under 12-year-old plantations of neem was used to grow food crops in north-eastern Nigeria. The results indicated that two months after planting, the crops produced five times higher biomass on the neem plantation soil than on the control. The trees had favourable effects on soil fertility and therefore improved crop yield (Verinumbe 1991).

The extensive root system of neem can extract nutrients from deep subsoils and enrich surface soils through litter. Thus, in northwest Nigeria, significantly higher total cations, cation exchange capacity, base saturation and pH were observed in soils under neem than on similar soils under fallow (Radwanski and Wickens 1981). Mulching sorghum (*Sorghum bicolor*) with neem leaves in Burkina Faso improved sorghum yields by up to 422% of the unmulched control (Tilander 1993).

The oil cake of neem is a good fertiliser and is effective in reducing the attack of agricultural crops by termites and diverse insects after fertilisation (Datta 1978). In northwest Nigeria, neem is used on degraded agricultural lands for soil amelioration to improve the pH value and to make available soil nutrients for commercial crops (Maydell 1990).

Medicinal Values

Many medicinal uses have been reported for neem. The bark, leaves, fruit, oil and sap reportedly cure various skin diseases, venereal diseases (syphilis), tuberculosis, etc. (Maydell 1990). Ahmed and Grainge (1985) and Hepburn (1989) also report that neem oil has contraceptive qualities. Undiluted neem oil showed strong spermicidal action and was 100% effective in preventing pregnancies in rhesus monkeys and human subjects (Sinha et al. 1984).

7.3.9. Pests and Diseases

Neem trees are generally pest-free, due perhaps to the presence of azadirachtin and other insecticidal compounds (Csurhes 2008). However, neem plantations have been badly damaged by a scale insect, *Aonidiella orientalis*, in Africa, and to a lesser extent in India (NRC 1992). Certain species of ants, moths and bugs are also known pests of neem (NRC 1992). Live neem trees are susceptible to borers and termites (Hearne 1975). Because neem has not been extensively studied in Ghana, there is no record of the pests and diseases that affect the trees in this environment.

7.3.10. Case Study: Assessing the Potential of Neem Plantations for Use in Agroforestry

An on-farm experiment to evaluate soil fertility status and crop productivity under neem plantations in an agroforestry setting in Northern Ghana is reported by Nanang and Asante (2000). Three neem plantations of different ages *viz*: 9, 7 and 5 years were selected. The status of soil nutrients under these plantations was determined, and sorghum and groundnuts were planted in the 7-year-old plantation. The treatments were: pruned, unpruned and control (adjacent plot without neem trees). Soil samples were collected from each plantation site at 0-20 cm and 20-40 cm depths. The experimental design was a randomised complete block with two replications.

Soil Analyses

Initial soil analysis before cropping indicated that pH under the neem plantation was significantly higher than pH on the control plot. Though differences in soil pH under plantation of different ages was not significant, a clear pattern of pH was observed, i.e., the older the plantation, the higher the pH. Generally pH below 20cm depth was lower than pH at 0-20cm depth. Soil organic carbon ranged between 0.340 – 0.550%. The older the plantation, the higher the organic carbon content obtained.

Total nitrogen followed the same trend as soil organic carbon and ranged from 0.019-0.046%. The control plot had lower soil total nitrogen than soils under the neem plantations. The difference between total N level at 0 – 20 cm and 20 – 40 cm was not significant. Soil available phosphorus ranged from 1.16 -5.55 mg/kg. No clear trend was observed with regards to either the depth of sampling or the age of the plantation. Soil exchangeable potassium was clearly higher at 20 – 40 cm than at 0 – 20 cm. It ranged between 37.88 – 177.7 mg/kg. The age of the plantation did not influence soil exchangeable potassium content.

Soil pH after harvesting food crops from the treatments with neem trees ranged from 5.29 to 5.53. The difference between control and soil under neem plantation was not significant. The higher soil pH under the plantation could be attributed to the higher amount of organic matter in the plantation produced by leaf fall as compared to the control plot with low vegetative cover.

Available soil phosphorus and exchangeable potassium may be more related to the soil than to the plant material addition. There was a reduction in soil organic matter, carbon, total nitrogen at harvest compared with initial levels. This could be attributed to mineralisation and uptake. Organic carbon after harvest was between 0.34 and 0.38%; total nitrogen ranged between 0.015 – 0.023%, soil available phosphorus 2.55 to 5.22 mg/kg and exchangeable potassium between 22.87 and 29.76 mg/kg. These soil chemical properties were not influenced by plantation, either pruned or unpruned.

Crop Yields

Sorghum stover yield ranged from 4.5 – 9.7 t/ha for all treatments. Both pruned and control treatments gave significantly higher stover yield than the unpruned treatment. The difference between the pruned and the control plot was however not significant. Grain yield ranged from 2.2 – 3.7 t/ha. Groundnuts vine yield was significantly influenced by treatment. Vine yield on control was more than twice that of the pruned area and more than six times that of the unpruned plot. Pod yields followed a similar trend as the vine ranging from 0.72 to 1.24 t/ha. The differences in yield were significant at 5% significance level. Grain yield ranged from 0.067 to 0.107 t/ha (67 to 107kg/ha). The pruned plot had higher grain yield than the control though the difference was not statistically significant. Both yields on the control and pruned treatments were significantly higher than the unpruned treatment.

The lower yields from the unpruned treatments for both crops were probably due to shading effect. The canopy of the plantation may have intercepted sun radiation of the intercrop reducing photosynthetic activities and hence a reduction in dry matter production. The lower yield observed from the unpruned plot could also be attributed to moisture stress. The tree with its broad canopy has a large surface area for transpiration and hence results in competing for soil moisture with the intercrop. Competition for moisture was probably lower on the pruned plot compared to the unpruned. This could have resulted in less moisture stress on the intercrop in the pruned plots, hence its better performance.

These results suggest that neem plantations can positively influence soil fertility and hence have a potential for use in agroforestry practices in Northern Ghana. However, adequate pruning of the trees at certain stages of growth to reduce shading of the companion crops and minimise moisture stress of this evergreen tree is required. The pruning should be incorporated into the soil to increase soil nutrients. This approach has been used in Burkina Faso, where neem branches and suckers are trimmed back at the beginning of the rainy season and left to enrich the soil (Bationo et al. 2004). This illustrates an integrated approach to manage neem on the farm: the crops benefit from the trees (organic matter and nutrients in the soil) and negative effects (shade) are minimised through management (Judd 2004). Further studies in Northern Ghana can quantify the influence of shading and moisture stress on the yield of companion food crops.

7.4. Critical Silvics of Gmelina Arborea

7.4.1. Gmelina in Ghana

According to FAO (2002a), there are about 6,350 ha of gmelina plantations in Ghana. The most extensive plantation of gmelina in Ghana is owned by the Subri Industrial Plantation Ltd. (SIPL), which is a parastatal organisation that was formed under the previous Forestry Department (FAO 2002a). An estimated 4,000 ha of gmelina was planted for the production of fibre for a proposed paper mill which was intended to be established at Daboase on the eastern banks of River Pra in the Western Region of Ghana (FAO 2002a). Plantation establishment at the SIPL site started in 1971 and ended in 1996. The pulp mill was never built due to the lack of financing, and this delay in the establishment of the pulp mill resulted in the addition of the production of sawlogs to the management objective in 1990. Exports of gmelina wood products in 2013 were about 4,459 m^3 (value of €508,681) in the form of air-dried lumber, billet and poles.

7.4.2. General Description

Gmelina arborea Roxb. is a tree species belonging to the family *Verbenaceae*. Gmelina is a fast growing tree frequently planted in plantations to produce wood for light construction, crafts, decorative veneers, pulp, fuel, and charcoal (Hossain 1999). The species is also planted in taungya systems with short-rotation crops and as a shade tree for coffee and cocoa (Hossain 1999). Gmelina is a deciduous, medium-sized tree that grows up to 40 m tall and 140 cm in diameter, but usually smaller than this (Jensen 1995). The tree form is fair to good, with 6–9 m of branchless, often crooked trunk and a large, low-branched crown (Hossain 1999). The leaves are simple, opposite, more or less heart-shaped, 10–25 cm long, and 5– 18 cm wide (Hossain 1999). Evans (1992) estimated that the number of seeds per kilogram varies from 700 to 2500. Gmelina begins to flower and set fruit at about 6 to 8 years of plantation age. The first flowers are borne 3-4 years after planting and, in nature, self-pollination is discouraged by the floral morphology (Orwa et al. 2009).

7.4.3. Distribution

Gmelina is native to the South and Southeast Asia from Pakistan and Sri Lanka to Myanmar (Hossain 1999). According to Jensen (1995), the species has been widely planted in Southeast Asian countries such as Bangladesh, Myanmar and Thailand. However, it has not been as widely planted in tropical African and Latin American countries (Evans 1992), although it has been introduced into many tropical countries, including the Philippines, Malaysia, Brazil, Gambia, Costa Rica, Burkina Faso, Ivory Coast, Nigeria, and Malawi (National Academy of Sciences 1980).

7.4.4. Site Requirements

Gmelina is found in rainforests as well as dry deciduous forests and tolerates a wide range of conditions from sea level to 1200 m elevation and annual rainfall from 750 to 4,500 mm (Orwa et al. 2009). It grows best in climates with mean annual temperature of 21–28°C (Jensen 1995). The best growth of gmelina is achieved on deep, well-drained, base-rich soils with pH between 5.0 and 8.0. The species can tolerate a 6–7-month dry season and grows on many soils, acidic laterites to calcareous loams, doing poorly on thin or poor soils with hardpan, dry sands, or heavily leached acidic soils, well-drained basic alluviums (Duke 1983).

7.4.5. Propagation

Gmelina can be propagated by seeds, cuttings, or stumps (Alam et al. 1985). Under natural conditions, germination takes place in the rainy season soon after fruits fall from the tree. The germination rate for fresh seed is 65–80% (Hossain 1999). Fresh seed can be stored at room temperature for about 6 months. Seed stored at 4°C will remain viable for about three years. The seed should be soaked in cold water for 24 hours before planting. Seeds should be planted in germination beds with a mixture of sand and loam and covered with a thin layer of sand or compost (Hossain 1999). Seeds germinate in 2–3 weeks and are ready for transplanting to polybags when the first pair of leaves appear. Root pruning and hardening off of the seedlings are beneficial for maximum field survival. Within six months, seedlings reach a height of 30–45 cm and are ready for planting in the field (Hossain 1999). Seedlings are usually ready for stump preparation in 7–8 months and should have a root collar diameter of at least 2.5 cm (Hossain 1999). The stem and roots of seedlings should be pruned back to 5 cm and 20 cm, respectively. Mortality rates for stump planting are usually as high as 50% (Hossain 1999).

7.4.6. Management

In plantations, an initial spacing of 2 x 2 m is commonly used, while a spacing of 4.5 x 4.5 m is used for agroforestry (Hossain 1999). Under favourable conditions, the growth of the seedlings is rapid, particularly from the second year onward (Duke 1983). Because gmelina is shade-intolerant and sensitive to competition, 3–4 weedings are required during the first two years of growth (Hossain 1999). If gmelina is grown for pulpwood and sawnwood, rotations of 6 and 10 years, respectively are commonly used, while for fuelwood, between 5–10 years are common. If 10-year rotations are used, thinning 50% of the stand at about half-way through the rotation is carried out, and another 50% at seven years (Hossain 1999). Gmelina lends itself to coppicing and the second rotation is usually produced using this method. If a third rotation is desired, seedlings and stumps are used (Duke 1983). In Ghana rotations of 6 years are used for pulpwood purposes, while, for sawnwood, the usual rotation is 10 years.

Gmelina has suitable characteristics for use in agroforestry systems, with its fast growth, ease of establishment, and relative freedom from pests outside its natural range (it can be repeatedly browsed without damage) (Orwa et al. 2009). It is an especially promising fuelwood species because it can be established easily, regenerates well from both sprouts and seeds, and is fast growing (Orwa et al. 2009). Although able to compete with weeds more successfully than many other species, it responds positively to weeding and also benefits from irrigation (Orwa et al. 2009). Planting gmelina with crops like maize and cassava has been found beneficial in increasing the simultaneous production of wood and food.

At the SIPL plantations, an initial spacing of 4 x 2 m (1250 trees/ha.) has been observed to be the best for gmelina for the production of pulpwood and sawlogs (FAO, 2002a). According to Cabaret and Nguessan (1988), due to difficulties of obtaining a straight bole and frequent problems of forking and multiple stems, a planting density of about 1,100 stems/ha should be used to ensure natural pruning and a sufficiently large stock for a good selection at the time of thinning.

7.4.7. Growth and Yield

Gmelina is a fast growing species, and on good sites, can reach 20 m height in 5 years, and can attain more than 30 m in height with about 60 cm DBH at maturity (Hossain 1999). The form of the tree is fair to good, with 6–9 m of clear bole (Duke 1983). Some trees can reach 3 m after a year from planting and 20 m after 4.5 years. In Nigeria, the yield of gmelina is 84 m^3/ha at age 12 in poor sandy soils, 210 m^3/ha at age 12 in clay or lateritic soils, and 252 m^3/ha at age 10 in favourable alluvial soils—all volumes are under bark to 7.5 cm top diameter (Adegbehin et al. 1988). The National Academy of Science (1980) reports annual increments greater than 30 m^3/ha on fertile sites. On the SIPL plantations in Ghana, MAI of about 19m^3/ha/yr have been recorded, with current annual increments of up to 22m^3/ha/yr (FAO 2002a). With proper management of the plantations, it is possible to obtain higher yields than these.

7.4.8 Uses of Gmelina

Davidson (1985) describes the wood as yellowish or greyish-white, even grained, and very useful for planking, panelling, carriages, furniture, and carpentry of all kinds. The specific gravity of gmelina wood is 0.42–0.64 (Davidson 1985). It is easily worked, readily takes paint or varnish, and is very durable under water. The wood is used for light construction and pulp as well as for fuelwood and charcoal. Fuelwood derived from gmelina provides 4,400–4,800 Kcal/kg (Davidson 1985). Graveyard tests indicate that the untreated timber may last 15 years in contact with the soil, with pulping properties superior to most hardwood pulps (Duke 1983). The wood makes a fairly good charcoal and according to Little (1983), the leaves are harvested for fodder for animals and silkworms; the bittersweet fruits were once consumed by humans.

7.4.9. Pests and Diseases

Numerous pests and diseases affect gmelina in its native range, but it is unknown what effect they may have on crop/production yields of native plant populations. Some fungal pathogens have been introduced into areas where the trees have been established as exotics Browne (1968). Among these, leaf spot caused by *Pseudocercospora ranjita* is most widespread although it has not caused any substantial damage (Wingfield and Robison 2004). Wingfield and Robison (2004) note that a serious vascular wilt disease caused by *Ceratocystis fimbriata* in Brazil resulted in the most significant failure of gmelina in plantations. Among the insect pests, the defoliator *Calopepla leayana* (Chrysomelidae) appears to be most important, although no serious insect pest problems have been recorded where gmelina is grown as an exotic (Wingfield and Robison 2004).

A bark disease (worm disease) that can girdle the base of the tree and cause dieback of branches in 2-year-old plantations is spread by *Griphosphaeria gmelinae* (Orwa et al. 2009). In Indonesia, one of the insects consistently associated with the species is a carpenter worm *Prionoxystus sp.*, which bores into stems of saplings, feeds from within and weakens them (Orwa et al. 2009). In Ghana, heart-rot in gmelina plantations at the age of 20 years has been observed at the SIPL plantations, which resulted in reduced coppicing ability (FAO, 2002a). In addition, grasshopper attacks on the leaves of gmelina have also been reported (FAO, 2002a). There are opportunities to reduce insect and disease problems through biological control of insects and integrated disease and pest management, especially using vegetative propagation breeding and selection for insect and pathogen tolerance will facilitate the propagation of healthy trees (Wingfield and Robison 2004).

7.5. Critical Silvics of Cedrela Odorata

7.5.1. Cedrela Odorata in Ghana

It is on record that *Cedrela odorata* was introduced into Ghana in 1898 (FAO, 2002a). Known plantations of cedrela in Ghana are: about 3,800 ha owned by the Forestry Commission within forest reserves, 40 ha planted under the Gwira Banso Project (FAO, 2002a), 1,000 ha at the SIPL site in Daboase and another 641 ha at the Fure River Forest Reserve near Prestea in the Western Region where a cedrela plantation was established in 1971. Due to its desirable wood characteristics, cedrela is one of the priority exotic species that is being promoted for use in plantation development programmes in Ghana. The Forestry Research Institute of Ghana has established a seed orchard for cedrela to produce good quality seeds for stakeholders. In 2009, about 2,000m^3 of cedrela lumber and 30,000m^2 of sliced veneer were exported from Ghana. The total volume of cedrela wood products exported in 2013 was about 3,019 m^3 in the form of lumber and veneer, earning about €1.9 million (Ghana Forestry Commission 2014).

7.5.2. General Description

Cedrela odorata belongs to the family *Meliaceae*. It is a deciduous tropical tree that grows to a maximum height of about 30 – 40 m. About two-thirds of the bole is clear of branches. The wood has a strong aromatic odour, and the fruit is a woody capsule, while the very small seeds are winged and spread when the ripen fruit splits open (Lamb 1968). The species is insect pollinated and has wind-dispersed seed (Cavers et al. 2004). Trees bear fruit from the age of 10 years according to Lamb (1968) or 15 years according to Lamprecht (1989). Flowering is annual, but good seed crops occur every 1-2 years (Orwa et al. 2009). Flowers appear early in the rainy season and fruits mature during the dry season when the leaves become deciduous. Seeds are wind-dispersed (James et al. 1998). Fruit development takes about 9 or 10 months, and fruits ripen during the next dry season. The fruit, a large woody capsule, is borne near branch tips. Fruits ripen, split, and shed seeds while still attached to the parent tree (Lamb 1968). Cedrela is listed by the International Union for Conservation of Nature (IUCN) category of species that face a high risk of extinction in the wild in the medium-term due to over-exploitation (CITES 2007).

7.5.3. Distribution

The native range of cedrela is the forests of moist and seasonally dry subtropical or tropical life zones (Holdridge 1976) from latitude 26° N on the Pacific coast of Mexico, throughout Central America and the West Indies, to the lowlands and foothills of most of South America up to 1200 m altitude, finding its southern limit at about latitude 28°S in Argentina (Chaplin 1980; Tosi 1960). Plantations of cedrela have been established both within and outside of its native range.

7.5.4. Site Requirements

Cedrela needs a plentiful supply of nutrients and is very intolerant of water logging (Cintron 1990; Lamb 1968). The species tolerates soils high in calcium and prefers fertile, free draining, weakly acidic soil (Orwa et al., 2009). It tolerates a long dry season but does not flourish in areas of rainfall greater than about 3000 mm or on sites with heavy or waterlogged soils (Beard, 1942; Malimbwi 1978). Cedrela develops best in seasonally dry climates, as reflected in its deciduous habit. It grows at altitudes from 0-1900 m, with a mean annual temperature range of 22-26 °C. It reaches the greatest prominence under an annual rainfall of 1,200 to 2,400 mm with a dry season 2 to 5 months long. The species can also survive in lower rainfall areas, to as low as about 1,000 mm of annual rainfall, although under these situations, it will grow slowly and become stunted (Miller et al., 1957; Wadsworth 1960). The common denominator appears to be drainage and aeration of the soil, not soil pH (Styles 1972; Holdridge 1972; Whitmore 1976).

7.5.5. Propagation

Transplanting of naturally regenerated seedlings or establishing branch and stem cuttings are the most common propagation methods. Grafting and budding methods have also been successful (World Agroforestry Centre 2009). The World Agroforestry Centre (2009) database on tropical forest species describes the propagation of cedrela as follows. Seeds may be broadcast or sown in lines in level nursery beds and lightly covered with soil, sand, sawdust or charcoal. Where there is adequate moisture, shade is not necessary; a shade increases the risk of damping-off. Germination takes 2-4 weeks. It is fastest at temperatures 30-35°C, but seed also germinates at 15°C. Seedlings grow very quickly and may attain 40-50 cm height after three months and 130-150 cm after 12 months. Stumps, striplings and container seedlings are used for planting; occasionally wildings may be used. Direct seeding is feasible, as the young plants develop very quickly; as trees seem to experience a rather severe planting shock; this method is recommended when there is no shortage of seed. Cedrela does not coppice readily nor produce root suckers, and is not fire resistant (Beard 1942).

7.5.6. Management

Although cedrela tolerates weed competition during the seedling stage (Whitmore 1976), it is classified as intolerant of weeds and shade at the sapling stage and beyond (Malimbwi 1978). Pruning is not required when cedrela is grown as a stand, but trees affected by *Hypsipyla* attack may need pruning to remove multiple leaders formed (Orwa et al. 2009). The root system of cedrela is superficial, and this makes the tree subject to wind-throw damage, especially after thinning. In mixed stands, it is realistic to raise only 10-20 high-quality trees/ha. Well-formed, straight stems are usual, except in trees grown in open places (Orwa et al. 2009).

7.5.7. Growth and Yield

Generally, cedrela is a very fast growing tree, once it passes the vulnerable early sapling stage, adding 2.5 cm or more in diameter and 2 m in height a year under good conditions (Omioyola 1972). During the first nine years in trial plantations of cedrela in Java, the mean annual increment was 17 m^3/ha at 650 m altitude and 28 m^3/ha at 800 m altitude. A 40-year-old plantation in Nigeria yielded a timber volume of 445 m^3/ha (Omioyola 1972). Cedrela shows potential for plantations, as it is fast growing and produces multipurpose timber (Orwa et al. 2009).

7.5.8. Uses of Cedrela

Cedrela is widely harvested for use as timber by virtue of its durability, excellent working qualities and appearance. It was reported to be perhaps the most important local timber for domestic use in tropical America (Rendle 1969). The wood is also in high demand in the American tropics because of its natural resistance to termites and rot (Cintron 1990). The aromatic wood is also used to make cigar boxes. The wood is suitable for making non-structural elements for exteriors and interiors, quality furniture and novelty and craft items (Anon., 2004; Echenique-Marique and Plumptre 1990). Cedrela trees have many low branches and spreading crowns, and this makes them suitable for use to provide shade and as a windbreak in courtyard gardens and cocoa and coffee plantations (Orwa et al. 2009). The species can also be planted for ornamental purposes along roads and in parks.

7.5.9. Pests and Diseases

Plantations of cedrela have suffered snail damage in Malaysia and Africa, while slugs killed some nursery stock of an exotic provenance in the Virgin Islands. Beetle damage is a problem in some plantations in Africa (Malimbwi, 1978; Omoyiola 1973). According to Holdridge (1976), the most serious insect pest of cedrela is the mahogany shoot borer *Hypsipyla grandella.* The larvae of this moth eat the pith just behind the growing tip of fast-growing shoots, causing death of the apical meristem, which slows seedling and sapling growth and may ruin tree form, since multiple leaders or bushiness often result from the insect attack (Holdridge 1976). Shoot borer attack may also contribute to seedling mortality, especially in already stressed populations (Allan et al. 1973; Grijpma 1976). Chaplin (1980) reports that provenance trials of cedrelas from a wide geographic range have shown that they may vary in response to the attack, and, therefore, careful selection may allow future development of tolerant strains. Because cedrela is highly susceptible to *Hypsipyla* attack, it is recommended that trees be planted in mixed plantations, for example with *Leucaena leucocephala*, or under the light shade of trees such as *Eucalyptus delgupta* (Orwa et al. 2009). Even though there is *Hypsilla alcate* in Ghana that attacks the *Khaya* species it has not been observed to attack cedrela yet (FAO 2002a).

7.6. Critical Silvics of Kapok (*Ceiba Pentandra*)

7.6.1. Ceiba Pentandra in Ghana

Ceiba pentandra (kapok or silk cotton tree) is an important tree species occurring in the natural forests of Ghana. It is exploited as a commercial timber species from the wild and exported in the form of lumber, plywood, sliced veneer or mouldings. In 2013, more than 52,443m^3 of ceiba was exported, mostly as lumber, sliced and rotary-cut veneer and plywood (Ghana Forestry Commission 2014). The total value of exports was about €18 million. The main problem with ceiba exports is the relatively low value of its wood in international markets. Local and national markets for ceiba timber are weak; the average log price in Ghana's domestic market was US$31–55/m^3 in 2005, while the average sawn wood price was US$ 53/m^3 (ITTO 2005c).

Although the potential of ceiba to be planted in plantations has been recognised and FORIG has been carrying out research on the artificial regeneration of the species including vegetative propagation, there are currently no large-scale plantations of ceiba in Ghana. The only known plantations of ceiba are two small ones in the savannah zone. These plantations are located near Tamale in the Northern Region and were planted to produce kapok and edible seeds (FAO 2002a). The first plantation was established in 1993 and is only about 0.8 ha in size and was planted at a spacing of 8 x 8 m (FAO 2002a). The second ceiba plantation is located at Bogunayili and was established on a 5 ha plot. This plantation is located on a sandy soil that was lying fallow and was planted at a spacing of 10 x 10 m. The planting is said to have been done at about 1960 (FAO 2002a). The average measured diameter of the trees was about 70 cm in 1997 (FAO 2002a).

In addition to its use as timber, ceiba is widely planted in Ghana as an ornamental tree along avenues and around buildings. Its seeds and leaves are popular as sources of food and are used to prepare local soups. The fibre is a main source of material for stuffing pillows and mattresses, and for insulation by local people across Ghana. Recent studies suggest that there is a potential to use new processing techniques to process the fibre in textile applications, as well as a biodegradable alternative to synthetic oil-sorbent materials, due to its hydrophobic-oleophilic properties (Duvall 2009). The hydrophobic-oleophilic characteristics of the kapok fibre could be attributed to its waxy surface, while its large lumen contributes to its excellent oil absorbency and retention capacity (Lim and Haung 2006).

7.6.2. General Description

Ceiba pentandra belongs to the family *Bombacaceae* and is a tall, deciduous tree bearing short, sharp prickles all along the trunk and branches; supported by pronounced buttresses at the base (Orwa et al. 2009). The trunk is cylindrical to slightly convex. The crown spherical to round, with bright green, open foliage; branches verticillate and abundant, sloping upwards; the bark is smooth to slightly fissured, pale grey, with horizontal rings, protruding lenticels and sharp prickles that are irregularly distributed on the upper part of the trunk (Jøker and Salazar 2000). It has a light crown and is leafless for a long period. The ceiba cultivated in West Africa and Asia is a medium-sized tree up to 30 m tall, bole unbranched, usually spineless, buttresses small or absent, branches horizontal or ascending, leaves intermediate, flowering annually after leaf-shedding.

The tree is obligately deciduous, losing its leaves for 10–14 weeks in the dry season, and it usually flowers annually in the leafless period. Leafing and flowering periods are more regular in drier parts of the distribution area; in moister areas, leafing and flowering periods are highly irregular (Duvall 2009). The flowers open at night and are senescent by midday; they are pollinated by bats, but are also visited by moths and bees. The fruits ripen 80–100 days after flowering, the dehiscent types releasing kapok with loosely embedded seeds that are wind-dispersed (Duvall 2009).

Leaves are alternate with slender green petioles. There are usually 5 leaflets in a mature form. The leaflets hang down on short stalks; short pointed at the base and apex, not toothed on edges, thin, bright to dark green above and dull green beneath (Orwa et al. 2009). Fruits are leathery, ellipsoid, pendulous capsule, 10-30 cm long, usually tapering at both ends, rarely dehiscing on the tree. White, pale yellow or grey floss originates from the inside wall of the fruit. Seed capsules split open along 5 lines. Each capsule releases 120-175 seeds rounded black seeds embedded in a mass of grey woolly hairs. Seeds are usually dark brown (Orwa et al. 2009).

The wood of ceiba is creamy white to yellow, often with greyish veins with a coarse texture and interlocked grain. The wood is light and not durable unless treated with appropriate preservatives. It is amongst the most vulnerable African timbers to termites, *Lyctus* beetles, and other boring insects (Duvall 2009). It is susceptible to white-rot and blue-stain fungi, but resistant to very resistant to brown rot. The wood is extremely vulnerable to decay when in contact with the soil; however, it readily absorbs preservatives: both vacuum-pressure and open tank systems give good penetration and absorption (Duvall 2009).

7.6.3. Distribution

Ceiba originated in the American tropics. Natural occurrence from 16°N in the United States, through Central America to 16°S in South America, but is cultivated widely in the tropics between 16°N and 16°S (Jøker and Salazar 2000). Its natural distribution has been obscured by its widespread introduction after about 1500 (Duvall 2009). Duvall (2009) argues that although the literature has often indicated that ceiba was introduced by humans into tropical Africa, there is no historical evidence of such introduction, and there is strong ecological, botanical and cytological evidence that the tree is native to Western and Central Africa as well. Ceiba is now cultivated all over the tropics, but mainly in South-East Asia, especially in Indonesia and Thailand. In addition, there are records of the species in 13 other countries in East and Southern Africa (including South Africa) and the Indian Ocean islands, but the tree has probably been planted in all other tropical African countries as well (Duvall 2009).

7.6.4. Site Requirements

According to Orwa et al. (2009) ceiba can be found in various types of moist evergreen and deciduous forests, as well as in dry forests and gallery forests. As a pioneer species, it mostly occurs in secondary forests. The preferred altitude is from sea level to 900m above sea level with a mean annual temperature range of 18-38°C. Ceiba requires abundant rainfall during the vegetative period and a drier period for flowering and fruiting (Duvall 2009). Optimal mean annual rainfall is 1500 mm, but does well in its area of natural distribution with mean annual rainfall of between 750-3,000 m. Ceiba can grow on a variety of soils, from sand to clay soils provided they are well-drained, but prefers alluvial soils, slightly acidic to neutral, with no water logging (Jøker and Salazar 2000). Assessments of soils in the dry semi-deciduous forests of Ghana showed that most of the upland soils with the exception of those with shallow, eroded, ironstone concretions, boulders and outcrops were suitable for the growth of ceiba (Adjei and Kyereh 1999).

7.6.5. Propagation

Ceiba is usually propagated by seed, although it can also be grown from cuttings (Orwa et al. 2009). Without any pre-treatment seeds germinate slowly (less than 10% one month after sowing) and germination may continue for 3–4 months. The seeds contain large amounts of oil that tend to go rancid quickly, and the viability diminishes rapidly (Jøker and Salazar 2000). Bushfires may cause simultaneous germination of seeds (Duvall 2009). The following description of propagation is adopted from Duvall (2009). The seeds of ceiba may be stored up to one year in glass or plastic containers at 4°C and 60% relative humidity. Fresh seeds have germination rates of 90–100% within 3–5 days after sowing when pre-treated by scoring lightly and soaking in water for 24 hours or by soaking in boiling water for 5 minutes. Germination is good in sandy soil with temperatures of 20–30°C (Duvall 2009). When the young plants are 12–15 cm tall, they can be exposed to full sunlight. Young plants can be grown in a nursery and be transplanted into the field when they are 4–10 months old. It is recommended, however, to sow directly on land that has been properly cleared for planting. *Ceiba pentandra* can also be easily propagated from cuttings, which should be taken from orthotropic branches. Trees raised from seeds root deeper than those raised from cuttings but develop slower (Duvall 2009).

7.6.6. Management

Ceiba is light-demanding, and growth is spindly and poor and mortality is high for seedlings and saplings in shaded locations, including small canopy gaps that close relatively quickly (Duvall 2009). The recommended initial planting distances should be 5 x 5 m. Plantations can be thinned to remove every other second row after 6 years of growth, to reach an optimal spacing of 10 m between rows (Duvall 2009). In the first 2 years after planting, vegetation must be cleared periodically around saplings. General, tending may be necessary for the following years, by cutting climbers and lianas, and removing dead and diseased trees. Plantations need not be thinned if planted with 7 x 7m spacing unless intercropped with smaller tree crops (Duvall 2009).

7.6.7. Growth and Yield

Growth of ceiba is relatively fast. Seedlings planted in Ghana were 29 cm tall 6 weeks after germination and 63 cm after 51 weeks (Duvall 2009). The annual increases in height and diameter during the first 10 years are about 1.2 m and 3–4 cm, respectively. In forest gaps, height growth may be 2 m/year (Duvall 2009). A tree 70 cm in diameter above the buttresses yields on average 4 m^3 of timber, and trees 100 and 150 cm in diameter above the buttresses yield 9.3 and 23 m^3 of timber, respectively (Duvall 2009). Under optimum conditions a full-grown plantation tree may yield 330–400 fruits per year, giving 15–18 kg fibre and about 30 kg seed. A satisfactory average annual fibre yield is about 450 kg/ha, whereas about 700 kg/ha will be considered very good yield (Duvall 2009).

7.6.8. Uses

Ceiba has many important uses around the world. The leaves, which contain about 26% protein is used as fodder to feed sheep, goats and cattle (Orwa et al. 2009). Supplementing goat feed with ceiba foliage increased growth rates of the animals and appeared to be a viable option to improve the nutritional status of goats during periods of the year when grazing is restricted (Kouch et at. 2006). In addition, the fibre from the inner wall of the fruit is unique in that it combines springiness and resilience and is resistant to vermin, to make it ideal for stuffing pillows, mattresses and cushions (Orwa et al. 2009). It is light, water repellent and buoyant, making it ideal for life jackets, lifeboats and other naval safety apparatus. It is an excellent material for insulating iceboxes, refrigerators, cold-storage plants, offices, theatres and aeroplanes (Orwa et al. 2009). It is a good sound absorber and is widely used for acoustic insulation; it is indispensable in hospitals, since mattresses can be dry sterilised without losing original quality (Orwa et al. 2009). The tree is an important source of honey and also suitable for soil erosion control and watershed protection. In agroforestry, it is grown with coffee, cocoa, and in Java as support for pepper trees. In India it is used in taungya systems (Jøker and Salazar 2000).

Traditionally, ceiba has been used to make canoes by hallowing out the entire trunks and for lightweight furniture, utensils, containers, musical instruments, mortars and carvings (Duvall 2009).The timber from ceiba, though not naturally durable can easily be machine planned and sanded and resistant to splitting when screwed. The wood is easy to peel for veneer. Reported uses of the wood include plywood, packaging, lumber core stock, light construction, pulp and paper products, canoes and rafts, farm implements, furniture and matches. Ceiba seed contains 20-25% non-drying oil, similar to cottonseed oil, used as a lubricant, in soap manufacturing and in cooking (Orwa et al. 2009). Orwa et al. (2009) report some medical properties of ceiba. For example, the compressed fresh leaves of ceiba are used as medicine against dizziness; decoction of the boiled roots is used to treat edema; gum is eaten to relieve stomach upset; tender shoot decoction is a contraceptive, and leaf infusion is taken orally against cough and hoarse throat. In Tamilnadu, India, the leaves are pounded together with fermented, boiled rice water and the extract is administered to cows orally as a remedy for reproductive problems. The dose is approximately 500 millilitres applied three times a day for three consecutive days (Orwa et al. 2009).

7.6.9. Pests and Diseases

Insect defoliators include *Ephyriades arcas, Eulepidotis modestula, Oiketicus kirbiyi* and *Pericalia ricini*. The tree is also a host to parasitic plants such as *Dendropthoe falcata* and *Loranthus spp*. (Orwa et al. 2009). High seedling and sapling mortality may occur in humid climates as a result of leaf spot, dieback, damping off and anthracnose. These infections are caused by various fungal pathogens (Duvall 2009). Ceiba is a host tree of Cocoa Swollen Shoot Virus (CSSV) causing swollen shoot disease in cocoa, a disease which has had a devastating effect on cocoa production in Ghana and neighbouring countries (Duvall 2009). Ceiba itself shows considerable resistance to this disease.

In Ghana, dieback of seedlings and saplings has been observed in the nursery and plantation. Studies by Apetorgbor et al. (2003) to identify possible pathogenic causes of damping off, leaf spot, anthracnose and dieback in ceiba seedlings and saplings showed that leaf spot and anthracnose were caused by *Colletotrichum capsici*, whereas under favourable conditions, *F. solani* and *L. theobromae* were found to be associated with dieback of stem in both the nursery and field. The authors found that Kocide at 6.6g/l and Aliette at 5g/l were the most effective fungicides to treat these infections but it retarded the growth of the ceiba seedlings. The fungicides were found to be most effective when applied as soon as seedlings emerged and continued fortnightly (Apetorgbor et al. 2003).

7.7. Critical Silvics of Wawa (*Triplochiton Scleroxylon*)

7.7.1. Wawa in Ghana

Triplochiton scleroxylon (known in Ghana as wawa) is a major commercial timber tree of West and Central Africa. In Ghana, it is found in the natural forests of the moist evergreen, moist semi-deciduous and the forest-savannah transition zones. The potential of wawa to be used in plantation development has been recognised, although there are no large scale plantations in Ghana. Wawa has been heavily exploited in Ghana for many decades. Available records show that in 1959 Ghana exported 650,000 m^3 of logs and 30,000 m^3 of sawn timber (Bosu and Krampah 2005). In 1992, wawa accounted for 64% of Ghana's log exports (Richards 1995). As a result, a ban on the exports of wawa in log form was implemented in 1993, which subsequently shifted the exports to secondary and tertiary wood products. In 2009, about 36,733 m^3 of wawa was exported in the form of kiln-dried lumber, earning about €10 million (Ghana Forestry Commission 2010) constituting 52% of all kiln-dried lumber exports and by far the number one exported timber species from Ghana in lumber form that year. In addition, smaller quantities of wawa were exported as sliced veneer, plywood, mouldings and kindling. Exports of wawa wood products in 2013 increased to about 43,274 m^3 mostly in kiln-dried lumber and moulding. These exports from wawa earned over €16 million (Ghana Forestry Commission 2014).

7.7.2. General Description

Triplochiton scleroxylon (Schum) belongs to the family *Sterculiaceae*. It is a large deciduous forest tree commonly attaining 45 m in height and 1.5 m in diameter. The boles of mature trees are often heavily buttressed but usually free from branches (Orwa et al. 2009). The timber is whitish to pale straw with no difference between heartwood and sapwood, while the texture is medium to coarse with grain typically interlocked that gives a striped figure. The wood has an unpleasant smell when green but usually does not persist after drying (USDA 2010). The heartwood is whitish to pale yellow, indistinctly demarcated from the sapwood, which is up to 15 cm thick (USDA 2010). Leaves are 10-20 cm long and broad, palmate with 5-7 lobes, cordate and 5-7 nerved at base, lobes broadly ovate, triangular or oblong, rounded or obtusely acuminate at the apex; glabrous; stalk 3-10 cm long (Orwa et al. 2009).

7.7.3. Distribution

Wawa is widely distributed in the West and Central African forest zones from Guinea east to the Central African Republic, and south to Gabon and the Democratic Republic of Congo. It is commonly planted in its natural area of distribution (e.g., in Côte d'Ivoire, Ghana and Nigeria), and occasionally elsewhere, such as in the Solomon Islands (Bosu and Krampah 2005).Within its natural limits, wawa is found mainly in forests at low and medium altitudes in the monsoon equatorial forest belt. Throughout its natural range, there is always a marked dry period between December and April. Wawa is referred to as a pioneer species, and it has been suggested that shifting cultivation in West Africa has influenced the natural distribution (Orwa et al. 2009). Hall and Bada (1979) note that where extensive forest disturbance results from human activity, wawa may invade areas where it was formerly rare or absent, becoming closely associated with a group of species quite different from that with which it grew originally.

7.7.4. Site Requirements

Wawa occurs up to 900 m altitude in regions with an annual rainfall of up to 3,000 mm, but is most abundant at 200–400 m altitude and in areas with an annual rainfall of 1,100–1,800 mm and 2 rainy seasons (Bosu and Krampah 2005). It prefers more fertile, well-drained, ferruginous soils with light or medium texture and acid to neutral pH (Bosu and Krampah 2005). It does not tolerate water logging, and, in general, avoids swamps. It is a light-demanding pioneer species and is unlikely that the species ever occurs naturally at elevations as high as 1000 m (Hall and Bada 1979).

7.7.5. Propagation

Due to its fast growth under plantation conditions, wawa is one of the indigenous species most favoured for artificial regeneration in the forest zone of Ghana (Hall and Bada 1979). However, problems of availability and storage of seed have in the past frustrated efforts to initiate large-scale planting schemes (Hall and Bada 1979). Wawa is known for its irregular flowering and erratic fruiting (Mackenzie, 1959; Jones 1974). The seeds are viable for only 2–3 weeks at room temperature, although they may be stored for up to 18 months if dried sufficiently (Howland et al. 1977). Seeds from mature fruits lose their viability after five years of subsequent storage at 0-5°C (Orwa et al. 2009). Vegetative propagation techniques have been developed for wawa to overcome the difficulties of seed supply, and to encourage reforestation efforts (Nketiah et al. 1998).

Most fruits collected from the ground have been attacked by insects. The fruits can be collected from the trees when still green just before maturation. Seeds start to germinate 1–2 weeks after sowing, but the germination rate is often low (Bosu and Krampah 2005). Germination rate and speed increase when the seeds are pre-treated by moistening between layers of damp cotton wool (Bosu and Krampah 2005).The erratic seed production is a major drawback for the establishment of plantations, but methods of relatively cheap vegetative propagation are possible and will offer great opportunities when superior germplasm becomes available (Bosu and Krampah 2005).

7.7.6. Management

In tests, wood material from plantation-grown trees was found to not be inferior to that from trees harvested in natural forests (Bosu and Krampah 2005). The high growth rates, allowing comparatively short cutting cycles, the generally good form of the boles, and the possibility of planting in mixtures with other timber species make wawa even more promising. The species has been used both for plantation establishment and in enrichment planting in Ghana and Nigeria (FAO, 2002a). From 1967 to 1995 about 3,000 ha was planted in Côte d'Ivoire, where they are grown on a cutting cycle of fewer than 40 years. In Nigeria, wawa is planted in agroforestry systems with cocoa (Bosu and Krampah 2005).

7.7.7. Growth and Yield

Under natural conditions, seedlings may reach 15 m tall and 15 cm in stem diameter after four years. Mean annual diameter increment in the forest averages 1 cm. In a 19-year-old plantation in Ghana with 600 stems/ha, the trees were on average 21.8 m tall with a bole diameter of 27 cm (Bosu and Krampah 2005). In Ghana, the total standing volume in natural forests was estimated at 39.30 m^3/ha in inventories in 2001, and an exploitable volume of 16.50 m^3/ha (Bosu and Krampah 2005). The final harvest in plantations in Côte d'Ivoire yields 200–250 m^3/ha of timber, of which 170–200 m^3 from the bole, with an annual volume increment of 8–13 m^3/ha (Bosu and Krampah 2005). In the moist evergreen forest zone of Ghana, the MAI was estimated to be 20m^3/ha/yr while it was about 12m^3/ha/yr in the moist semi-deciduous zone (FAO 2002a).

7.7.8. Uses

The most important value of wawa in Ghana is for the production of lumber for both domestic uses and for exports. Besides, veneers, plywood and mouldings are manufactured from the timber of wawa. The wood is also used for fibreboard and particle board, block board, boat and ship building, boxes and crates, cabinet making, plywood, furniture components, marquetry, mouldings, bedroom suites, building materials, casks, chests, cutting surfaces, excelsior, furniture, interior construction, radio, stereo and TV cabinets (Orwa et al. 2009). The leaves are prepared as a cooked vegetable or sauce in traditional cuisine in Côte d'Ivoire and Benin, while the bark is used to cover the roof and walls of huts in southern Ghana (Bosu and Krampah 2005).

7.7.9. Pests and Diseases

In Ghana, the moth *Anaphe venata* causes extensive defoliation, and *Trachyostus ghanaensis* (the wawa borer) weakens the wood by causing tunnels in standing trees (Orwa et al. 2009). Heartwood is not resistant to attack by termites and other insects, and it is susceptible to attack by pinhole borers, long-horn beetles and sap-stain fungi (Orwa et al. 2009). Sapwood is prone to powder-post beetle attack. A die-back fungus, *Botryodiplodia theobromae*, which reduces most mechanical properties, is sometimes present in this species.

The roots are very sensitive to fungal rot. In Nigeria, the cricket *Gymnogryllus lucens*, the grasshopper *Zonocerus variegatus* and the psyllid *Diclidophlebia* sp. can cause serious damage to seedlings, and the wood borers *Eulophonotus obesus* and *Trachyostus ghanaensis* also cause damage to adult trees (Bosu and Krampah 2005).

7.8. Matching Species to Vegetation Zones

Based on the critical silvics of the six tree species, it is possible to make some general recommendations regarding the suitability of these species for use in plantation forestry across the different vegetation zones in Ghana.

Table 1 summarises the site and climatic characteristics of the major vegetation zones in Ghana. Based on these characteristics, Table 2 then summarises the site and climatic requirements of the six plantation species discussed in the previous sections. All six species can be grown in all parts of Ghana, although as noted in Table 2, the species should be grown in areas where their productivity can be optimised, especially if large-scale commercial plantations are envisaged. For example, gmelina has the potential to grow well in northern Ghana based on its site requirements, although the species has not been widely promoted. Cedrela, on the other hand, will not do very well in the Sudan savannah zone, due to the limited rainfall of about 1000 mm/year.

Table 1. Site and climatic characteristics of the major vegetation zones in Ghana

Vegetation zone	Rainfall (mm)	Mean annual temperature (°C)	Soils
Wet Evergreen (WE)	1750-2200	26-29	Highly leached and acidic. pH 4-5.5. They are low in cation exchange capacity, available phosphorus, nitrogen and organic matter
Moist Evergreen (ME)	1500-1750		
Moist semi-deciduous (MSD)	1250-1700		
Dry semi-deciduous (DSD)	1250-1500		
Guinea savannah (GS)	960-1200	26-32	Well-drained, friable, porous loams; sandy or silty loams. Slightly acidic with pH 4-6
Sudan savannah (SS)	960-1000		

Source: FAO (2002a); Nanang and Asante (2000).

Table 2. Summary of site requirements for the six selected plantation species

Species	Optimum site requirements				Uses	Suitable Ecozone
	Rainfall (mm)	Elevation (m)	Temp (°C)	Soil		
Teak	Optimum 1200-1600 Tolerates: 500-5000	0-1300 m	13-40	Well-drained and fertile soils Neutral to acidic pH of 6.5-7.5	-Timber production - Electric and telephone transmission poles -Fuelwood	All vegetation zones. However, productivity will be lower in the GS and SS due to reduced rainfall
Neem	Optimum: 450 – 750 Tolerates: 450-2000	50-1500	0 – 44	Well drained soils, neutral to alkaline pH<6	• Building construction • Fuelwood • Windbreaks • Pest control Soil improvement • Medicinal uses	All vegetation zones, but better suited to the GS
Gmelina	Optimum: 750 – 4000 Tolerates: 6-7 months of dry season	0-1200	21-28	Deep, well-drained soils, pH of 5-8	• Timber production • Fuelwood Pulp	All vegetation zones

Table 2. (Continued)

Species	Optimum site requirements				Uses	Suitable Ecozone
	Rainfall (mm)	Elevation (m)	Temp (°C)	Soil		
Cedrela	Optimum: 1200-2400 Tolerates: 1000 – 3000	0-1900	22-26	Fertile, free-draining soils that are weakly acidic	• Timber • Construction • Shade • Windbreaks • Ornamental tree	All vegetation zones, but less suited to the GSZ due to rainfall limitations
Ceiba	Optimum: 1500 Tolerates: 750-3000	0 – 900	18 – 38	Deep permeable volcanic sandy to loamy soils that are free from water logging. Slightly acidic to neutral	• ornamental • timber • plywood, • veneer • fodder • fibre • medicines	All vegetation zones, but less suited to the evergreen and moist-semi-deciduous zones
Wawa	Optimum: 1000 - 3000 Tolerates: 4 -6 months of dry season	0 – 900	24 – 27	Fertile, well-drained, ferruginous soils with light or medium texture and acid to neutral pH	• furniture components • timber • plywood • boxes and crates particle and fibre board • artificial limbs	All vegetation zones, but less suited to the savannah zones due to limited rainfall

7.13. Conclusion

Successful plantation establishment and management depend on several factors, most of which are decided by the manager. However, information on the critical silvics is also important for matching species to their most productive site. The critical silvics information can also be used to produce maps of the best ecological zones for the target species. The information on the six species in this chapter contributes to that effort.

Chapter 8

PLANTATION ESTABLISHMENT AND PROTECTION

This chapter provides an overview of plantation forestry practices related to the establishment and protection of young trees that include seed collection and storage, germination and the types of seedling stocks raised in nurseries. It also describes the methods of site preparation, planting techniques and methods of protecting plantations from damage by biological and physical agents.

8.1. SEED COLLECTION

8.1.1. Methods of Seed Collection

Most plantations are regenerated from seed, and hence it is important that seeds that are used for plantation establishment are chosen carefully. Ideally, seeds should be collected from parent trees of superior phenotypes that show vigorous growth and high yields, physical attributes suited to the intended purpose of the plantation, adaptable to the site to be planted, and resistant to local pests and diseases (Nyland 2007). Understanding the reproductive biology and ecology of the tree species is central to proper seed collection, handling, germination and subsequent nursery practices (Evans and Turnbull 2004). The two methods used to collect seeds are: from the ground and from standing trees. The factors that determine the seed collection method include the location of the seed source, the size of the trees, and the type of seed or fruits and their maturity (Evans and Turnbull 2004).

Species that produce large, heavy fruits or seeds that usually fall to the ground can be collected by spreading polythene sheets. This method is frequently used for collecting the seeds of teak, gmelina, and wawa (Evans and Turnbull 2004). Seed collection must be timed according to the characteristics of the species and can be combined with other forestry operations in order to ensure efficiency and minimise costs of the operations. Collecting seeds from standing trees can either be accomplished by using long-handled tools, manual shaking of the trees, or climbing the trees to collect the seeds (Evans and Turnbull 2004). This method has been used for collecting the seeds of various species of pines and eucalyptus. Before commencing seed collection from standing trees, it is critical to verify that the seed has reached maturity.

To maximise seed viability and genetic quality, seeds should be collected from a stand of trees rather than isolated individuals; single trees tend to self-pollinate and produce seed that is less viable and genetically inferior (Rudolf 1974). The importance of using the best seeds and the need to produce seed in large quantities to satisfy plantation development programmes suggest that seed selection for plantations be carried out by those with the requisite expertise. The traditional practice in Ghana has been for forestry technical officers and forest guards to collect seeds from parent trees for propagation in tree nurseries. This practice of not confining seed collection to superior phenotypes may result in poor quality seeds being collected, problems with seed germination and poor quality trees in the plantations.

8.1.2. Seed Storage

The timing of seed collection and effective storage methods are critical components of achieving planting objectives. Some seeds lose their viability soon after falling off the tree and hence seed collection should rely on a good knowledge and understanding of the ecology of the tree species. For tree species that lose their viability quickly, the seeds should be collected before the fruits fall off the trees. Neem seeds, for example, are short-lived and do not retain their viability for long periods (Troup 1921), with a two-week upper limit. The seeds begin to germinate as soon as they fall from the trees (Evans 1992). Some species such as wawa (*Triplochiton scleroxylon*) fruit irregularly and their seeds may need to be collected and stored only in years that seeds are produced.

Seed storage can prolong the viability of seeds up to ten times longer under controlled than in field conditions (Nyland 1996). Unfortunately, the only controlled conditions for storing tree seed in Ghana at the FORIG can only contain 20kg of seed (FAO, 2002a). In addition, storage facilities need to incorporate the ability to monitor the seeds to ensure that they do not deteriorate over time. To meet demand for seeds at a national level, the FORIG has established seed orchards of teak, wawa, cedrela and gmelina. Two private companies, Dupaul Wood Treatment and Subri Industrial Plantation Ltd. (SIPL) also have seed orchards of teak and gmelina, respectively (FAO 2002a).

The FORIG has a Tree Improvement and Seed Technology Laboratory, but this is ill-equipped and there are no control environments for the proper testing of seeds. FAO (2002a) reports that the estimated average annual production of seeds from FORIG is about 1000kg of teak, 20kg of cedrela, and 20-30kg of indigenous species. It is obvious that this level of seed production is woefully inadequate to meet any large-scale plantation development needs in Ghana. To successfully implement a national plantation programme, it is imperative that more research be undertaken to develop improved seeds, and increase storage and testing facilities. Furthermore, tree species that are adapted to many parts of Ghana can produce desired wood products, and are resistant to local pests and diseases should be developed.

8.1.3. Quantity of Seed Required for Plantations

Large-scale national level plantation development programmes will require large quantities of genetically superior seeds to be successful. Given the long process involved in a plantation establishment, it means that adequate seeds have to be procured at least one year ahead of the season in which planting is expected to occur. Forestry authorities, therefore, need to be aware of the seed requirements and whether they can be obtained locally, and if not, make arrangements to import them.

Evans and Turnbull (2004) provide a formula for computing the weight of seeds required for plantation programmes. This formula is given in Equation (1) as:

Weight of seed needed in kg =

$$W = \frac{H \times N_s}{S \times PS} \tag{1}$$

where:

H = the annual planting target in ha

N_s = intended number of seedlings to be planted per ha

S = estimated survival rate as a decimal

PS = estimated number of plantable seedlings per kg of seed

While data on H, N_s and S are usually available at the local level, the number of plantable seedlings (PS) is more difficult to obtain, and can be estimated from Equation (2) as:

$$PS = N \times G \times P \times (1 - M) \tag{2}$$

where:

N = number of seeds per kg

G = germination percentage as a decimal

P = purity percentage, as a decimal (the proportion of seed that is not other matter)

M = estimate of seedling mortality during nursery, as a decimal

The planting target under the NFPDP in Ghana was set at 20,000 ha per year. Given that most of the planting is teak, we can estimate the quantity of teak seeds required per year to plant that area. The following assumptions are made: the number of seedlings per kg (N) is an average of 1500 (Evans and Turnbull, 2004 provide a range of 1000-1900 seeds/kg for teak). Given that most of the plantations are planted out at 2 x 2 m initial spacing, N_s is 2500. Survival (S) is estimated at 0.80, P is set at 0.90; M is 0.20, and G is set at 0.70. Based on these assumptions, the total quantity of seed required to support a national plantation programme to meet the annual planting target is 82,672 kg per year. The quantity of seeds actually required for the various species used in the plantation programme can be calculated using this method once the exact areas to be planted with each species are known.

8.2. SEEDLING PRODUCTION

Producing the growing stock for forest plantations involves several stages including preparing the nursery site, sowing the seeds, tending, lifting, transplanting, and packaging for storage or transport. The exact requirements will depend on the stock type being produced (container, bare-root or stumps), but in each case special care must be taken at each stage to produce good quality stock that has high chances of survival and growth rates after out planting. Nursery practices should be geared towards producing the best quality and quantity of seedlings to meet demand.

8.2.1. Pre-Sowing Treatment

The seeds of some tree species germinate slowly or incompletely because they are completely or partially dormant (Evans and Turnbull 2004). These types of seeds may require pre-sowing treatment to overcome such dormancy problems and improve germination results. One of the causes of seed germination failure is that the seed may need to be "after-ripened" before they can germinate. *Stratification* is the process where the seed mixed with a moist medium (such as sand or saw-dust) and placed in a storage container and stored at a controlled temperature to help "ripen" the seed and break dormancy. Other methods of breaking dormancy are by soaking or treatments with laser beams. For example, in an experiment in Tamale, pre-sowing treatment significantly improved germination performance of dawadawa (*Parkia biglobosa*) seeds. Seeds soaked in cold water for 12 hours had the highest germination (91.3%), compared with untreated seeds, which had 80.0% germination rate (Korankye-Gyamera 1997). For neem seeds, no pre-treatment is usually required, although pre-soaking the seeds in water for 24 hours and removing the endocarp or cutting the seed coat at the round end with a sharp knife also increases the germination capacity. Seeds that have hard shells may have to be scarified by soaking in water, acid or by mechanical means before stratification. *Scarification* is an artificial way to prepare seeds with hard coats for germination. Other reasons for treatment are to protect seeds from destruction by fungi, insects and rodents or preserve their viability.

A key consideration in nursery practices is to decide what type of seedlings to produce. When seedlings are raised in the nursery, they are usually of four main types: container stock, bare-root stock, striplings and stumps, and vegetative propagation. Sometimes unrooted cuttings are also produced.

8.2.2. Bare-Root Stock

The bare-root method is the oldest seedling production method and was used before the advent of motorised vehicles mainly to reduce the cost of transporting seedlings. This method has been the traditional way of raising planting stock in the Sahel-Sudan zone of Africa though in dry years survival is often poor (Evans and Turnbull 2004). It is usual when producing bare-root stock, to first broadcast the seeds on a germination bed before lifting them out unto transplant beds about a week or more after germination. The seedlings are allowed to grow to the desired size and age on the germination bed, where they are then lifted up (bare of any soil on the roots) to the field for transplanting. Under the RAP in the 1990s, nursing of seedlings began in the dry season (December/January) for out planting in the next rainy season (May-September). The most popular method of planting neem in northern Ghana is as bare-root stock (Nanang 1996). Other species that are suited for planting as bare-root stock are cedrela and mahogany species such as *Swientenia macrophylla*. The main advantages of bare-root stock are (after Nyland 1996): lower costs of production and transportation; easy adaptability for size and condition; and easy handling during field operations. Usually, bare-root stock takes longer to grow in the nursery before transplanting compared to container stock.

8.2.3. Container Stock

The usual practice in Ghana when producing container stock is first to broadcast the seeds on a germination bed. Once the seeds germinate, they are lifted into polythene bags after about a week (Nanang 1996). The seedlings grow to the desired size in the polythene or biodegradable container before being transported to the planting site. The most common container type is polythene tubes of various sizes. If polythene tubes are used, these are removed just before the seedling is put into the soil during transplanting. However, the use of biodegradable containers (such as paper tubes) would allow the seedling and container to be put into the planting hole, thereby speeding up the planting operation. In the RAP nursery in Tamale, container seedlings were produced using one-litre polythene bags filled with 3/6 soil, 2/6 rice chaff and 1/6 composted cow dung. When raising container seedlings, care has to be taken to avoid poor drainage and over-watering as this often results in die-back of the seedlings. Poor drainage was a major problem observed by the first author in some nurseries in northern Ghana.

Container-grown seedlings have the following advantages: the containers protect the roots from injury and desiccation until planted; the rooting medium around the seedling provides a microsite that favours root-soil contact; easy to plant without twisting roots; and more adaptable to mechanised handling (Nyland 1996). These advantages increase the survival rates of container seedlings over those of bare-root stock. For example, in an experiment with dawadawa in Tamale, survival of seedlings after three months of out planting was found to be highest in container seedlings (83.0%) than bare-root (40.0%). Only 20% of stumped seedlings survived after the same period (Korankye-Gyamera 1997). Despite these advantages, container stock need more logistical support to produce and transport and hence is more costly than bare-root stock.

8.2.4. Striplings and Stump Stocks

Most seedlings of tree species in the family *Miliaceae*, such as cedrela and *Khaya* spp. are usually stripped off most, if not all, of their leaves before planting. The main objective of using striplings is to reduce the amount of transpiration stress following out planting of the seedlings. This practice is used in places where animal damage is common, where weed competition and stem borers are a problem (Evans and Turnbull 2004). This type of seedling stock is recommended for transplanting in semi-arid conditions where the seedlings are likely to be subject to desiccating weather conditions as pertains in the savannah vegetation zones of Ghana.

Figure 1. Tree nursery near Bontanga in the Northern Region. Foreground: container seedlings of neem. Background: bare-root seedlings of teak. In the picture are Mr. James Amaligo (left), Prof. Robert Day (second from left) and some nursery labourers 1995.

Figure 2. Teak seedlings in a nursery in Tamale - 1995.

The term stump is used to describe a type of cutting applied to a seedling, which is usually 15-25 cm long with about 80% root and 20% shoot with the shoot stripped of leaves and the lateral roots trimmed off (Evans and Turnbull 2004). This method is particularly suited for tropical species that are capable of producing new roots and shoots when in contact with moist soil. Stump planting is especially suitable for taproot dominated species and is frequently used when establishing teak, gmelina and a number of other important tropical genera (e.g., *Afzelia*, *Cassia*, *Chlorophora*, *Khaya*, *Lovoa*, *Pterooarpus*, *Terminalia*, *Triploohiton* and many Leguminosae) (Chapman and Allan 1978). Plantation-grown teak is established using stump plants rather than direct sowing of teak seeds, which has been found to give less satisfactory results (Borota 1991). Neem seedlings can also be planted as stumps by trimming the tap and lateral roots, pruning the stem to leave 20-30 cm of the stem above the root collar, and any side branches and leaves are cut off as close as possible to the stem, without damage to the stem. This leaves a sufficient number of buds for sprouting. Leaving a longer stem may enable the wind to whip and loosen the plants. In India, as little as 2.5 cm of the stem is left with apparently good results (FAO 1995).

8.2.5. Vegetative Propagation

Vegetative propagation is an asexual reproduction in which a vegetative part of the original plant is used in the reproduction process, rather than the seed. This is slightly different from the three methods of raising planting stock discussed above as it does not necessarily use seedlings in a nursery. The three main types of vegetative propagation used in forestry are grafting, air-layering and cuttings. FAO (1993) describes the three main types as follows. Propagation by cuttings is the most convenient and cheapest method and usually preferred when possible. Air layering is a variation of propagation by stem cuttings in which root formation is initiated before the plant part is separated from the mother tree. In grafting, the shoot (scion) of the desired tree is joined with a root (stock or root stock) of different genetic origin. The use of unrooted cuttings is currently the subject of research in Ghana by FORIG, which is researching into the vegetative propagation of *Triplochiton scleroxylon, Milicia excelsa, Khaya* spp., *Ceiba pentandra, Terminalia superba, Baphia* spp. and *Ficus* spp.

8.3. Site Preparation

Before planting, it is essential that the site be prepared to a high standard. Site preparation refers to preparing the land before regeneration (planting trees in the case of plantations). It is any measure that makes the physical environment suitable for germination, and later survival and growth of seedlings (Nyland 1996). Site preparation may be needed to reduce or redistribute slash, reduce competition from existing vegetation, prepare the soil, and facilitate and reduce the cost of regeneration and future management operations (Nyland, 2007; Day 1996). Benefits such as ease of planting, increased tree survival, better weed control and improved growth rates may far outweigh the time and effort involved in site preparation. The following factors affect the type of site preparation that may be required: the amount, size and species distribution of existing vegetation; the method of harvesting existing forests; and the nature of the stand desired in the next rotation (Day 1996).

In situations where no active measures are taken to prepare the site (known as passive site preparation) foresters rely on conditions created by previous forest management practices such as logging to achieve sufficient regeneration. In many cases however, active site preparation is required when existing conditions favour less-desirable species, and the vegetation or soil conditions might limit the regeneration of the desired species (Nyland 2007). Despite the advantages of site preparation, it should be noted that site degradation and loss of nutrients and organic matter could result from the practice (Nyland 1996). Loss of nutrients varies with the nature of site preparation and management practices and can have a dramatic effect on the growth of the next stand. Nutrients lost by removal or leaching can be replaced by fertilisation, but best results are usually obtained when fertilisation is combined with other practices such as weed control or site preparation (Tiarks et al. 1998). There are generally three methods of site preparation used by foresters: mechanical, chemical applications, and the use of prescribed burning.

8.3.1. Mechanical Site Preparation

Mechanical site preparation practices affect the soil surface layers, either to modify the seedbed conditions or to alter the physical attributes of the rooting zone (Nyland 1996). This generally includes scarification that loosens the upper soil or breaks up the organic layer; removes undecomposed litter and humus to expose mineral soil; or mixes surface organic materials with mineral layers beneath them (Ford-Robertson 1971). There is the potential to increase soil compaction with mechanical equipment, erosion, or siltation to nearby waters (Williams 1989). Furthermore, removing and burning slash and other organic materials prior to planting may cause some nutrients to leach, and reduce the amount available for tree growth. With the exception of large-scale industrial plantations, smallholder and community plantations may not be able to afford the cost of mechanical site preparation in Ghana.

8.3.2. Chemical Treatment

Chemical site preparation involves the use of chemical applications, mainly herbicides, to kill competing vegetation. Herbicide use in forestry is limited to not more than a couple of applications per rotation. Herbicide use is regulated by health and environmental authorities, and hence only authorised chemicals can be applied. Herbicides are used to kill dense and unwanted vegetation, control harmful insects and animals, alter the habitats for insects and animals, prevent the development of understory vegetation, inhibit germination of seeds for weeds, and supplement available nutrients by fertilisation (Lowery and Gjerstad 1991). The general technique for applying herbicides is by aerial spraying, but injection into bark of woody plants (as was done in the application of the tropical shelterwood system in Ghana), or incorporating it into the soil have also been used. The main advantages of herbicides are that they are generally cost effective, control a broad range of weeds and prevent sprouting from stumps and root systems, while not interfering with the productive capacity of soils.

8.3.3. Prescribed Burning

Prescribed burning in forestry refers to any burning that is carried out intentionally for a specified purpose. Prescribed burning is used to prepare seed beds for planting, reduce fuel loads, control natural succession, improve grazing, manage wildlife, and improve sanitation on a site. In plantation establishment, bare-root stock or container stock are usually used, and unimpeded access to the site is often required. Furthermore, enough of the surface organic matter has to be removed to enable the planters to reach the mineral soil with their tools. In most cases, burning can achieve this cheaper than mechanical treatment (Day 1996). Moreover, in Ghana, fire has been used for generations to prepare land for agricultural purposes, and hence its use is already widespread and could easily be applied to the site preparation for plantations. Burning, if used successively, removes the organic materials that would have been recycled into the soil, and hence reduces the long-term sustainability of the plantation. The main risk associated with using prescribed fire is the safety of workers and the possibility that the fire might get out of control and burn areas that were not initially targeted.

8.3.4. Examples of Site Preparation in Ghana

In Ghana, site preparation can take several forms depending on the scale of the plantation and the nature of the land to be prepared. For small-scale farmers, site preparation would involve clearing the land with simple farm implements such as hoes, cutlasses and ploughing manually or with animal traction or the use of fire. Industrial plantation investors will more likely rely on mechanical or chemical methods to prepare the sites. For example, sites of the plantations belonging to Bonsu Vonberg Farms Ltd. were site prepared by stumping and ploughing. It was observed that the plantations on such treated sites performed better (by as much as one and a half times) than those that did not receive similar treatment at the beginning of the project (FAO, 2002a). A new technique of site preparation in Ghana is the Subri conversion technique whereby degraded natural forests are cleared, and lines are cut through the debris and planted up without any burning. This conserves the humus layer and the extra debris on the forest floor (Nwoboshi 1994). In Northern Ghana, plantations are commonly established on either farmlands or abandoned lands that are out of cultivation mostly as part of the cycle of shifting cultivation. In either case, lands are prepared in March/April prior to the rains. Land preparation is often done by the farmer with his family or the community in 'communal labour' programmes in the case of community plantations.

8.4. Out Planting

The planting exercise should be planned to give seedlings the best chance of survival. Seedling quality directly affects the survival of the trees. Experience has shown that seedlings and cuttings of poor quality would not survive and grow well even under the best site conditions. Also, the initial spacing and stocking of the plantation is determined at this stage, taking into consideration the management regime envisaged for the plantation enterprise. The tree-planting exercise should be well thought out ahead of time. Planting should be carried out after rain when the soil is moist. It is advisable to soak seedlings in their pots prior to planting out. Several factors to be considered include species selection, stock type, spacing, the arrangement of trees, site preparation, seedling handling, and other operational considerations. The techniques for planting differ between the container and bare-root stocks. Bare-root seedlings are planted by digging a small (20-25 cm deep) planting pit or open up a slit with a hoe or spade and insert seedlings up to the root collar level with the ground, using one hand. Soil is carefully filled back with the other hand and pressed with the foot around the seedling (Evans and Turnbull 2004). For container seedlings, the planting pit should be slightly bigger than the container. Planting is carried out by removing the container and positioning the seedling in the pit with the root collar level with ground; the soil is filled back and firmed with the foot around the seedling (Evans and Turnbull 2004).

8.4.1. Stock Type

Foresters have a choice between direct seeding, transplanting seedlings, and vegetative reproduction when regenerating forests. The method chosen depends on which one gives the best results for the chosen species and site. The stock type also has implications for when the trees can be planted out and the cost of the operation. For example, direct seeding does not require nursing the seedlings in a nursery, and hence can be carried out more readily and cheaply compared to raised seedlings. Direct seeding means artificially spreading seed over an area where a landowner wishes to regenerate a new forest stand artificially (Ford-Robertson 1971). The most important factors affecting the success of direct seeding are the rate of seed applied, the degree of site preparation, and the weather. The seeding methods include broadcast sowing by hand or aerial sowing; spot sowing with or without covering; drilling by hand or with small mechanical seeders (Day 1996). Direct seeding is unsuitable in droughty conditions or on seedbeds that dry up quickly, on erodible surfaces where seed can easily be washed off, on soils that become inundated during the rainy season, and on sites where seed-eating organisms can easily consume the seed (Nyland 2007). Direct seeding by spot sowing is used by farmers in Ghana to plant most fruit trees on farms and home gardens. However, for large areas, this method would be labour intensive and costly, except where the main objective is to create jobs.

8.4.2. Planting Season

Out planting of seedlings must take into consideration the season of the year and the rainfall pattern in the location. In northern Ghana, there is a single rainfall season, followed by a long, dry season when there is no rain and dry desiccating conditions that are exacerbated by rampant bushfires. In southern Ghana, there are two rainfall seasons, a major and a minor one. The temperatures are moderated by the vegetation cover, and the impact of the dry winds is less severe on the soil. The season of planting would, therefore, have important ramifications for survival.

Out planting of seedlings in the savannah zones is done exclusively during the rainy season [from May to the end of August]. The beginning of the planting season each year, however, depends on when the rains start and, therefore, varies from year to year. In most years, planting starts before the end of May. Sometimes planting is carried on into September but this practice is not recommended since the newly transplanted seedlings do not have sufficient rainwater to establish themselves to survive the long, dry season starting from October to March. Late planting (after September) should be avoided whenever possible. There is more flexibility in when to plant in southern Ghana, but the considerations are identical.

Table 1. Plantation spacing and stocking levels in English and metric systems

English system		Metric system	
Feet	Trees per acre	Metres	Trees per hectare
2 x 2	10890	0.3 x 0.3	111,111
3 x 3	4840	0.5 x 0.5	20,000
4 x 4	2723	1 x 1	10,000
5 x 5	1742	2 x 2	2,500
6 x 6	1210	3 x 3	1,111
7 x 7	888	4 x 4	625
8 x 8	680	5 x 5	400

8.4.3. Initial Spacing

One of the critical decisions in plantation establishment is the initial spacing at which the trees will be planted. The ability to determine initial spacing and hence the stocking levels of the plantations is a major advantage of plantations over natural forests. Initial spacing is a function of the chosen species' growth rate and form, its habitat and shade tolerance, site fertility, planned time to first treatment, the planned silviculture, intended products from the plantation, and costs considerations. Other considerations include the need to plant trees in a way that facilitates subsequent operations and enables general access to different parts of the plantation. Initial spacing also affects the costs of seedlings, area to be planted and labour and machinery costs of out planting seedlings. Costs for plants and labour tend to increase with decreasing planting distances, but on the other hand, costs of weeding tend to increase with wider spacing (Chapman and Allan 1978). Closer-spaced plantations produce biomass in less time, but wider-spaced plantations can produce larger trees in less time. Therefore, in general, if the intended purpose of the plantation is to produce fuelwood within a short time, closer spacing should be used, while wider spacing should be used if bigger diameter trees are required for sawlogs or electric transmission poles.

The growth rate of the species planted should influence the initial spacing. In general, slower-growing species tend to be planted at closer spacings than faster-growing species, and for this reason spacing in the tropics tend to be greater than in temperate regions (Chapman and Allan 1978). Furthermore, species that branch heavily need to be planted closely to promote the formation of a well-defined leading stem. Most tropical species are self-pruning (e.g., *Terminalia superba*) and can, therefore, be planted more widely apart than temperate trees (Chapman and Allan 1978). If mechanised management operations such as weeding are intended, then the spacing at which the trees are planted should take this into consideration. A distance of 2.8 m between rows is considered a minimum spacing where weeding is mechanised (Chapman and Allan 1978). Wider spacing is used on poorer, less fertile, and arid soils to allow more room for trees to develop roots to access nutrients.

The initial spacing determines the stand density (number of trees per ha) of the plantation. Closer initial spacing promotes slower diameter growth, compared to wider spacing. Table 1 gives examples of various spacing in English and metric systems and their corresponding stocking levels in acres and hectares. In the metric system, stocking levels of trees/ha are determined as 10,000/ (spacing in m)^2, while in the English system, trees/acre is calculated as 43,560/(spacing in ft.)^2.

For many decades, the initial spacing for plantation grown trees in Ghana has been 2 x 2 m apart, giving 2500 trees/ha. Most species are planted out as pure stands of exotic species such as teak, neem, or lueceana. Some farmers grow food crops such as maize and millet until canopy closure. Shade tolerant food crops (e.g., pepper) are also grown under older trees on farmlands. Since almost all plantations are established in proximity to villages, intensive silviculture is not only feasible but is beneficial to the communities. For the purposes of producing fuelwood and poles, the initial spacing of 2 x 2 m is considered wide and can be reduced to 1.75 x 1.75 m to give an initial planting density of 3265 stems/ha (Nanang 1996). A higher density than this will lead to increased plantation costs and also likely produce inferior poles and rafters since there will be a proliferation of many small stems and branches rather than increases in diameter of individual trees. The adoption of closer initial spacing and thinning regimes will circumvent the current problem of wood removal by villagers (for poles and rafters) which interferes with the development of fully stocked plantations (Nanang 1996).

8.5. Protecting Planted Trees against Damage

Protecting forest plantations from physical damage, the weather and other biological damage is one of the most difficult tasks of managing a plantation forest, especially when the trees are still young. Most communities and forestry agencies have no problems planting trees; it's a lot more difficult to protect them to survive the first few years. In this section, we will discuss the methods that can be used to help planted trees survive, especially when planted around villages.

8.5.1. Fire

In many dry parts of Ghana, especially in the savannah regions where it is often dry with low relative humidity, fire remains a serious threat to forest plantations. Even in the moist south, the threat is not completely eliminated as there are warm and dry periods that could increase the risk of fire. It is absolutely critical that all forest plantation development planning include a fire management plan. This should detail the level of risk that the plantation is exposed to, how these will be mitigated, and in the event of a fire, the fire control measures to be taken to limit damage to lives and property, and clearly outline responsibility for implementing the plan. A plantation fire protection plan aims to prevent outside fires from getting to the plantation, prevent fires from being started inside the plantation and limiting the spread of fire within the plantation once it is started (Evans 1992). Most fires in forest plantations in Ghana will originate from outside of the plantation and the key will be to eliminate the chances of fires from reaching into the plantation. The most effective way is to create a firebreak of at least 20 m between the plantation and adjacent land uses. This space should be free of all vegetation during the dry season when fires are most likely to occur.

Although fires can be caused by nature, such as lightning, most fires in Ghana are caused by human activities either started within the plantation or as a result of fires spreading from the surrounding farmlands or savannah woodlands. The risk of fire is directly related to the availability of fuel (dead combustible organic debris at the ground level) and weather conditions. Ways to minimise fire risk include:

Preventing fire from starting within the plantation

It is very important to reduce fire hazard within plantations through:

- Reduction or complete elimination of dead combustible materials on the forest floor, through burning in a controlled environment (annual prescribed burn), weeding, or cleaning. When a prescribed burn is used, the fire should be continuously monitored by trained firefighting personnel to ensure that the fire does not spread beyond its planned size or boundaries;
- Choosing species that are generally less susceptible to fire. Some tree species such as teak and gmelina produce leaf litter on the forest floor that is a serious fire hazard. These hazards should be minimised through fuel load reduction strategies;
- Restricting access to the plantation and restrictions on activities within the plantations, such as not smoking or starting fires; and
- Use public education on the risk and dangers of fire and involve local people in forestry.

Preventing outside fires from getting into the plantation

- Plantations should be designed to include effective fuel breaks surrounding them (i.e., a band which may be heavily grazed or slashed as bare-earth strip) and within the plantation. For example, a band of 20 m between the plantation and the adjacent land use is effective in reducing the risk of outside fires reaching the plantation. At the end of the rainy season, these firebreaks should be weeded and completely cleared of dry vegetation. There should also be fire breaks within the plantation itself, such as roads to restrict the spread of fire once started.
- Adequate monitoring and warning systems to immediately detect, communicate and suppress fires before they reach the plantation. If plantations are located close to grasslands, for example, which are prone to annual bushfires, then every effort should be made to plan for how to prevent these grass fires from spreading into the plantation, through the use of appropriate firebreaks and weeding to eliminate fuel loads within the plantation during the dry season.
- Fire management for plantations should be planned within the context of the overall land use of the area within which the plantation is located. Regular prescribed burning of grassland or other vegetation bordering the plantation may be required to reduce the risk of fires.
- Strictly enforce the Bushfire law in Ghana that was enacted to help reduce indiscriminate wildfires.

Limiting and suppressing fires after they have started

- Community participation in fire prevention, detection and suppression should integrate both fire and people into sustainable land use and vegetation management systems. This can be achieved through community fire brigades or volunteers who could be mobilised to patrol plantations and respond immediately to any fires that are detected. This will limit the spread of the fire and hence the damage that would have been caused if the fire were left to burn.
- Once a fire has started, suppression will include the use of water, beaters and physical construction of fire breaks. Fire burns in the presence of oxygen, heat and fuel (combustible material) and any fire suppression activities must target to eliminate one or more of these variables.
- Commercial plantations in particular should have fire tenders and plan roads within the plantation to provide access for forest management and fire control purposes. These fire tenders should be used to suppress any fires. Bulldozers and other earthmoving equipment should be used to construct physical barriers/firebreaks during suppression activities to reduce the spread of the fire to other parts of the plantation.

8.5.2. Animals

When plantations are located in close proximity to villages or trees planted around homes in home gardens and compounds, there is a risk that the trees may be grazed or browsed by domestic animals such as sheep, goats and cattle and can damage young plantations. Animals damage plantations by killing young planted trees and natural regeneration by repeatedly browsing or breaking them, or by stripping off the bark; removing a lot of the foliage so that growth is reduced; making young trees bushy or crooked by taking out the leading shoot; damaging food crops; and taking fruits and seeds that have already been harvested (FAO 1995).

The most effective way to protect trees in this case is to use hedges and fences to prevent intrusion by domestic animals. In cases where fencing is too expensive, guards can be hired to protect the trees from the animals. Although minimal browsing is beneficial to a plantation due to spreading manure, seed distribution and weed reduction, one must avoid having too many animals for the particular site to reduce damage. Other techniques include tethering or watch animals and using *pollarding* (cutting fodder trees at 2 – 4 m above ground) to discourage over-browsing.

The planting of tree species not browsed by domestic animals around houses is another option. In this case, one needs only worry about potential physical damage, rather than damage caused by browsing and grazing.

If plantations are located away from human habitation, damage to forest plantations by wild animals may be the concern. This usually involves tree browsing or de-barking. The most common wild animals responsible for damage include rodents (rats, mice, and moles and squirrels), rabbits, antelopes, pigs and buffaloes. The principal methods of controlling damage by wild animals involve the use of fences, hedges or ditches, trapping and removal, and poison baits (FAO 1989).

8.5.3. Humans

In some circumstances, plantations need to be protected from humans as well. There are several reasons for this. Humans may cause damage to plantations such as starting fires, driving through plantations to cause damage to trees, or stealing fruits/firewood and illegal logging or grazing animals.

In areas where there is acute shortage of wood for fuelwood, there is a real chance that wood will be stolen from plantations for cooking and heating. This will be more difficult to control if the plantation is located far away from homes, such as on farms and woodlots. It is difficult to control theft for firewood, outside of fencing the whole plantation and keeping it under key and lock or hiring guards to patrol the plantation. Promoting community participation in woodlot establishment and management would help create communal ownership of property and reduce theft. The more effective solution, however, is to have supply of wood outstrip demand by creating more plantations and reducing fuelwood demand through providing affordable alternative energy sources.

When humans allow animals to graze in young plantations, these animals can cause severe damage, as the animals feed on the leaves and could also physically destroy the seedlings. It is important that plantation establishment consider protection against human who allow grazing animals into the plantation. This problem is often associated with land tenure arrangements in the community and who has rights to the land on which the plantation is located. Effective control includes having shepherds control the animals, weeding to keep plantation clean of grazing materials and fencing around the plantation.

8.5.4. Pests and Diseases

Pests and diseases occur in every forest environment and plantations are no exception. However, in their natural environments, trees and shrubs normally attain a state of equilibrium with indigenous pests (FAO 1989). When exotic trees and shrubs are planted, exotic pests can also be introduced as a result. It is also quite possible that exotic species may not have diseases in their new location, but if they do, then the natural enemies of these pests and diseases will be absent, making any infestations more disastrous. To minimise the impacts of pests and diseases the following precautions should be taken (FAO, 1989; Evans 1992):

- Plant tree species that are suitable to the climatic and soil conditions of the site, and use surveys of indigenous pests to ensure that none are among the known forms to which the selected species is susceptible;
- Maintain healthy vigorous young trees to help make a plantation more resistant to insects and fungi;

- Promptly remove and destroy infested trees and shrubs as an effective way to prevent the spread of the pest attacks to the rest of the plantation;
- Use mixed species to reduce the risk of pest and disease infestations and spread;
- Use appropriate chemical insecticides or fungicides to control pests and diseases. Ensure that only previously tested and environmentally sound insecticides and fungicides are used;
- Biological control of insects has been employed with success in some situations; in most instances, the introduction of a parasite to control the insects is required; and
- Mechanical control, either by physically removing and destroying the pests or by eliminating the alternative hosts, can be effective.

8.5.5. Climatic Factors

Irrespective of the location of the plantation, it is difficult to predict weather patterns that can potentially damage the trees. Since these events cannot be controlled, it is impossible to completely protect plantations from these weather events. Climatic events such as windstorms, flooding, drought, extreme heat, etc., can all have negative impacts on either the growth and productivity of the plantation or destroy the trees altogether. A related factor is that once trees are under stress, they are more susceptible to insect and disease attacks.

The best advice on weather-related damage is to grow tree species known to be resistant to the detrimental effects of local weather patterns, or locating the stands of trees in sheltered areas (FAO 1989). In addition, some silvicultural practices can reduce the impacts of climate-related damage such as timing of planting and using moisture-conserving site preparation methods in dry areas or designing thinning carefully to minimise wind throw (Evans 1992).

8.6. Conclusion

Plantation establishment begins with a clear plan on seed procurement, nursery activities, out planting and a management plan to protect the newly planted trees from biological and inorganic damage. Planning should also carefully include timing to ensure that seedlings are in the best condition and ready for out planting at the targeted time. The quality of seedlings has important implications for the survival and subsequent productivity of the plantation. Therefore, nursery practices, out planting activities and protection measures for the seedlings should be taken seriously.

APPENDIX A. SOME PLANTATION SPECIES SUITABLE FOR THE HIGH FOREST ZONES OF GHANA

English/ Common name	Botanical name	Key uses
Indegineous species		
Ofram	*Terminalia superba*	Veneer, Plywood, Furniture
Otie	*Pycanthus angolensis*	Plywood, Veneer, Boards, Furniture
Rosewood	*Dalbergia sissoo*	Furniture, Veneer, Plywood
Rosewood	*Dalbergia retusa*	Furniture, Veneer, Plywood
Makore	*Tieghemella heckelii*	Plywood, Veneer, Boards, Furniture
Koto	*Pterygota macrocarpa*	Plywood, Veneer, Boards, Furniture
African mahogany	*Khaya ivorensis, Khaya grandifolia, Khaya angolensis*	Plywood, Veneer, Boards, Furniture
Wawa	*Triplochiton scleroxylon*	Plywood, Boards, Particle Board
Edinam	*Entandrophragma angolensis*	Plywood, Veneer, Boards, Furniture
Sapele	*Entandrophragma cylindricum*	Plywood, Veneer, Boards, Furniture
Kosipo	*Entandrophragma candolli*	Plywood, Veneer, Boards, Furniture
Utile	*Entandrophragma utile*	Plywood, Veneer, Boards, Furniture
Ceiba/kapok	*Ceiba pentandra*	Veneer, Plywood, Boards
Kokrodua	*Afromosia elata*	Furniture, Parquet, Veneer
Emire	*Terminalia ivorensis*	Veneer, plywood
Dahoma	*Piptadeniastrum africanum*	Mining Timber, Sleeppers, Furniture
Exotic species		
Mahogany	*Swietenia macrophylla*	Plywood, Veneer, Boards, Furniture
Cedrela	*Cedrela odorata*	Furniture, Veneer, Plywood
Teak	*Tectona grandis*	Veneer, Furniture, Parquet, Transmission poles
Red Meranti	*Shorea spp*	Veneer, Furniture, Parquet, Plywood, Panelling
Gmelina	*Gmelina arborea*	Veneer, Plywood, Furniture, Parquet, Pulp
Paulownia	*Paulownia elongata*	Plywood, Pulp, Parquets
Eucalyptus spp	*Eucalyptus*	Plywood, Veneer, Boards, Paper, Electric transmission poles
Paulownia	*Paulownia tomentosa*	Plywood, Veneer
Paulownia	*Paulownia fortunei*	Plywood, Veneer
Aucoumea	*Aucoumea klaineana*	Plywood, Veneer, Panelling,
Sentang	*Azadiracta excelsa*	Veneer, Furniture, Panelling
Bamboo	*Bambusa spp*	Furniture
Pines	*Pinus spp*	Furniture, Window frames, Panelling, Floors and Roofing, Turpentine

APPENDIX B. SOME PLANTATION SPECIES SUITABLE FOR THE TRANSITION ZONE OF GHANA

English/Common name	Botanical name	Key uses
Indegineous species		
Ceiba/kapok	Ceiba pendandra	Veneer, Plywood, Boards
African Mahogany	*Khaya ivorensis, Khaya grandifolia, Khaya angolensis*	Plywood, Veneer, Boards, Furniture
Wawa	*Triplochiton scleroxylon*	Plywood, Boards, Particle Board
Ofram	*Terminalia superba*	Veneer, Plywood, Furniture
Emire	*Terminalia ivorensis*	Veneer, Plywood
Kusia	*Nauclea diderrichii*	Furniture, Parquets, panelling
Mansonia	*Mansonia altissima*	Furniture, Parquets, Veneer
Odum	*Milicia excelsa*	Furniture, Parquet
Kokrodua	*Afromosia elata*	Furniture, Parquet, Veneer
Dahoma	*Piptadeniastrum africanum*	Mining Timber, Sleeppers, Furniture
Asanfena	*Aningeria spp.*	Veneer, Plywood
African Walnut	*Lovoa klaineana*	Veneer, Plywood
African Rosewood	*Pterocarpus erinaceous*	Furniture, parquet
Exotic species		
Teak	*Tectona grandis*	Veneer, Furniture, Parquet, Transmission poles
Sentang	*Azadiracta excelsa*	Veneer, Furniture, Panelling
Cedrela	*Cedrela odorata*	Furniture, Veneer, Plywood
Acacia	*Acacia mangium*	Energy, Charcoal, Paper
Cassia	*Senna siamea*	Agroforestry, live fences, windbreak, shade, fuelwood, posts, rafters
Eucalyptus	*Eucalyptus spp*	Plywood, Veneer, Boards, Paper, Transmission Energy
Bamboo	*Bambusa spp*	Furniture
Pines	*Pinus spp*	Furniture, Window frames, Panelling, Floors and Roofing, Turpentine
Neem	*Azadirachta indica*	Fuel wood, shade, agroforestry medicines
Albida	*Acacia albida*	Wood, posts, fences, agroforestry, windbreaks, shade, food, fodder
Gmelina	*Gmelina arborea*	Fuelwood, rafters, live fences, windbreaks
Leucaena	*Leucaena leocelapholus*	Agroforestry, fuelwood, windbreaks, soil conservation, food (seed), fodder
Quickstick	*Gliricidia sepium*	Fuelwood, agroforestry, soil conservation, posts, rafters
Lebbek	*Albizia lebbeck*	Fuelwood, posts, rafters, some oils and dyes,
Moringa	*Moringa oleifera*	Live fences, windbreaks, shade, dyes, wood, food (leaves), all parts used for medicines

APPENDIX C. SOME PLANTATION SPECIES SUITABLE FOR THE NORTHERN AND COASTAL SAVANNAH ZONES OF GHANA

English/Common name	Botanical name	Key uses
Indegineous species		
Ceiba/kapok	*Ceiba pendandra*	Veneer, Plywood, Boards, food
African Mahogany	*Khaya senegalenses*	Boards, Furniture
Akee apple	*Blighia sapida*	Charcoal, fuelwood, dyes, shade
Anogeissus	*Anogeissus leiocarpus*	Fuelwood, posts, poles, local tools
Cashew	*Anacardium occidentale*	Fruits, fuelwood, rafters, posts, fodder
Dawadawa/Locust bean	*Parkia Biglogbosa*	Condiments, rafters, posts, fuelwood, nitrogen fixing, agroforestry
Desert date	*Balanites aegyptiaca*	Fuelwood, rafters, fruits, soil improvement, oils
Ebony	*Diospyros mespiliformis*	Fuelwood, posts, tools, glues, fruits, medicines, windbreaks
Mango	*Mangifera indica*	Fruits, agroforestry, fuelwood, timber, shade, windbreak, fences
Pawpaw	*Carica papaya*	Latex, fruits, medicines
Tamarind	*Tamarindus indica*	Fruits, fuelwood, rafters, tools, some dyes, agroforestry, windbreaks, shade
Shea	*Butyrospermum parkii*	Fruits, oils, agroforestry, shade, fuelwood, posts, rafters
Rosewood	*Dalbergia sissoo*	Furniture, Veneer, Plywood, windbreaks, soil improvement
Exotic species		
Acacia	*Acacia mangium*	Fuel wood
Millettia	*Millettia thonningii*	Fuel wood
Neem	*Azadirachta indica*	Fuel wood, shade, agroforestry medicines
Teak	*Tectona grandis*	Veneer, Furniture, Parquet, panelling Transmission poles
Eucalyptus	*Eucalyptus camaldulensis*	Veneer, Plywood, Boards, Energy
Eucalyptus	*Eucalyptus spp*	Plywood, Veneer, Boards, Energy
Albida	*Acacia albida*	Wood, posts, fences, agroforestry, windbreaks, shade, food, fodder
Cassia	*Senna Siamea*	Agroforestry, live fences, windbreak, shade, fuelwood, posts, rafters
Gmelina	*Gmelina arborea*	Fuelwood, rafters, live fences, windbreaks
Leucaena	*Leucaena leocelapholus*	Agroforestry, fuelwood, windbreaks, soil conservation, food (seed), fodder
Quickstick	*Gliricidia sepium*	Fuelwood, agroforestry, soil conservation, posts, rafters
Lebbek	*Albizia lebbeck*	Fuelwood, posts, rafters, some oils and dyes,
Moringa	*Moringa oleifera*	Live fences, windbreaks, shade, dyes, wood, food (leaves), all parts used for medicines

Chapter 9

PLANTATION MANAGEMENT OPERATIONS

This chapter discusses plantation management tending operations (weed control, pruning, and thinning). Stand density management is one of the most important tools used to ensure that the optimum potential of forest sites is achieved. The different types of stand density indexes are described, and a case study that develops a stand density diagrams for teak is presented.

9.1. WEED CONTROL

9.1.1. The Concept of Weed Competition in Forestry

Successful plantation establishment requires that planted trees are protected from undesirable competing vegetation. Competing vegetation is defined as unwanted or undesirable vegetation which suppresses or inhibits the growth and survival of desirable tree crops (Coates and Haeussler 1986). One objective of silviculture, both from biological and economic viewpoints is to restrict the composition of forest stands to those species that are best adapted to a particular site (Smith et al. 1997). Notably, sites that are most favourable to tree growth tend to be also favourable for competing vegetation, especially in the early successional stages (Day 1996). This vegetation often competes with tree crops for growth resources such as water, light and nutrients and hence directly interferes with growth of the desired species. Undesirable vegetation can also negatively interfere with crop trees by directly competing with seedlings or out plants, causing physical injury to seedlings, smothering seedlings, producing or exuding toxins that limit the growth of seedlings, increasing the fire potential of the site, and providing a favourable habitat for biological pests which have the potential to damage or kill seedlings or trees (Day, 1996; Bell, 1991; Ross and Walsted, 1986; Sutton 1985).

On some sites, the tree crop will eventually grow through the weeds, dominate the site and become established; and on such sites the main function of weeding is to increase crop uniformity and speed up the process of establishment, while on other sites, the type or density of the weed growth is such that in the early stage of a plantation it will suppress and kill some or all of the planted trees, and in such areas the main purpose of weeding is to reduce mortality and maintain an adequate stocking of trees to establishment (Chapman and Allan

1978). Weed control provides the most benefits when it is done before competing vegetation has had an opportunity to impede tree growth, i.e., before crop trees display visible signs of suppression, as the response of trees to weed control is rather slow once they have already been suppressed. Weed control is not always necessary. Competing vegetation does not always have a level of impact on crop tree survival or growth that warrants an investment in weed control.

There are three opportunities for vegetation control: during site preparation; release or cleaning operations after the trees have established, and during pre-commercial thinning. Characteristics of competing vegetation include the ability to produce abundantly and at frequent intervals, shade tolerance, grow faster than tree crops, and have the ability to survive various kinds of natural disturbances (Nyland 1996).

The timing and frequency of weeding depend on the site preparation that was carried out as well as the tree species, initial and subsequent spacing, the nature of the competing vegetation and climate. An effective site preparation controls the existing competing vegetation and ensures that the trees have a good start ahead of weeds. Therefore, weed control could be delayed until later. Without site preparation, weeding may need to be undertaken immediately to avoid suppression of the trees by weeds. Sites that have previously been forested tend to be invaded by pioneer species and have more woody regrowth than sites that did not previously contain a forest. Also, some tree species are more competitive and tolerate competition better than others. Moreover, some species achieve crown closure faster than others and can suppress weeds much sooner. For example, for successful eucalyptus establishment, clean weeding is required for good early growth (Evans and Turnbull 2004). The timing and frequency of weeding also depend on the amount of rainfall in the area. Weeds, just like other plants, require adequate rainfall, and hence in years of high rainfall, more frequent weeding may be required compared to drier years.

In northern Ghana, most small-scale farmers only weed their plantations if they are intercropped with food crops. However, weeding is used more as a fire management tool. Most farmers and communities make fire belts around their plantations to exclude fire from the trees using hoes and cutlasses to clear brush in the form of belts one to two metres wide. Fire belts are often created at the end of the rainy season (October). Despite these preventive measures, some plantations especially those established on unfarmed lands are still burnt by annual bushfires during the dry season.

9.1.2. Weed Competition in Ghana's Forested Ecosystems

Weed invasion into forest and savannah ecosystems presents a serious threat to successful plantation establishment. However, information on the diversity, modes of introduction, spread and impacts of invasive weeds in Ghana is limited (Anning and Yeboah-Gyan 2006). Anning and Yeboah-Gyan (2006) examined the diversity and distribution of invasive weed species in the humid forest environment of Ashanti Region, Ghana and found a total of 43 species belonging to 19 families, 41 genera and six life/growth forms in cultivated, degraded, aquatic, ruderal and forested ecosystems. The dominant invasive weeds were *Chromolaena odorata* (L.) (12.71%), *Centrosema pubescens* Benth. (10.42%) and *Rottboellia cochinchinensis* (Lour.) Clayton (6.39%).

Generally after forest clearance, tropical forests are dominated by pioneer species that are light-demanding with fast growth rates. For example, in Ghana, forest landscapes are invaded by *Chromolaena odorata* Linn (known locally as *Acheampong*), a highly competitive weed species noted as the primary cause for poor natural regeneration (Honu and Dang 2000). *Chromolaena odorata* forms a dense canopy, which induces severe competition for below ground growth resources such as water and nutrients. The grass cover also suppresses and prevents tree seedlings from growing through the canopy (Honu and Dang 2000; 2002). Competition from *C. odorata* may account for 64% mortality in tropical tree species (Honu and Dang 2000). Thus, seedling mortality is high when shade-intolerant pioneers regenerate under *C. odorata* canopy (Richards 1996).

In many sites across Ghana, grass competition is also intense from such species as *Imperata cylindrica* (speargrass) which invades after forest and woodland clearing and must be controlled to re-establish tree species. Speargrass was once ranked as the world's seventh worst weed (Holm et al. 1977). The persistent and aggressive rhizomes of speargrass are the main mechanisms of spread, and, coupled with their resilience, speargrass is very difficult to control (Bolfrey-Arku et al. 2006). Managing competition from competitive vegetation may be required to increase natural regeneration success or before and after establishing forest plantations. For example, removal of 50% of *C. odorata* cover reduced competitive effects, increased seedling height growth and number of leaves (photosynthetic area will enhance CO_2 capture) three fold in released relative to control plots in Ghana (Honu and Dang 2000).

9.1.3. Methods of Weed Control

Even after adequate site preparation, undesirable vegetation will still likely compete with desirable crop trees, and there may be a need to control this vegetation to optimise the productivity of the planted trees. Several kinds of grasses and shrubs that are abundant in both plantations and natural forest ecosystems will tend to pose threats to trees. In general, four methods of weed control are used in forestry: manual, mechanical, chemical and biological controls. Any of these methods could be applied as spot treatments around individual seedlings or in the case of chemical treatments, can be broadcast across the entire stand of trees (Nyland 2007).

A. Manual Control

This refers to cutting back or removing weeds using manual implements such as hoes and cutlasses. This is the most common weed control method, and is identical to weed control for agricultural crops in Ghana. It is suitable for smallholder plantations, and in agroforestry systems. Labour is usually supplied by the farmer or his household or relatives. Occasionally, hired labour may be used such as in the modified taungya system where the government provides incentives and pays for labour to undertake weeding of planted trees. In most cases, cutlasses are used to cut off the competing vegetation, while hoeing, which is a more effective method, is used to remove the weeds and also helps aerate the soil. In addition to increased soil aeration, hoeing is especially important in the savannah areas that have a long, dry season, where it can increase rainfall percolation and reduce evaporation from the soil. Manual weeding is cheap, but more labour intensive and ineffective where the weeds are cut off, as the weeds may regrow following the weeding.

B. Mechanical Control

Mechanical weed control involves the use of machines to remove weeds and cultivate the ground between trees by harrowing, rotovating and shallow ploughing (Evans and Turnbull 2004). Since machines are used, plantations must have been planted at a spacing that makes this feasible. Machines are more effective than manual weeding but are also more expensive and beyond the financial abilities of average small-scale farmers. At the BVFL plantations in Somanya in the Eastern Region of Ghana, mechanical weeding is carried out during the first two years before the dry season using a tractor and slasher attached to it. However, this type of weeding does not usually affect the weed growth at the base of the trees, which are the crucial ones with respect to competition and hence such weeds are removed manually. Mechanical weed control with large machinery carries risks associated with harm to wildlife, potential soil compaction, increased erosion and also excessive burning of fossil fuels.

C. Chemical Control

Chemical weed control involves the use of herbicides to reduce, eliminate or suppress competition. Chemicals can be applied through aerial spraying, or through on-ground treatments involving vehicle-mounted equipment, backpack sprayers, or other hand application tools. Weed control by chemical methods is more effective than other weed control techniques because the chemicals kill the weeds (Evans and Turnbull 2004).

The use of chemicals to control weeds is widespread in agriculture in Ghana, while its use in forestry is still limited. For example, the use of *glyphosate* to control water hyacinth in areas with heavy infestations is accepted and approved in Ghana (Madsen and Streibig 2003). Glyphosate is used around the world in forestry in countries such as Canada. It is a non-selective, systemic herbicide, that translocates (or moves throughout) plants very effectively once it penetrates the waxy cuticle of plant leaves or stems. As such, it is particularly useful for control of weedy plant species that re-sprout from roots, rhizomes or cut stumps and it exhibits a high degree of effectiveness on most of the key competitive species in Canadian forest regeneration sites (Thompson 2009). Herbicide use includes the risk that weeds could become resistant to the active ingredients used. It is good practice to use spot application and apply herbicides to the as minimum areas as are required to achieve weed control.

The risks of herbicide use are generally associated with the potential for direct or indirect effects on wildlife species or to humans that may be inadvertently exposed to herbicide residues. However, such risks are significantly mitigated by the extensive scientific research that enhances and defines biological effects thresholds for herbicides (Thompson 2009). Furthermore, in Canada, operational practices have put into place to reduce the probability that actual exposures will exceed such thresholds (e.g., buffer zones, signage, use of minimum effective rates, advanced application technologies to optimise targeting and reduce drift potential, etc.) (Thompson 2009). Economic evaluations have shown that the returns to small-scale farmers could be considerably increased by the use of *glyphosate* for weed control in agriculture in Ghana (Darkwa et al. 2001), although there is no similar evaluation for forestry use yet.

D. Biological Control

Biological control involves the use of biological agents such as the natural enemies of the weeds or cover crops to control weeds. In Ghana using biological agents to control weeds in plantations of cedrela is an option that may help manage competitive effects of the species

(Timbilla and Braimah 2000), but this approach must be implemented with caution. The oil palm and para rubber industries in Ghana and elsewhere have used and continue to use the creeper plant *Peuraria phaseoloides* for weed control and soil improvement. But forestry in Ghana has not applied this technique (FAO, 2002a). In Ivory Coast, the performance of the leaf mulches of *Leucaena leucocephala, Gliricidia sepium* and *Flemingia macrophylla* in weed control has been tested by Budelman (1988). Of the three mulch materials only that of *F. macrophylla* showed promise in retarding weed development. The effective lifespan of a mulch layer of 3 tonnes was found to be between 12 and 13 weeks, while that of treatments 6 and 9 tonnes had effective life-spans of over 14 weeks. The value of mulching in weed control is limited to the control of weed species that multiply by seed since regrowth originating from roots or stumps from former vegetation are unlikely to be checked by a mulch layer (Budelman 1988).

9.2. PRUNING

9.2.1. Definition and Importance

Pruning is a silvicultural activity that removes branches to improve tree form or wood quality, health and vitality of a forest. It is used almost exclusively for conifer plantations. As trees grow, branches become incorporated into the wood of the tree trunk and form knots. When trees are widely spaced, the lower branches stay alive longer, thereby producing larger diameters, but also increasing the number of knots, the rotation age and the time to produce clearwood (knot-free timber) through natural pruning (Nyland 1996). Branches may have to be artificially removed if there is a desire to obtain clear wood sooner than it would otherwise occur under natural conditions. Clearwood is valuable for producing veneers, which are obtained by cutting thin layers of timber from logs. Veneers improve the finished appearance and value of lower quality timber products.

Pruned timber will result in a higher proportion of clean, knot-free timber that is easier to plane and can be put to high-value end-uses. Knots cause structural and visual defects that can lower the value of the timber; a correctly pruned tree can potentially be worth 15%-50% more than the value of an un-pruned tree (Rodney-Bowman 2007). The timing of pruning is important to the overall management of the stand; premature pruning wastes time and effort while delayed pruning results in large loses of valuable clearwood (CQFA 2009). Pruning also serves to reduce the chances of ground fires reaching the crowns and to facilitate access to the stand. It is a costly operation, which should be perceived as an investment to improve the quality of the final product. It is, therefore, justifiable only when the extra revenues involved outbalance the costs (Centeno 2009)

The frequency, season and standard of pruning vary between species, growth rates and the length of clearwood to be produced (CQFA 2009). In general, removing live branches of trees also removes part of the photosynthetic capacity of the plant, and therefore, severe pruning will decrease the ability of the trees to produce carbohydrates and decrease growth (B. C. Ministry of Forests 1995). However, removing live branches in a pruning would not affect the radial increment unless more than 50% of the crown or the live crown ratio is reduced to less than 40% (Daniel et al., 1979; Smith et al., 1997). Pruning only the lowest branches of a tree will have little effect on tree growth since these branches produce few carbohydrates (B. C. Ministry of Forests 1995). With regards to timing of pruning, dead branches can be removed at any time, but it is advisable to remove live branches during dormant periods, such as during the dry season in order to reduce bark damage resulting from peeling.

If the spacing is appropriate, hardwood species normally tend to self-prune (Nyland 1996). However, for small-scale plantation growers in the tropics, biomass from pruning could serve as fuelwood to households. In agroforestry practices, lower branches of trees are pruned to grow high quality, high-value timber, provide fuelwood, reduce shade and promote pasture growth, improve access and visibility, reduce fire hazards and increase tree stability in windy areas (Evans 1992). Pruned material can be used as mulch in an agroforestry system, although the biomass on the forest floor could increase fuel loads and hence the risk of fires. From an ecological point of view, pruning removes lower branches that serve as perching and nesting sites for some birds and contributes to the structural diversity of the forest (Hunter 1990).

Pruning is currently not a major forestry operation in Ghana. The plantations established by the Forestry Department were neither pruned nor thinned until recently when a market was established for teak poles and cedrela logs (FAO, 2002a). Centeno (2009) recommends pruning teak trees to remove branches up to a desired height near the time of canopy closure. For cost effectiveness, pruning should be done selectively, coordinated with the intended method of thinning, to clear 2 to 3 metres of the stem at a time. Centeno (2009) notes that teak has the propensity to produce adventitious branches and epicormic shoots next to the scars caused by pruning; therefore a balance is necessary between the need to produce knot-free timber, the stem length pruned at a time, and the need to prevent a slowing down of growth due to an excessive reduction of the crown.

9.2.2. Methods of Pruning

There are two basic types of pruning: form pruning - the selective removal of branches to produce a single straight trunk and aids future pruning by controlling tree form; and clearwood pruning - the removal of lower branches in such a way as to grow the maximum amount of knot-free wood (CQFA 2009). Pruning is further divided into low and high pruning, depending on the height at which pruning takes place. Low pruning is the removal of branches up to 2m up the stem of the tree, or just after canopy closure in plantations (Evans 1992). It is usually carried out by hand or a chainsaw to provide easy access to the plantations, reduce the fire hazards, facilitate felling and markings for thinning and produce knot-free timber at the base of the tree (Evans 1992). Not all tree species would require low pruning. For example, in teak plantations, low pruning is not required except the removal of shoots during weeding or cleaning (Evans 1992).

Pruning can be carried out manually, using saws, axes, cutlasses, shears, chainsaws, pole loppers, etc., or by mechanical methods using powered hydraulic shears and mobile mechanical pruners. For taller trees, these simple implements can be mounted on long poles to reach branches from the ground. Pruning is labour intensive, and even when machines are used, it is slow. For example, analyses of the time required for high mechanical pruning various species of eucalyptus plantations in Australia show that about 15-22 trees per hour were pruned with experienced contractors (Rhodey-Bowman 2007). This suggests that a one-hectare plantation with 2500 stems would need between 14-21 days per ha, working at 8 hrs per day. The time requirements would even be higher with manual pruning.

Pruning should first remove structurally weak and dead branches. Those branches that appear dead or have leaves that appear unhealthy and distressed should also be removed. Pruning cuts should be made just outside the branch collar and nearly, but not completely, flush to the trunk, so as to provide viable growing branch bark that would encourage early closure of the wound (Rodney-Bowman 2007). Generally, pruning for the form should start in the second or third year, depending on the species and other site characteristics. Forked stems should be reduced to only one dominant trunk to minimise future pruning requirements and encourage healthy stems. Due to the potential for disease infestation following pruning, permanent branches should be shortened by pruning them back to a lateral branch or bud where an immediate growth response will be initiated (Rodney-Bowman 2007).

There is no consensus in the literature regarding the profitability of pruning as a silvicultural exercise. Because pruning is labour intensive, it is an expensive operation. For it to be profitable, the benefits must outweigh the costs of doing so. The benefits include the increase in the value of the pruned trees relative to unpruned stands and the value of the pruned biomass if used for fodder, fuelwood, or other non-commercial purposes.

The main factors affecting the costs of pruning are the number of branches to be removed per tree, the number of trees to be pruned, and the height at which pruning is done (low versus high pruning). The economic basis of pruning is to produce a higher proportion of clearwood and hence increase the value of the log. However, in addition to a knot-free log, other factors affect the value of the log as well, such as the size and shape of the log. The value of clear lumber produced from a pruning treatment can be three to five times greater than comparable knotty boards (Forest Practices Code 1995). However, when viewed at the stand level, the net present value (NPV), measured in dollars per hectare, of a pruned stand were found to be lower for lower site qualities in British Columbia, Canada. There is an optimum stand density to obtain maximum NPV for a properly pruned stand; which is a balance between sacrificing the size of the clear wood shell and the overall stand volume density (B. C. Ministry of Forests 1995). It is reasonable to conclude that pruning will only be profitable under limited circumstances. Given the analyses above, it is advisable to use planting densities that will minimise or completely eliminate the need for pruning, except in situations (e.g., agroforestry applications) where there is an explicit goal to use the pruned material.

9.3. Stand Density Management

Conceptually, stand density management is the process of controlling the level of growing stock through initial spacing or subsequent thinning to achieve specific management objectives (Newton, 1997; Newton et al. 2005). Manipulation of stand density with thinning affects stand structure, yield, tree size and rotation length (Kumar et al. 1995). Determination of appropriate levels of growing stock at the stand level is a complex process involving biological, technological and economic factors specific to a particular management situation (Castedo-Dorado 2009). The process requires selection of upper and lower limits of growing stock while taking into account the fact that the upper limit is chosen to ensure site occupancy, to maintain adequate stand-level yields and to produce trees of a desired form, whereas the lower limit is chosen to promote individual-tree growth rates (Dean and Baldwin 1996).

Once the initial spacing decision has been made, the only remaining tool to control the density of forest stands is through thinning. Thinning is the removal of some stems in an immature forest stand in order to give the residual trees improved conditions for growth and increase wood quality, improve plantation health, or remove product prior to mortality due to crowding or natural events. The main purpose of thinning, therefore, is to allow fewer stems to utilise the full potential of the growing site and ensure that volume increment is put on residual stems of high quality. Thinning also ensures the utilisation of all the merchantable material produced by the stand during the rotation.

Thinning is not new, it has been practiced in forests since the first century; the first recorded thinning regimes date to the 14th Century and have been progressively developed since that time (Day and Nanang 1997). The development of the principles of modern thinning since the 1920s by Craib (1939), MarMoller (1954), Assmann (1970) and many others, now provide a basis for manipulating the growth and development of natural and managed forest stands to favour almost all objects of management. Three very important principles for optimum timber production are (after Day and Nanang 1997):

1) determining the optimal density range for the species to be grown throughout the proposed rotation;
2) thinning natural regeneration of the species to, or out planting it at, spacings that minimise intra-specific competition and maximise growth to the time of first thinning. The time of first thinning is then scheduled to occur when the species reaches the maximum permissible density; and
3) implementing subsequent thinnings to maintain the species in an optimum band of density that maximises growth and stem quality through the remainder of the rotation.

9.3.1. Objectives of Thinning

Thinning is expected to result in the following bioeconomic benefits (Day and Nanang 1997):

(i) increases yield per ha by harvesting trees that would otherwise die because of intense competition;
(ii) favours the trees with the best growth potential and higher economic return;
(iii) shortens the time needed to reach a specified diameter or log volume;
(iv) extends the time to reach peak current annual increment;
(v) improves the quality and value of the timber by concentrating growth on the most valuable trees;
(vi) maintains high tree vigour and permits the removal of trees susceptible to insects and disease;
(vii) generates income early in the rotation, which can defray investments in silviculture and often results in higher economic returns;
(viii) reduces mortality almost to zero;

(ix) addresses age-class imbalances and shortening technical rotations, thereby alleviating impending wood-supply problems;
(x) creates employment opportunities thereby contributing to community stability;
(xi) minimises the inoculum potential of facultative parasites, which often cause disease in over-dense natural stands; and
(xii) strengthens the bole and branches of residual trees, making them resistant to breakage and windthrow.

Despite the many advantages of thinning, there are some subtle disadvantages associated with the practice. First, intervention may break the canopy and render the stand more vulnerable to windthrow, or harvesting may cause physical damage to and initiate biological deterioration of residual stands (Price 1989). If stands are over- thinned, it reduces the amount of photosynthesis and also forces the branches of the residual trees to elongate, which creates substantially weakened branches that may break easily during storms. Furthermore, if the thinnings have no economic value, thinning occurs at a net cost to the forest manager, especially if it occurs in unroaded areas of the forest. Finally, since thinning generally leaves large trees, volume increment due to thinning may do little to enhance the value of the volume to which it is added, since large trees may have already reached the plateau of the price size relationship (Price 1989). Due to these disadvantages, no general case can be made for or against the practice, since thinning may be viable and justified for a particular circumstance and may not be in other cases.

9.3.2. Types of Thinning

Foresters traditionally distinguish between five types of thinning (Smith, 1986, Ford-Robertson 1971): thinning from below, thinning from above, selection (or improvement thinning), mechanical or systematic thinning and free thinning. Three very important factors in thinning are the timing of the practice, the intensity (how much volume to remove), and the frequency.

Low thinning (German thinning) or thinning from below is the removal of smaller, weaker, and deformed trees whose crowns are in the lower portion of the stand canopy to ensure the right planting density is maintained and to allow resources to be concentrated on the better trees. This type of thinning leaves trees that are larger and more vigorous, with the most developed crowns in the stand (Nyland 1996). Low thinning is the oldest thinning technique and is not widely used in commercial forestry because the removal of these lower trees provides little extra growing space for the dominant trees in the stand. Thinning from below is identical to pre-commercial thinning, where the thinnings would not have a commercial value and is done at a net cost to the landowner. There is evidence that hardwood species on poorly drained soils do not respond to low thinning (Tepper and Bamforf 1959). Unlike in natural forests where trees often regenerate at high densities outside the control of the forest manager, in plantations, the initial spacing can be chosen so as to eliminate the need for low thinning – and in fact, this should be done. Unless the low thinnings have value as fuel, fodder, or mulch in an agroforestry system, it would make little economic sense to plant trees at high densities and costs, only to thin them out a few years later at an additional cost to the landowner.

Crown thinning is also known as the "French" method or "thinning from above," and is targeted at reducing crowding within the main canopy and involves the removal of intermediate and upper level trees (co-dominant and dominant trees). By removing the overstory trees, light and air can penetrate to the lower and middle crown classes. The increased light and air circulation within the crowns often reduce the incidence of diseases and pests. Tree selection is based on health and growth potential, with healthy and higher potential trees left in the stand. Weak or diseased trees in the lower canopy are also removed at this stage for convenience. Crown thinning improves the diameter growth of upper-crown residual trees (Nyland 1996).

Selection, or improvement thinning, removes trees based on size, quality, and their position in the canopy. Usually, dominant trees are cut to release trees in the lower canopy. This is not a widely used practice and is generally used in plantations with some shade-tolerant species. In agroforestry or community plantations, this method of thinning can be applied by pre-determining a size limit for trees. Once trees reach that size, they are harvested. This practice was found to be common in Northern Ghana on neem plantations planted on individual or community plantations, where trees that reached pole sizes were harvested leaving only the smaller trees (Nanang 1996). Evans (1992) also reports that this method is applied on eucalyptus plantations in Ethiopia once the trees have reached 12 cm diameter at breast height (Evans 1992).

Mechanical (systematic or line thinning) is a method in which trees are thinned based on an objective and systematic procedure, where individual tree quality is not considered (Evans and Turnbull 2004). Trees may be removed within a fixed spacing interval or by strips within fixed distances between them (Ford-Robertson, 1971; Smith et al., 1997). In natural forests, mechanical thinning is applied as spacing thinning, while, in plantations, it is applied as row thinning (Nyland 2007). In plantation applications, the landowner decides on the interval for the rows that will remain, and then harvest all trees between those rows. For example, the landowner could decide to remove every second or third row in the plantation. The thinning intensity desired will inform how many rows are removed. This practice fits well with mechanised operations such as weeding, thinning, and logging.

The final method of thinning is free thinning, which releases selected crop trees, without regard to the position of the trees in the crown. It may require a combination of thinning methods to achieve this. Once the crop trees have been selected based on a set criteria, thinning is undertaken to favour those crop trees, leaving the remainder of the stand unthinned (Smith et al., 1997). Irrespective of the criteria used for selecting crop trees, landowners usually mark out those trees that interfere with the designated crop tree and cut them out. This practice allows the landowner to release only a minimum number of choice trees per unit area without investing in the removal of undesirable trees as well (Nyland 2007).

9.3.3. Effects of Thinning on Stand Development and Volume Production

All major textbooks of silviculture (e.g., Nyland, 2007; Smith et al., 1997) agree that merchantable timber volume can be optimised by periodic reductions in stand density that maximise the diameter and volume growth of residual trees without sacrificing gross volume production. If plantations are planted at close densities, they would be subject to severe intraspecific competition and wasteful mortality, and hence thinned stands often produce more gross growth in volume than unthinned stands. Thinning is complex because of the many variables that determine the success and outcome of the practice – frequency, intensity, type, method and timing. Therefore, the growth and yield of individual stands and their ability to respond to thinning depends primarily on the species, the productivity of the site, the stand's age, density and historical development, and the nature and intensity of the thinning. The main benefits of thinning is to increase the size of individual trees, and redistribute the growth potential of the stand to optimum advantage, which ultimately enhances the value of the final crop, while lowering its harvest and milling costs (Johnstone 1997).

Thinning provides more space for the growth of the residual trees. It opens the canopy for improved photosynthesis and the soil for improved absorption of water and nutrients. In the years after thinning, residual trees form more and larger buds and extend new foliage and branches filling the canopy, and roots filling the soil space (Nyland, 2007; Smith et al. 1997). The current annual increment per hectare declines immediately after thinning and then increases as the residuals extend. The loss in increment that occurs immediately after thinning is soon recovered as the residuals extend and fill the growing space.

Over the life of the rotation, landowners can recover more usable volume by periodically thinning to control stand density and by selling the excess and potentially mortality trees (Nyland 1996). In Figure 1, the upper line depicts a yield curve and shows the gross volume produced during the life of the stand. This volume includes the volume currently in the stand, plus the amount removed through thinnings. Three thinning regimes are shown in Figure 1 by the bottom curve. It shows that thinning removes some of the standing volume, and then the stand is allowed to regrow before the next thinning. The gap between these two curves shows the amount of wood recovered in thinnings. If the stand had not been thinned, this volume would have been lost to wasteful mortality. It is important to point out that although the final harvest volume is less when the stand is thinned than unthinned, the cumulative volume is actually higher for the thinned stands. In other words, the combined volume of thinnings and final harvest in the thinned stand is greater than the final harvest in the unthinned stand.

9.5.4. Determining Optimal Stand Density

When trees are young, they need less space to grow than when they are older. As they grow older, the increased space requirements per tree lead to competition among the trees for resources below ground (water, nutrients, etc.) and above ground (light and air) compared to younger trees. Competition can occur between plants of the same species or of different species. Intraspecific competition is the negative interaction that occurs between plants of the same species while interspecific competition occurs between plants of different species (Radosevich and Osteryoung 1987). For example, in a monoculture plantation of gmelina, only intraspecific competition will be observed, since all the trees are of the same species, while in a mixed stand of gmelina and teak, there will be both intraspecific (e.g., among gmelina trees) and interspecific competition among the gmelina and teak trees.

The amount and timing of thinning older trees depends on growth rates and initial spacing. Crowding of the tree crowns gives a rough indication of when thinning is necessary. Deciding the optimum stand density for managing any tree species remains one of the most controversial issues in forest management. The question of how many trees per ha constitutes the ideal number for any tree species is continuously debated by foresters. This debate has led to the development of indices to assess the densities of forest stands and predict when stands should be thinned to reduce or eliminate competition.

Measures of Stand Density

Several measures of stand density have been used in forestry. These include: a) trees per ha, b) volume, c) basal area, d) crown competition, e) stand density index, and f) relative stand density index. Density measures are used to estimate economic timber production and evaluate or estimate total biomass, design management regimes and evaluate other non-timber values such as carbon sequestration, wildlife habitats, etc. (Davis et al. 2001).

With regards to stand density and relative stand density indices, Kira et al. (1953) and Shinozaki and Kira (1956) studied the effect of density (number per area) on the mean dry weight of plants in competing populations. These Japanese population biologists have shown that there is a linear relationship between the reciprocal of mean plant dry weight (1/w) and the density of a wide range of plant species. This relationship is referred to as the "Reciprocal Yield Law" (Harper 1977). In other words:

$$\frac{1}{w} = Ad + B \tag{1}$$

where w = mean plant dry weight
d = plant density (number per unit area)
A and B = species dependent constants

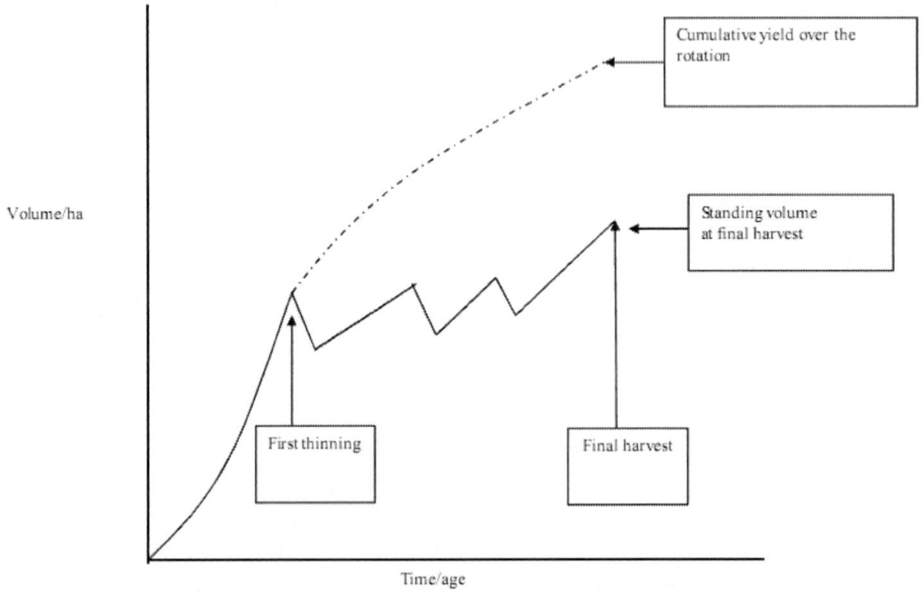

Figure 1. Hypothetical relationship between volume thinned, net productivity and gross productivity of a forest stand following thinning.

Kira et al. (1953) and Shinozaki and Kira (1956) further showed that when plant populations are grown at high densities, mortality causes self-thinning in accordance with the reciprocal yield law. From this law, it has been demonstrated that when the logarithm of the mean dry weight of the survivors in a competing plant population are plotted against the logarithm of their density, there is a linear relationship with a slope of -3/2 or -1.5 (Harper, 1977, Yoda et al., 1963). This relationship is defined as the -3/2 Power law. In forestry, the Power law often defines the self-thinning line, which has proven valuable in understanding the dynamics and development of forest stands and their projection through time (Solomon and Zhang 2002). The slope of the self-thinning line represents the largest number of trees of a given size that can occupy a hectare at any one time.

Several density indices have been used in forestry to describe measures of stand density: the stand density index (Reineke 1933), the self-thinning rule (Yoda et al. 1963), the relative density index (Drew and Flewelling 1977; 1979) and the relative spacing index (RS) (Wilson 1946). The advantage of these density indices is that they are independent of site quality and stand age (Long 1985; McCarter and Long 1986). The self-thinning line has been commonly expressed by the following mathematical equations:

$$\ln V = k - 1.5 \ln N \tag{2}$$

(Drew and Flewelling 1977)

$$\ln W = k - 0.5 \ln N \tag{3}$$

(Zeide 1987)

$$\ln D_q = k - 0.625 \ln N \tag{4}$$

(Jack and Long 1996)

$$\ln N = k - 1.605 \ln D_q \tag{5}$$

(Reineke 1933)

where:

V = mean tree volume
W= total tree volume or biomass
N=number of trees per unit area
D_q = quadratic mean diameter
k= species specific constants (note: the value of k is not the same across equations)

In general, tree species that are very intolerant will have their Reineke's stand density index (RSDI) in the 1000 to 2000 range, while very tolerant species will range in RSDI of between 8000 and 10,000 (Day 1996).

The relative spacing index (expressed as Spacing Factor %) is defined as the mean distance between trees in a uniform canopy crop expressed as a percentage of the stand's top height (Day 1996). Mathematically, this relationship is expressed as:

$$N = k\, Top\, Ht^{-2.0} \tag{6}$$

or

$$\log N = k - [2.0 \log(Top\, Ht)] \tag{7}$$

where k is calculated as:

$$k = \log N + [2.0(Top\, Ht)] \tag{8}$$

In Wilson's SF% relationship, the number of trees per ha is plotted over the logarithm of the top height of the stand in metres on a Wilson's graph, with the slope of the SF% lines being -2.0. Tree species that are very intolerant will have SF% values of between 30 and 20% while very tolerant species will have SF% from 15-10% (Day 1996).

Stand Density Management Diagrams

Stand density indices are the basis for developing stand density management diagrams (SDMDs). These have greatly facilitated stand density management decision making (Castedo-Dorado 2009). SDMDs are average stand-level models that graphically illustrate the relationships among yield, density and density-dependant mortality at all stages of stand development (Newton and Weetman 1994). Their utility has been largely limited to evaluating density management outcomes in terms of mean tree size and stand-level volumetric yields (Newton 1997). In addition to volumetric yield, maximisation of product value is becoming an important management objective, which requires estimation of the underlying diameter distribution, given the inherent relationship between monetary value and tree size (Castedo-Dorado 2009). Newton et al. (2005) have made recent innovations to include structural yield prediction in the development of SDMDs and of traditional stocking guides which enable forest managers to estimate the number of trees in each diameter class at any point during stand development (Castedo-Dorado 2009).

SDMDs are constructed by characterising the growing stock with indices that relate the average tree size (e.g., mean weight, volume, height or diameter) to density (e.g., the number of trees per ha). While stand density management diagrams have been developed for some Japanese and North American tree species (e.g., Drew and Flewelling, 1979; McCarter and Long, 1986; Newton and Weetman 1994), there are no such diagrams for tropical species, with the exception of teak in India (Kumar et al. 1995).

9.3.5. Profitability of Thinning Operations

There is evidence in the forestry literature regarding the ability to increase gross timber production through optimal stand density management (e.g., Frank 1973, Van Cleve and

Zasada 1976). From an economic perspective, however, there is still controversy over the economic profitability of intensive forest management practices such as thinning. For example, Anderson (1992) argues that the net benefits of silvicultural investments are at best marginal. The economic consideration in applying enhanced silviculture is very critical because it determines whether firms or landowners involved in forest management will invest in thinning activities. Forest managers, researchers, industry, and governments are interested in knowing whether thinning makes economic sense, and if so under what conditions.

The profitability of thinning operations is seldom evaluated in isolation from the overall profitability of the plantation enterprise. That is, instead of assessing whether the revenues received from thinning operations are greater than the costs, it is often the case that the analysis is done at the plantation enterprise scale, to determine whether the overall profitability of the plantation is higher with and without the thinning operation. If the thinning operation is profitable in isolation, it increases the chances that the whole plantation enterprise would be profitable at the rotation age. Conversely, if the thinning results in a net cost to the landowner, the chances that the operation would be judged profitable when evaluated at the final harvest are reduced. The methods for undertaking this kind of analysis are presented in the chapter on the economics of plantations (Chapter 11).

In cases where thinning has been shown to be profitable (e.g., Day and Nanang, 1997; Ghebremichael et al., 2005), the increase in profitability of the thinned over the unthinned stands was mainly due to the increase in total merchantable volume (thinnings plus final harvest) of the thinned stands. Secondly, the bigger trees (greater volume/tree) at the end of the rotation resulted in reduced final harvesting costs. Also, intermediate revenue from thinnings offset the cost of regeneration early in the life of the plantations and, therefore, reduced the interest on capital. The profitability of thinning, therefore, depends on a combination of site productivity, final timber price, whether the thinnings have a market value, timing of the thinning operation, the rotation length, and site stability. Thinning increases revenues but incurs a risk in terms of windthrow on unstable sites (Phillips 2004).

It has been argued (e.g., Braathe 1957) that even in cases where the thinnings do not give a direct profit, it *may* pay to carry them out because the first thinning in a stand may be more than repaid by a better development of the stand in the immediate future. This increase in value results from reduced harvesting and transportation costs, and increased value due to the bigger trees at final felling. The financial viability of thinning is also determined in part by the distribution of fixed and variable costs of the operations. Thinning involves relatively large fixed costs, such as movement of equipment to treatment sites. In this regard, there are economies of scale in the application of thinning treatments.

9.3.6. Case Study: Stand Density Management for Teak

Experience from India and Vietnam

This section begins with examples of stand density management from Asia, where they have had more experience managing teak than we have in Ghana. It has been found that the initial planting spacing and the timing of thinning strongly affects the pattern of growth and yield of teak plantations in India (Krishnapillay 2000). Wider planting spacing and/or early heavy thinning may result in heavy side branching and epicormic shoots resulting in reduced volume yield. On the other hand, closer spacing and /or late light thinning may result in a decline in growth rates (Krishnapillay 2000). Kumar et al. (1995) provide guidance on the maximum and minimum levels of the RSDI for teak in India as 1200 and 720, respectively. The lower limit for full site occupancy to occur is 420 trees/ha while the lower limit for the onset of competition is 300 trees /ha. An initial planting density of between 1,200 and 1, 600 trees per hectare is suggested for teak in India (i.e., an initial spacing of about 8.3 m x 8.3 and 6.25 x 6.25 m, respectively). These suggested spacing for teak are far wider than those typically used in Ghana of 2 x 2m (2500 trees/ha).

Krishnapillay (2000) describes the thinning regimes in India for teak to consist of three thinnings. The timing of the first thinning is often determined by the height of the trees and is commonly carried out when the trees reach 9.0 to 9.5 m. The second thinning may be carried out when the trees reach 17 to 18 m (Krishnapillay 2000).The mean basal area is often allowed to reach 20 to 22 m^2/ha after the second thinning. A third thinning is then carried out to reduce the mean basal area to between 13 and 15 m^2 per hectare. A final stocking of about 300 trees per ha would be the ideal (Krishnapilla 2000). In Vietnam, the thinning schedule for teak includes three thinnings at plantation ages 6, 12, and 20 years, leaving 800, 400, and 200 trees/ha on good sites, and 1200, 600 and 300 trees/ha on medium sites (FAO 1998).

Stand Density Management for Teak in Ghana

Stand density management for teak in Ghana is illustrated here using data from Nunifu (1997). The data for these analyses were collected through a field study in the Guinea Savannah zone of Ghana described in Section 6.10.

Using this information, the stand characteristics, as well as the RSDI and the SF%, were calculated and presented in Table 1. The calculations were done using the following equations:

$$RSDI = TPH(D_q / 25)^{1.605} \tag{9}$$

$$D_q = \sqrt{(40000/\pi)(G/TPH)} \tag{10}$$

$$SF\% = \frac{SI}{TH} \times 100 \tag{11}$$

$$SI = \sqrt{\frac{10000}{TPH}} \tag{12}$$

where: D_q = the quadratic mean diameter in centimetres

G = stand basal area in m^2/ha

TPH = the number of trees per ha

TH = top height in m

SI = spacing interval (average distance between trees)

From Table 1, the maximum RSDI for teak in the savannah vegetation zone was about 1200, while that in the high forest zone was 373. The SF% was much larger in the HFZ than the savannah due mainly to the lower densities of trees in the HFZ.

Table 1. Summary of stand parameters of teak plantations in the savannah and high forest zones

Stand age (years)	Average DBH (cm)	Top Height (m)	Density (trees/ha)	Basal area (m^2/ha)	Gross volume (m^3/ha)	RSDI	Spacing factor %
Savannah Zone							
3	2.76	4.10	1510	0.99	4.13	47	62.77
4	3.25	4.82	1613	1.50	5.49	67	51.66
6	5.95	7.36	2100	6.87	29.49	239	29.65
7	7.67	9.45	1765	8.34	44.33	270	25.19
8	7.69	9.62	1705	7.94	43.12	258	25.17
9	7.82	9.33	1591	8.22	38.09	261	26.87
17	10.78	10.29	1488	14.21	73.74	400	25.19
26	18.21	16.82	1413	37.89	339.42	870	15.82
31	21.06	17.19	1396	50.14	472.28	1086	15.57
38	23.58	24.53	1299	58.83	776.72	1218	11.31
40	19.53	18.37	1429	43.83	220.33	980	14.40
High Forest zone							
13	21.00	16.80	291	10.38	84.00	225	34.89
14	27.00	19.90	265	13.03	114.00	265	30.87
15	22.50	17.45	311	11.71	91.50	251	32.50
16	27.33	21.03	276	15.07	136.67	301	28.62
17	29.33	21.22	243	15.57	142.17	301	30.23
18	28.00	20.65	246	14.08	126.75	278	30.88
19	30.00	19.40	177	11.91	100.00	228	38.74
20	29.40	20.82	264	16.23	148.00	316	29.56
21	30.50	23.10	243	16.79	164.50	320	27.77
22	31.00	24.40	271	19.83	204.50	373	24.90

The RSDI and /or the SF% can be used to develop stand density management regimes for teak. A generalisation is that about 60% of a species' maximum RSDI approximates the lower limit of the zone of imminent competition mortality (Drew and Flewelling 1977). This means that below 60% of maximum RSDI, a stand experiences little or no density-related suppression, while, above 60%, self-thinning would be expected to occur. For the savannah zone, the lower limit for growing teak is a RSDI of 720 (60% of 1200) and the upper limit is an RSDI of 1200 in order to avoid competition that leads to self-thinning. This information can now be used to set upper and lower limits of RSDI for thinning. Stands should be thinned as soon as their RSDI reaches 1200, and reduced to an RSDI of 720, and allowed to grow back up until 1200 again, and thinned until the rotation is reached. It is also usual to assume that 35% of the maximum RSDI represents the lower limit for full site occupancy (i.e., the density that captures the most growth potential of the site). For teak, this would be about

RSDI of 420 in the savannah zone. If however, the emphasis of management is on individual tree growth at the expense of some potential yield, then it might involve limiting the upper density to about 25% of the maximum RSDI (Kumar et al., 1995). In the case of teak, this will be an RSDI of 300.

A Thinning Schedule for Teak in the Savannah Zone of Ghana

Below, we describe a simple schedule and a SDMD for thinning teak plantations in the savannah zone based only on the RSDI (henceforth SDI for short) and quadratic mean diameters calculated in Table 1. It was not possible to develop a similar diagram for teak in southern Ghana because the stand density data shows that these stands are already understocked. However, the same techniques can be used if stand-level data for fully stocked plantations in southern Ghana were available. Similar schedules can be developed using the SF% or any of the other stand indices previously described. In practice, more complicated SDMDs are developed using three or more of these indices at the same time (e.g., Kumar et al, 1995; Castedo-Dorado et al. 2009; Newton et al. 2005 and Newton and Weetman 1994). The example below is simplified for illustrative purposes.

The thinning schedule is developed for teak using the information on the upper and lower limits on the SDI and the diameter of the tree at the rotation age. In this example, two scenarios are evaluated. First, the maximum SDI of 1200 is used as the upper limit for managing teak. However, other authors have used lower percentages of the maximum SDI as upper and lower limits. For example, Kumar et al. (1995) used 35 and 20% of maximum SDI as the upper limit and lower limits, respectively in developing a SDMD for teak in India. Dean and Baldwin (1993) used 45 and 30% of maximum RSDI as the upper and lower limits, respectively for a SDMD for loblolly pine plantations in the United States. The limits are usually determined based on experience with the species in the location where it is grown.

To illustrate how this is done with upper limit of maximum SDI =1200 and lower limit of SDI=720, we assume that the plantation is grown for sawlogs and the desired diameter (D_q) at the rotation is 40 cm. It is further assumed that no thinning will take place until the trees reach a minimum diameter of 10 cm. From Figure 2, the point at which the 40 cm D_q line intersects the 100% RSDI line (1200) represents the point at which final harvesting should take place (rotation). To find the sequence of thinnings to get to the final point, we stair-step backwards between the upper and lower limits until D_q falls below 10 cm. The thinning is represented by the horizontal portion of the stair-step and approximates the number of trees removed in the operation. The vertical portion of the stair-step represents the post-thinning growth phase of the operation.

To determine the timing of thinning and the final harvest, we use the site index curves and a relationship between D_q and the top height of the plantations. For this data, top height can be predicted from D_q using:

$$TH = 7.3159\ln(D_q) + 0.3487 \qquad R^2 = 0.73 \tag{13}$$

We know that all thinning occurs when the SDI reaches the maximum of 1200, and these occur at points G, E, C and A (final harvest) in Figure 2. At these points, the corresponding D_q's are 6, 18, 29, and 40 cm. Using these D_q's, we predict the corresponding top heights of the plantations using Equation (13) as 14, 22, 25 and 27m, respectively. The plantation ages to thin are then the ages on the site index curves that correspond to these top heights in Figure 3 (using a site index of 20).

From Figure 3, the first thinning should occur at about four years after planting, followed by a second thinning at about 15 years and a third thinning at about 34 years. The final harvest should occur in year 50. We can determine that the initial stocking will be about 2100 trees/ha (at an initial spacing of about 2.2 x 2.2 m). The number of trees to be removed at each thinning is determined in Figure 2 and are 400, 420 and 450 in the first, second and third thinning, respectively. Assuming no natural mortality, about 600 trees will remain for the final harvest. The information is summarised in Table 2.

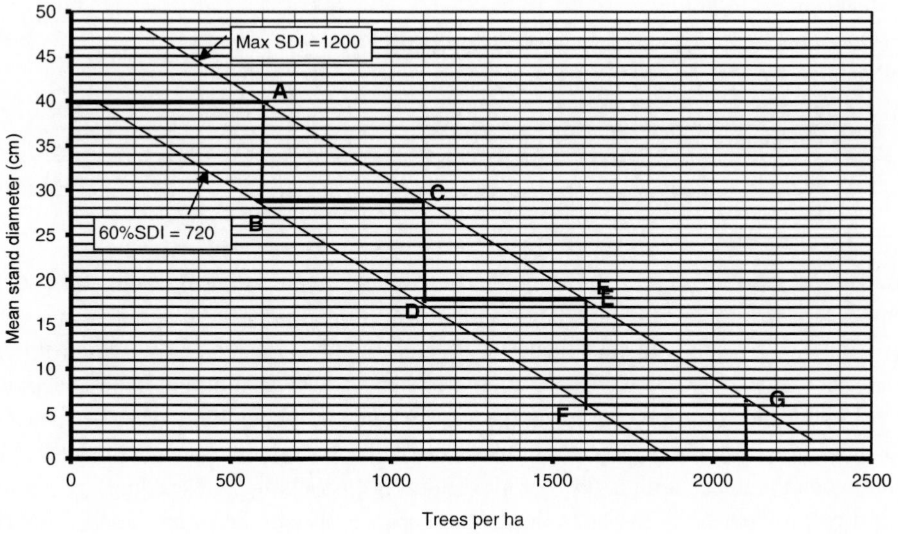

Figure 2. Stand density management diagram for teak in the savannah zone of Ghana.

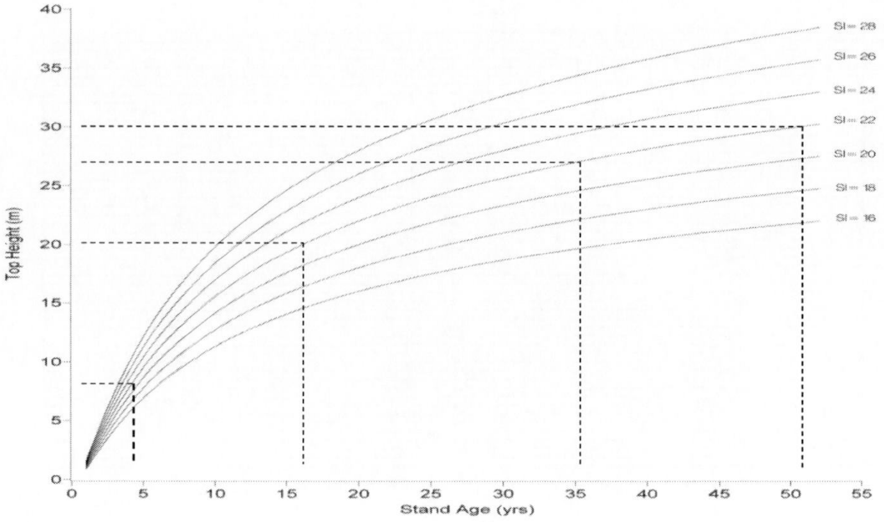

Figure 3. Site index curves for teak plantations in the Savannah Zone of Ghana.

Table 2. A comparison of two density management regimes for plantation teak grown for sawlogs in the savannah zone of Ghana

Operation	Age (year)	Top height (m)	TPH		No. of trees Removed /ha
			Before	After	
Upper limit of maximum SDI =1200; lower limit of 60%SDI=720					
Planting	1	-	2100	2100	-
First thinning	4	14	2100	1600	500
Second thinning	15	22	1600	1100	500
Third thinning	34	25	1100	600	500
Final Harvest (rotation)	50	27	600	0	600
Upper limit of 60% SDI =720; lower limit of 30%SDI=360					
Planting	1	-	1600	1600	-
First thinning	4	12	1600	1100	500
Second thinning	15	21	1100	600	500
Third thinning	34	25	600	100	500
Final Harvest (rotation)	50	27	100	0	100

Figure 4 presents a regime for stand density management for teak using 60% of the maximum SDI as the upper limit and 30% as the lower limit. Under these circumstances, self-thinning would be avoided. In this scenario, the minimum diameter below which thinning would not take place is 5 cm. The plantation ages at which thinning take place were determined using the same procedure as the case where the upper limit was the maximum SDI (but not shown in Figure 3). A comparison of the two density management regimes shows that when the upper limit is 60% of maximum SDI, stocking levels reduce accordingly, and more growth is put on individual trees at the expense of overall volume yield. This is an obvious balance that forest managers need to keep in mind when setting density management regimes.

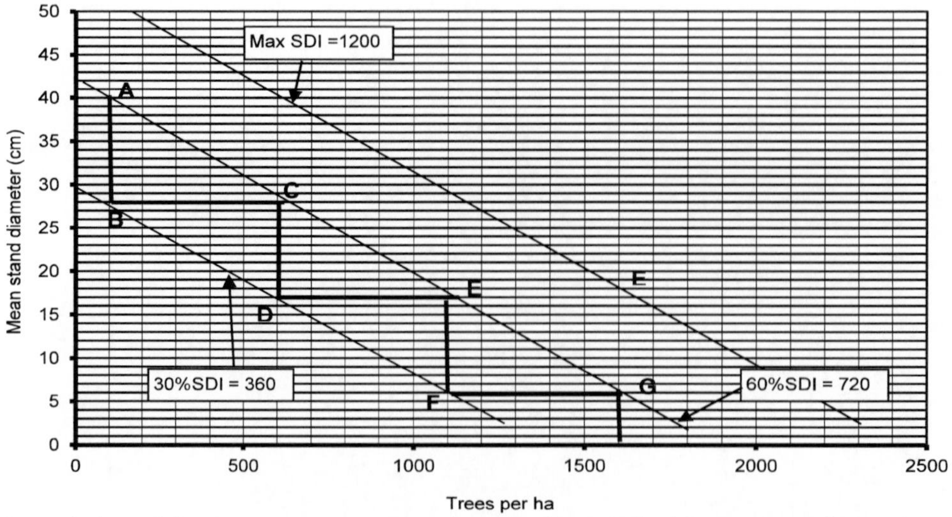

Figure 4. A stand density management regime for teak using a 60% of the maximum SDI as the upper density limit.

9.4. CONCLUSION

This chapter described the common practices in plantation silviculture and management – weed control, pruning and stand density management. Understanding and implementing these practices are essential to optimising the growing potential of forestry sites and improve the productivity and the economic profitability of the plantations. The concepts outlined here are of general application, and, therefore, the specific practices will depend on the tree species under consideration. Hence, some of the knowledge of critical silvics discussed in Chapter 4 will be crucial to determining the management regime that should be pursued.

Chapter 10

RURAL DEVELOPMENT AND ENVIRONMENTAL PLANTATION FORESTRY

Environmental plantation forestry refers to growing plantations or forests for environmental purposes. The purpose of this Chapter is to discuss in detail the development and management of forest plantations to achieve non-timber or environmental objectives. In many rural communities, tree planting is often undertaken to provide wood products for the individual, community or the wider market; however, for others, the reason may be less about wood products. These non-timber products and services provided by plantations are often considered minor though they can be major in certain circumstances. This Chapter discusses the role of trees in providing these products/services and in particular what species are most suitable for each use, the planting methods required and the management strategies that can be applied to achieve the intended results.

10.1. RURAL DEVELOPMENT AND POVERTY REDUCTION

10.1.1. Role of Trees in the Rural Economy

Trees play several important functions in the rural economies around the world. In Ghana, forest resources play a significant role in rural livelihoods of a large number of people. The role of plantations include soil stabilisation, fodder for livestock, shelter for humans and animals, windbreaks, erosion control, foodstuff for the population, fuelwood for cooking and heating, wood for building construction and medicinal values. Although wood and wood products are important internationally traded commodities, the importance of trees to meet the wood (energy, building construction, etc.) and non-timber forest products (NTFPs) needs of local communities continue to exist. Forest plantations can be a powerful tool to assist the poor households across Ghana to establish productive and environmentally sustainable income generating opportunities, meet their wood supply needs and provide critical inputs into local industries, while strengthening their capacity for local self-reliance.

There are a number of reasons for the widespread interest in the use of NTFPs for poverty reduction and biodiversity conservation (Pfund and Robinson 2005). The first is that the demand for many NTFPs is growing fast (e.g., medicinal plants) and their habitats and populations are increasingly threatened by ecosystem degradation. Secondly, economically viable NTFP harvesting may be less detrimental to forest cover and biodiversity than timber harvesting. Third, sustainable incomes from NTFP harvesting and marketing can provide sufficient incentives for forest and other natural habitat conservation. Finally, the contribution of NTFPs to the livelihoods of the poor is often high compared to other measures (Pfund and Robinson 2005). It must be noted, however, that given the often biologically less diverse structures of plantations, their potential use for NTFPs may be less than for more heterogeneous natural forests.

The ability to develop plantations in a location of the planter's choosing allows governments, non-governmental organisations (NGOs), and the private sector to use forest plantation initiatives as rural or economic development tools and to improve food security and alleviate poverty. There are several ways in which plantations contribute to improved rural livelihoods and poverty reduction: employment in the establishment and management of plantations, employment in wood processing plants, revenue from the sale of timber and non-timber forest products, use of forest products as inputs into rural arts and crafts industries, payment for ecosystem services or carbon trading schemes, and eco-tourism. To achieve poverty reduction objectives, poverty alleviation should be incorporated as a major objective of both government- and private sector-led plantation development efforts. The benefits accruing from plantations must reach rural communities for their livelihoods to improve.

In many rural areas without natural forests, the major forest products need is often for wood energy and wood for building construction. Community and small-holder individual plantations can be used to serve these needs of rural dwellers. Some of the forest products can also be sold to supplement household incomes. Also, government and NGO-sponsored plantation projects offer job opportunities to rural people, where there is usually limited access to alternative employment opportunities. Employment opportunities are created especially if industrial plantations are established with wood processing facilities that can employ a diversity of professionals, and bring along the development of much-needed infrastructure to the communities. These projects allow communities to participate meaningfully in the management of their resources and also build their capacity in project management. For example, the World Bank-sponsored Rural Afforestation Project that was undertaken in the savannah regions of Ghana from 1989 to 1995 employed several hundreds of technical and manual labourers to produce, distribute and transplant seedlings for interested farmers, communities and organisations.

In fact, the provision of employment to rural dwellers and re-stocking of commercial species in degraded forest reserves were the main objectives of the plantation development programmes by the then Forestry Department in the 1960s (FAO 2002a). Again, in 2001, when the Government of Ghana launched the National Forest Plantation Development Programme (NFPDP) with a target planting of 20,000 ha/annum, a secondary objective of the programme was to provide employment for the youth in rural areas. Job opportunities created under the NFPDP were either full time or casual by day jobs. The full- time jobs were in the form of farming opportunities granted to peasant farmers from forest fringe communities. The casual by day jobs were offered mainly for activities such as site preparation, peg cutting, pegging, seedling production, and planting (Ghana Forestry Commission 2006b). In 2005, the programme recorded an estimated 31,500 full-time jobs, and 600,905 man-days of casual by day jobs (Ghana Forestry Commission 2006b). When the government re-launched the NFPDP in 2009, job creation for rural dwellers was again at the centre of its focus, promising to create 30,000 jobs by the end of 2011 (Ghana News Agency 2009). As noted in Chapter 1, job creation for the youth was also a key goal of the SADA afforestation project.

Non-timber forest products already form the backbone of many communities across the country. These range from fruits, gums, thatch for roofing, fuelwood, wildlife (e.g., bush meat and snails), and medicinal plants to arts and crafts. There are some other NTFPs that are in their infancy and have the potential to be developed into significant income generating activities in Ghana. These include eco-tourism, payment for ecosystem services, and other climate change-related programmes such as REDD+ and carbon trading schemes.

Despite the interest and potential for use of NTFPs to help alleviate poverty and improve rural economies, Pfund and Robinson (2005) point out a number of issues concerning NTFPs that make them a difficult group of products to frame in terms of their characteristics and actual potential contribution to poverty alleviation, economic development in general and to sustainable natural resource conservation. These include:

- Resource assessment of NTFPs is usually complicated for both plant and animal products. For plant products, unlike timber, few standard inventory methods can be applied. Species-specific population inventory techniques need to be adapted and combined with appropriate yield assessment techniques to arrive at production figures for such diverse products as roots, tubers, leaves, fruits, sap, bark, etc.;
- Sustainable management and harvesting recommendations can, therefore, be difficult to develop as they may need to be species specific;
- Quality assessment of the resource is difficult when the valuable ingredient(s) of the NTFP requires complex chemical analysis (e.g., medicinal properties);
- The end products of many NTFPs are often the outcome of a series of successive, varied and sometimes complex processing measures;
- It is much easier to smuggle valuable/trade-banned NTFP products under another species/product name than for timber; and
- Certification and fair trade requirements can be particularly difficult and/or costly to introduce, given the large possibilities to hide the true sources of the products.

10.1.2. Choice of Species

The choice of species for planting when the main objective is rural development and poverty alleviation is limited to those species that can clearly improve on rural livelihoods.. Recall the several ways in which plantations contribute to improved rural livelihoods and reduce poverty in the previous section. If the objective is to grow wood for timber products for sale to raise revenue, then the choice of species will focus on those that have the potential to grow into merchantable timber sizes or electric transmission pole sizes, such as teak, kapok, mahogany, etc.

On the other hand, species that are suitable for revenue generation through the sale of their products are wide-ranging and include fruit trees such as mango, dawadawa, sheanut, cashew, guava, etc. In the Northern Region, the Integrated Tamale Fruit Company (ITFC) has been successful in developing a large number of mango plantations through out-grower programmes that produce and export organic mangoes. An out-grower programme is an excellent example of how fruit plantations can be used to generate revenue and reduce poverty in rural areas. With the advent of carbon trading schemes, plantations could be developed to sequester carbon as part of an offset trading scheme. In this case, the ideal species will be those that are fast growing and would produce large amounts of biomass (and hence carbon) within a short period.

Local arts and crafts typically depend on the natural forests for their wood needs, and this may likely be the best option unless there is a specific industry that has commercial demand to make growing plantations of those species economically viable. Another option would be to explore the use of wood from plantations species that are being grown for other uses for arts and craft production.

10.1.3. Management

Planting methods and management practices for plantations established to promote rural development objectives will depend on the development initiative itself. Readers may refer to the subsequent sections for specific management approaches that may apply to an initiative of their interest.

10.2. Wood Energy Production

10.2.1. Role of Trees in Wood Energy Production

The terms *fuelwood* or *woodfuel* describe wood that is used as fuel. Fuelwood usually includes firewood, charcoal, chips, pellets and sawdust. Firewood, on the other hand, is any woody material used as fuel, usually in unprocessed log form. Wood has been used as a source of energy for millennia and remains the most important source of energy in Ghana today. The majority of the plantations established around the world are intended to produce wood products for domestic and commercial purposes. One of the most important roles that trees play in Ghana is the provision of biomass for use in bioenergy applications. This includes the use of tree biomass for firewood and charcoal production for domestic uses in cooking and heating. Firewood and charcoal remain the largest sources of energy for Ghana's population, especially the rural dwellers. This means that the majority of the population relies on wood from the natural forests, savannah woodlands, and plantations to meet their wood needs.

In 2000, the annual production of wood in Ghana was about 30 million tonnes of which about 18 million tonnes was available and accessible for woodfuels (Energy Commission, undated). According to the FAO (2014), Ghana produced 42 million (m^3) of woodfuel and 1.8 million metric tonnes of wood charcoal in 2012. Figure 1 shows the trends in woodfuel production in Ghana from 1961 to 2012. This graph shows a steady increase from 1961 until about 1991, and then a very steep increase thereafter.The bulk of woodfuels amounting to 90 percent is obtained directly from the natural forest. The remaining 10 percent is from wood waste, i.e., logging and sawmill residue and planted forests.

There is currently an enormous amount of pressure on natural forests to produce fuelwood to meet the needs of both rural and urban populations in Ghana and across the globe. There are several reasons for this. First, increasing population requires more wood to meet their needs. Cities have been expanding, in part due to the general increase in population and partly because of rural-urban migration. According to the World Bank (2010), the rural-urban split in Ghana is now 49/51%, which reflects decades of rural movements into urban areas in search of job opportunities. A shift in the rural/urban population also reflects a shift in energy use patterns between fuelwood and charcoal given that urban populations rely heavily on charcoal for cooking and heating while the rural population relies on fuelwood. A second major reason for fuelwood shortages is the loss of forest cover, through deforestation and forest degradation as discussed in Chapter 2. The third factor is the issue of land and/or tree tenure rights of people to harvest wood for fuelwood. Related to tenure rights is the fact that in many parts of Ghana, farmers who cultivate lands for which they are not owners do not have rights to own the trees they grow, and hence no incentive to manage these trees for the long term.

Given the state of the natural forests and savannah woodlands in Ghana, it is clear that plantation forests can play a significant role in providing the woody biomass needed for bioenergy. The ability to plant trees within compounds, close to homes or integrate them in agroforestry practices provides an additional advantage for forest plantations to be used as a reliable source of fuelwood supply. Other reasons, why forest plantations are well-suited for producing wood energy, are (State Government of Victoria 2009): a) Under favourable growing conditions, the average household could grow enough firewood to be self-sufficient on less than a hectare of land; b) It is relatively quick and easy to grow and requires few specialist skills and little management, compared to growing sawlogs, for instance; and c) fuelwood can be harvested as a profitable by-product of other farm forestry regimes such as sawlog production or agroforestry practices.

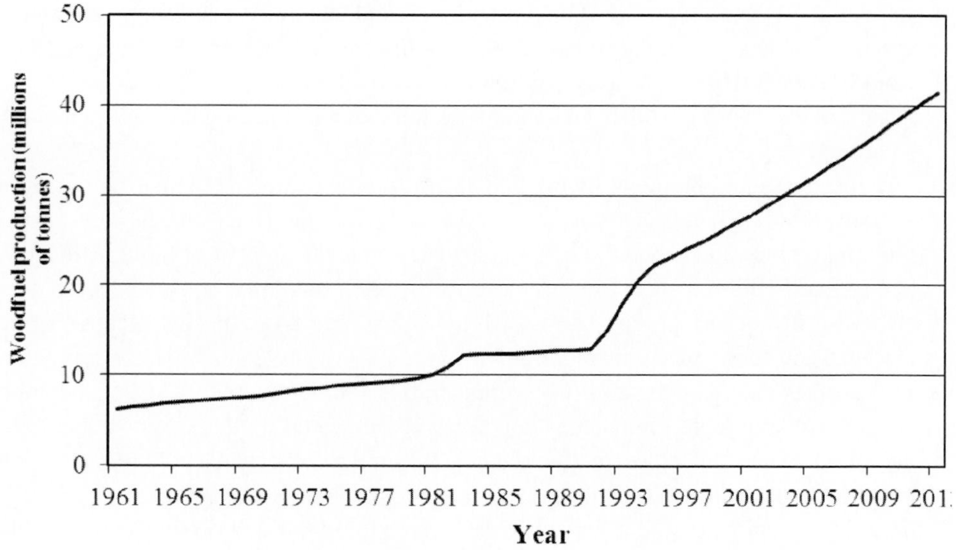

Figure 1. Woodfuel production in Ghana from 1961 to 2013.

10.2.2. Choice of Species

Species chosen for production of fuelwood should, first of all, be able to grow in the climatic conditions of the area where the plantation will be sited. One simple way to determine this is to scan the environment to see which tree species are growing well on the site. Other important characteristics of the species include:

i) Availability of seed and ease of cultivation. Tree species suitable for fuelwood production should ideally have locally available sources of seed for propagation. These species should be easy to grow and manage within the confines of the home compound or farm. Species that are easy to establish with minimal technical supervision would be preferred. The resistance of the species to browsing by animals will make it easy to cultivate around homes. Ease of cultivation is also related to the level of technical skills or expertise required to manage the species. Most farmers that would be involved in managing trees for fuelwood will have less education and technical skills to learn the management of specific tree species. Even then, it is still more advantageous to grow a species that is simple to manage than one that requires trained expertise.

ii) Good survival under adverse conditions: In order to produce biomass needed to meet household and commercial demands, tree species selected for fuelwood must have good survival rates when planted and must also be able to survive adverse conditions of the areas in which they are planted. For example, in northern Ghana, trees must survive high temperatures during the dry season, poor soil conditions, desiccating heat and winds, low relative humidity, frequent annual bushfires, etc. Hence, trees that are well-adapted to these conditions will be more suitable than those that cannot survive these conditions. If it is possible to find naturally occurring native tree species that have acclimatised to these conditions over several generations, they would be ideal. However, consideration also needs to be given to the fact that these local species may have other characteristics not suitable for growing in a plantation environment.

iii) The species should be fast-growing and able to produce large amounts of biomass in a short period. When species are grown for fuelwood, the shapes of the trees are of little importance and hence the major consideration is the amount of biomass that can be produced within a short time. As a result, species should be chosen because they are fast-growing and produce a lot of biomass. Some exotic species such as neem, leucaena and cassia meet this requirement.

iv) Tree species that have the ability to coppice offer additional options for regeneration beyond growing from seed. Trees coppice when they can produce new shoots from the stump when cut close to the ground. A similar concept is pollarding, which involves cutting the tree at the crown to produce new shoots. For trees that are browsed by animals, pollarding provides protection against browsing compared to coppicing. Coppicing can be an effective tool for regenerating plantations for fuelwood as they have the potential to produce wood in a shorter time than regenerating from seed.

v) In addition to the above, there are important characteristics of the wood itself that may determine its suitability for use as fuelwood. Fires produced from wood should be easy to start and should produce a large quantity of heat in a short time. The amount of heat produced by any fire depends on the wood density, resin, ash and moisture content of the wood. As a result, the following characteristics of wood are also important considerations in the choice of species: quick drying, ease of burning, amount of smoke produced, whether it throws sparks while burning, high calorific values, freedom from smells, resins and spitting when burnt, moderate wood density for consistent burning, and low ash content (Evans 1992).

A list of several examples of exotic plantation species that are suitable for use in wood energy production in Ghana include: *Leucaena leucocephala*, *Cassia siamea*, *Eucalyptus spp.*, *Azadirachta indica*, *Gmelina arborea*, *Albizia lebbeck*, *Anogeissus latifolia*, *Terminalia catappa*, *Gliricidia sepium,* and *Acacia spp.*

10.2.3. Management

There many different ways in which plantations can be grown and managed to produce fuelwood. The management approach used will depend on the scale of production required. For example, large-scale fuelwood plantations may be required to supply urban areas, including industrial operations, while for rural areas, which are under less population pressure, small-scale forestry activities may be sufficient to meet the demand for fuelwood (FAO 1989).

Dedicated Woodlots

A woodlot usually refers to trees grown purposely to produce wood products for fuel or building materials. These are usually, but not necessarily, grown close to homes, where they can be easily managed. Woodlots in Ghana can either be individual (private) or communal in nature. Woodlots have been a main feature of afforestation and reforestation initiatives in Ghana. For example, under the Rural Afforestation Programme in the 1990s, woodlots accounted for about 90% of the seedlings while the remaining 10% were aimed at agroforestry, boundary planting, windbreaks and home gardens (FAO 2002a).

There are several advantages to woodlots:

i) Given that most of the wood used as fuel in Ghana still comes from natural forests, the planting of woodlots has a potential to satisfy local fuelwood needs and reduce the pressure on natural forests, and subsequently, reduce deforestation;
ii) Woodlots located close to homes reduce the amount of time needed to collect fuelwood. As the local supply of fuelwood continues to decline, the time required to collect fuelwood from the wild has been increasing proportionately. The time saved can be used for other productive ventures in the community such as farming or running a business;
iii) Woodlots provide a reliable source of fuel, so that local communities do not have to rely on the use of animal dung and other agricultural residues for fuel, which could otherwise be used to improve soil fertility; and
iv) Because products from woodlots can be sold, they can provide supplemental income or in some cases the only source of income for farmers.

Despite these advantages, the promotion of woodlots in plantation forestry development initiatives in Ghana has not always been successful. At the local level, problems are usually associated with disagreements on land ownership and the commitment of the community to share responsibilities in the management of, and benefits resulting from, the woodlot. Another persistent issue with woodlots is the fact that most were established with incentives from the central government, and in some cases, the private sector. Once the incentives were removed, these initiatives were also abandoned. Some of the problems with woodlots were identified in Section 2.3.2.

Woodlots are often planted on land that has been taken out of agricultural production. While this does not necessarily have to be the case, it is important that lands for woodlots be chosen in consideration of the larger land-use context of the area, so as not to disadvantage other important land uses, such as food production. Nursery practices and timing of planting for woodlots follow identical practices for the chosen tree species as for any other purpose. Initial spacing will depend on a few considerations such as the final size of the product desired, planned operational methods to be used (manual or machinery), length of rotation, whether it will be thinned or not, the productivity of the site, etc. When the main reason for the woodlot is for fuelwood, the objective is to produce a large amount of biomass in a short period with small diameter trees that can easily be cut and handled. In this case, closer spacings may be used. Closer initial spacing is also suitable where the operational management relies on manual labour, rather than machinery. The literature recommends 1x1m spacing where the main objective is for fuelwood (Evans 1992). Depending on site productivity and species, closer or wider spacing may be used. When there is a need to produce poles, then wider initial spacing (e.g., 2x2 m), followed by management practices that ensure the straightness of the boles such as thinning that is appropriate for the species are recommended.

Trees planted for fuelwood are often harvested much sooner than for poles or commercial purposes. Though the rotation age depends on species and site productivity, it is common to observe trees being harvested for fuelwood as early as three years after planting. In northern Ghana, it was observed that neem plantations were thinned by age three for fuelwood (Nanang 1996).

Land Area Requirements for Growing Fuelwood

The amount of land area needed to grow fuelwood depends on the amount of fuelwood consumed. Simply, by dividing the amount of fuelwood needed by the growth rate of the species, it is possible to estimate how much area (in ha) that will be required to plant the desired quantity of wood. Table 1 below provides a simple example, but the same formula can be applied to determine other values outside the range used in the table below. Because growth rates depend on site productivity, it means that less land area will be required to produce the same amount of wood on more productive sites than on less productive sites. We can use the information presented in Table 1 of Chapter 2 on the growth rates of neem on good and poor sites to illustrate this concept. At a spacing of 2x2 m, the growth rate of neem in the Guinea Savannah zone is $12m^3$/ha/year, while it is $4m^3$/ha/year on poor sites. This information is in the first two rows under growth rates of Table 1. It is clear that on a poor site ($4m^3$/ha/year), and a fuelwood use of $15m^3$/year, 3.75 ha of land will be needed, while on a good site ($12m^3$/ha/year) only 1.25 ha will be required to yield the same volume of fuelwood. This information provides some insights into site selection for growing wood.

Table 1. Approximate area of land required (ha) for given growth rates and firewood consumption

Growth rate (m^3/ha/yr)	Quantity of fuelwood used per year (m^3/yr)			
	5	**10**	**15**	**20**
4	1.25	2.50	3.75	5.00
12	0.42	0.84	1.25	1.68
15	0.33	0.67	1.00	1.33
20	0.25	0.50	0.75	1.00

There is a tricky component to this analysis. Recall that the spacing of trees in the plantation affects the number of trees planted and hence the amount of wood that can be produced in the same amount of time. So in analysing the area requirements, a natural question is, at what spacing are these area estimates valid? That is, would one need the same area irrespective of the spacing at which the trees are planted? Note that the focus in determining the area required is on the amount of wood produced in a year (or productivity). Therefore, the answer is not tricky at all. The most important consideration is to ensure that whatever spacing is used results in the volume growth used in the calculation, i.e., the area estimates are valid for all spacings that achieve the volume growth rate used in the calculation. Depending on the species and site, it is possible for a woodlot planted at a spacing of 1x1m to produce the same amount of wood in a year as another planted at a spacing of 2x2m. In this case, the same one-ha area will be needed to meet the wood requirements, even though the two woodlots are planted at different initial spacings.

According to the FAO (2013), the production of woodfuel in Ghana in 2013 was about 40 million m^3 from all forest and woodland areas. If this amount of wood were to be grown in a woodlot, it would require about 3.33 million ha of land for neem on a good site in the Guinea savannah zone (mean annual increment of $12m^3$/ha/year) or 2 million ha of *Gmelina arborea* grown in the Moist evergreen forest zone, with a mean annual increment of $20m^3$/ha/year.

In Section 1.8.4, the Forestry Commission estimated the fuelwood needs of Ghana to be about 16.8 million cubic metres. Assuming this demand was all produced from forest plantations, we can estimate how much land will be required. If we assume that the productivity of the plantations is $20m^3$/ha/year, then 840,000 ha would be required. On the other hand, if the productivity is much lower at $4m^3$/ha/year, then 4.2 million ha will be needed to produce the wood to meet this demand.

Integrated Fuelwood and Multipurpose Plantation

Besides dedicating a piece of land solely for the production of wood, fuelwood can be produced as part of a multi-purpose tree plantation. A plantation can be set up to produce timber, fuelwood and other domestic products such as fruits, fodder for animals, shelter, habitat protection, land rehabilitation or other environmental purposes.

In a case where the main purpose is to produce timber, fuelwood is produced as part of pre-commercial or commercial thinning and during the final harvest, whereby small diameter trees are selected to be used as fuelwood. It is also possible to use branches and twigs from pruning for fuelwood. It is important to note, however, that the timing of these operations to produce fuelwood must be guided by the management plan for timber production, as fuelwood is a by-product of the operations, rather than the main reason for establishing the plantation. Thinning or pruning should not be delayed unnecessarily to the detriment of sawlog production to produce fuelwood. An advantage of growing fuelwood as part of a timber production process is that because sawlogs are more valuable economically than fuelwood, the investor can obtain revenue from timber, which essentially means that sawlog production is subsidising the fuelwood production.

Producing fuelwood as part of other multipurpose plantations such as fruits, fodder, shelter etc., also has benefits in the sense that though there may be minimal commercial benefit accruing to the landowner, the joint production of several outputs with largely the same inputs (joint costs) minimises the cost of producing each output. This method is also attractive in the sense of the larger context of land-use in a community, where land is scarce and each family may be limited to a small parcel of land where they can produce all their domestic needs. In this case, again, the management plan has to follow the other uses for which the plantation was set up, and consider fuelwood as a by-product of the process, unless of course, the initial intent was to optimise all the outputs at the same time.

Coppicing and Pollarding

Coppicing and pollarding can be used as effective tools when managing plantations for fuelwood. Coppice ability is a function of tree species and the condition of the stump. Many exotic plantation tree species, such cassia, leucaena, acacia and albizzia, Gmelina and teak species coppice well (Evans 1992). The reader may refer to section 2.4.4 for advantages of coppice systems compared to other silvicultural systems.

Two main disadvantages of the coppice system is that it draws heavily on the store of nutrients in the soil, particularly if the rotation is short, since it consists largely of vigorous young shoots and branches, which require more nutrients than older wood. This can be overcome by redistributing logging debris through the plantation or fertiliser application (State Government of Victoria 2011). Secondly, the material produced from coppice is of comparatively small size. There is, therefore, a limit to its general utility from the viewpoint of timber production, and its financial success depends on the existence of a special demand for the produce yielded by it (State Government of Victoria 2011).

10.3. Soil Improvement and Erosion Control

10.3.1. Role of Trees in Soil Improvement and Erosion Control

Soil is an important resource that supports vegetation and should be protected and maintained to prevent the decline in its productivity. Erosion is one of a number of forms of soil degradation; soil conservation efforts should address not only erosion, but also other forms of physical, chemical, and biological deterioration of soil (Nair 1993). Soil erosion, together with other soil degradation factors such as the loss of soil organic matter and plant nutrients can lead to a serious decline in the ability of the soil to sustain vegetation whether agricultural or forestry.

With increasing deforestation in the high forest zones and loss of tree vegetation in the savannah woodlands, most areas of Ghana are rapidly losing forest cover. This loss in vegetation cover can easily lead to soil erosion and reductions in soil fertility. Soil erosion has three main effects: loss of top soil for cultivation; loss of ground stability on steep slopes; and siltation of water bodies, which may lead to blockage of dams and irrigation channels (Evans 1992). Trees play an important role in soil conservation and improvement through reducing soil erosion, increasing soil organic matter, improving soil structure, and assisting in nutrient cycling.

Soil erosion can be categorised into natural and man-made (human-caused) erosion. Natural erosion is a very slowly occurring natural process caused by the influences of nature such as the weathering of mountains, hills, etc. On the other hand, when people cause the soil to become susceptible to be carried away by rain or the wind through practices that destroy the natural protection of the soil, man-made erosion is said to occur. Soil erosion occurs in several forms, such as gully, rill, wind, sheet and water erosion. In general, however, soil erosion occurs because of lack of vegetation cover due to over clearing of the land, overgrazing, strong winds and heavy rains.

Trees help to prevent or minimise soil erosion in five main ways: intercepting rainfall; providing ground cover; reducing wind speed; helping to retain moisture in soils; and through the binding action of roots (Evans 1992). Through interception of rainfall by their crowns, trees reduce the force with which rain drops hit the ground, thereby the quantity of, and the rate at which, the rain reaches the ground. With regards to moisture retention, the litter and humus layers under trees help to maintain soil moisture by soaking up moisture in the soil and thereby reducing infiltration. Finally, tree roots are capable of binding to the soil through which these roots penetrate to hold the soil together to reduce erosion. These five factors help to reduce surface soil erosion and improve soil productivity. Reducing soil erosion is one direct way to maintain or improve soil fertility. Trees and their falling leaves/debris provide protection to the ground/soil from erosion agents such as wind and water and also act to reduce wind speed that would otherwise be strong enough to blow loose soil away to cause soil erosion. The effectiveness of trees to reduce soil erosion will depend on the species, the shapes of the crown, the amount of litter produced, etc.

Through litter fall (dead and falling leaves, twigs, branches, fruits, etc.), dead trees and other biological processes, trees add organic matter to the soil, which improves soil nutrient content and structure. The amount of organic matter contributed by trees is affected by the volume of foliage dropped, the quality of foliage, additions from the roots and the tree species (Faleyimu and Akinyemi 2010). The rate at which organic matter is added to the soil is determined by the rate of litter decomposition, which in turn is affected by the ratio of carbon to nitrogen in the soil. Normally, the higher the level of nitrogen in the leaves, the greater the contribution of litter to organic matter in the soil. Tree litter low in nitrogen may immobilise fertiliser. Nitrogen from foliage is much less prone to leaching than that from commercial fertilisers. Trees vary greatly in type, volume and effectiveness of nutrients they recycle (Faleyimu and Akinyemi 2010).

The deep and spreading root systems of trees allow them to recycle nutrients by taking them up from deep down the soil and depositing them on the soil surface as litter, which then decomposes to form soil organic matter. If the trees are leguminous, they would usually add more nitrogen to the soil than other tree types. On the other hand, if the trees are low in nitrogen, they may immobilise fertiliser. Some *Acacia* species seem to produce litter that decomposes very slowly. The actual amount of nitrogen added depends on tree species and local conditions. Some trees such as *Eucalyptus gummifera* increase the availability of phosphorus by secreting root exudates. Others grow in association with mycorrhyzae, which also increase the availability of nutrients (e.g., *Pinus radiate* and *Eucalyptus marginata*). Establishment of such plants can be a step in the rehabilitation of degraded land. Nutrient recycling of trees on crop and pasture growth can be most effective when trees are planted in alley farming system. Tree canopies can trap significant amounts of nutrients, a source of free fertiliser that is washed from the leaves to the soil by rain (Faleyimu and Akinyemi 2010).

Increasing soil organic matter and litter from trees and other vegetation help improve soil structure and increase infiltration. The better the soil structure, the more spaces there are for soil infiltration and storage. Similarly, the more soil organic matter, the greater the activity of soil organisms and the more channels they make for water entry and storage. All vegetation forms root channels which promote the entry of water. The more water that enters the soil, the less there is to run-off and carry the soil away. Because the roots can penetrate compact soils, they can be particularly effective in increasing water infiltration in areas of high soil compaction. When planting trees to overcome soil compaction/water infiltration problems species should be chosen carefully as not all trees are suitable for planting on compact soils (Ludwig et al. 2002).

There is no doubt that incorporating trees into the landscape has important benefits to the soil as described above. Despite these beneficial effects of trees on soils, there are a few negative effects that trees could have on soils. Trees, both as individual plants and when grown in association with herbaceous plants, can have adverse effects on soils. The main soil-related problems are (Nair 1993):

- *Loss of organic matter and nutrients through tree harvesting*: trees accumulate large quantities of nutrients while growing, and when these are harvested, the nutrients are lost from the site if the whole tree is taken away. The amount of nutrient loss depends on the type of harvesting used. Whole tree-harvesting leads to larger quantities of nutrients lost than methods that allow branches, twigs and other litter to decay on site;
- *Nutrient competition between trees and crops:* Trees planted on the same piece of land as crops will compete with the crops for nutrients and water. This problem is mitigated due to the different rooting levels of trees and crops (i.e., trees draw their nutrients and water from deeper down the soil, given their extensive rooting systems, compared to crops);
- *Moisture competition between trees and crops:* In areas of water scarcity such as the savannah zones, the competition for water between trees and crops can be very severe; and
- *Production of substances that inhibit germination or growth:* Some tree species such as *Eucalyptus* produces toxins (allelochemicals) which can inhibit the germination or growth of some annual herbs (Poore and Fries 1985). Other examples of trees that have been reported to have allelopathic effects are *Gliricidia sepium* on maize and rice seedlings (Akobundu 1986) and *Leucaena leucocephala* on cowpea, sorghum, rice and maize seedlings (Suresh and Rai 1987).

10.3.2. Choice of Species

It is obvious that while trees can help reduce or control soil erosion, not all tree species will be effective in doing so. Growth rates and number of stems/ha (stocking) are key factors in determining how effective trees protect the soil from erosion. The larger the tree size, the smaller the number of stems/ha required to control soil erosion. Studies in New Zealand have shown that for a 5 cm DBH, 12,000 stems/ha will be required to control erosion, while for 60 cm DBH, the corresponding number of trees will be 45 stems/ha (New Zealand Farm Forestry Association 2014). Of course, the effectiveness also depends on the slope of the terrain (steep hill, gentle slope or flat terrain).

The tree species that are growing in the natural environment always provides an indication of species that will grow and survive in such locations. Kunkle (1978) and Evans (1992) list the following desirable characteristics of trees for use in erosion control:

- Good survival and growth on impoverished sites. This factor is the most important as most areas that are prone to erosion are usually poor in soil fertility.
- Ability to produce a large amount of litter. Litter is important for increasing the organic matter content of soils, and in areas where soils are impoverished, the ability of tree species to produce large quantities of litter will be important to increase soil organic matter content.
- Strong and wide-spreading root system with numerous fibrous roots and deep roots in landslide zones. A large number of fibrous roots enable the trees to hold large quantities of soil together and reduce erosion of the soil from the water. Deep roots help with nutrient cycling and reduce competition with food crops that draw their nutrients from the upper levels of the soil.
- Ease of establishment and need for little maintenance. It is advantageous if the species can be established from cuttings. When planting on poor soils, species that establish easily will colonize the site quickly and provide badly needed vegetative cover for the soil to prevent erosion.
- Capacity to form a dense crown and retain foliage all year round or at least through the rainy season. A dense crown is more effective in reducing erosion from rainfall than a less dense crown, and so the former is preferable.
- Resistant to insects, disease and browsing by animals. Most tree species are susceptible to one form of disease or pest or subject to browsing by animals. Damage caused by these agents will reduce the vigour and development of the planted trees. Choosing species to minimise damage by biological agents will help the growth of the trees.
- Ability to improve soil, especially through nitrification. For soil improvement, choosing nitrogen-fixing tree species is ideal.
- Ability to provide economic returns such as fruits, fodder nuts or beverage products. In most communities, access to employment opportunities are often non-existent, and so if tree species chosen for soil improvement can also provide economic benefits to the landowner, then these should be preferred.

Some tree species that can be used in erosion control include (Kuypers et al. 2005):

- *Moringa oleifera* can function as windbreaks for erosion control, live fences, as an ornamental or intercropped to provide semi-shade to species requiring less direct sunlight.
- *Acacia spp. (e.g., albida, nilotica).* The most adaptable and widely grown species in semi-arid regions and suitable for growing in difficult conditions.
- *Gliricidia sepium:* This species is fast growing and the easiest way to propagate it is from branch cutting. It is suitable to grow in dry places in sub-humid lowlands, and it is a N-fixer. This tree is not edible as it is poisonous to humans.
- *Leucaena leucocephala:* Genetically very variable; suitable for ground stabilisation in semi-arid to humid areas. Easy to multiply by seed and cuttings, fast growing, strong root system, good cover crop and a N-fixer. It has many uses, such as firewood, fodder crop.
- *Mimosa pigra:* This is a shrub with many long thorny branches that can form a dense thicket. Used for stabilisation of riverbanks.
- *Parkinsonia aculeate:* This shrub can survive and grow in very dry regions. It can be used for land reclamation after mining and road construction activities.
- *Pinus sp.* This species has favourable properties for erosion control. It is grown on dry, well drained, poor soils. It can withstand competition from grasses; forms a good litter; in later stage undergrowth is often formed; thick bark gives tree resistance to bush fires.
- *Pithecolobium sp.* This is a large tree with extensive and deep root system. Good for planting alongside river banks.
- *Prosopis chilensis:* In semi-arid climates: fast growing, also does well in difficult situations. It is deep rooting and has high regeneration potential.

10.3.3. Planting Methods

Soil Erosion

In all cases, trees should be planted early in the rainy season to give seedlings a good chance to establish themselves before the arrival of the dry season. Seedlings that are healthy, sturdy and container grown should be used to ensure high survival rates and minimise the gap between planting and establishment. Where soil erosion is a problem, trees can be planted to minimise it. However, it is important to note that the ability of trees to prevent erosion depends on their capacity to provide cover to the soil and the ability of the roots to hold the soil together. If land is completely bare, planting trees may not provide immediate protection of the soil from rain or wind, as it takes time for the trees to grow and develop crowns and litter fall to provide this protection. This means that tree planting should be used in combination with other land management practices that maintain vegetative cover such as growing cover crops, contour alignment practices or terracing that can provide immediate vegetative cover to the soil.

To maintain the integrity of the soil, planting for erosion control will involve trees intercropped with grasses and/or mulch. On a hill, planting should be done at intervals down the slopes complemented by legumes and grasses in-between the bands of trees to reduce surface run-off. On areas with natural vegetative cover, bushfires, overgrazing and practices that remove vegetative cover should be avoided.

Planting to control soil erosion can take several forms such as:

- Spaced planting is used on flat or less steep terrain. The trees can be planted as part of an agroforestry practice.
- Pair planting is used to control gully erosion around streams and other water bodies. Trees can be planted about 2–3 metres apart on either side of the gully floor. The roots have the ability to create a damming effect that slows the water flow, stabilises the gully sides, and allows other plants to regrow (Gregg 2012).
- Stream bank planting occurs when trees are planted at the edge of streams and rivers to stem the erosion of the banks. Traps or small dams may also be used to catch material dislodged from the edge.
- Riparian strip: Erosion is particularly frequent around riparian areas, given the sloping nature of the river banks. To minimise this, trees can be planted in a strip (10 or more metres wide) near a stream or river. The objective is to trap sediment, and nitrogen and phosphorus in water run-off, from nearby slopes before it enters waterways. Trees may be planted with a mix of trees, shrubs and pasture.
- Shelter belts are usually in one or more rows of trees planted at right angles to the prevailing wind, which help to reduce wind erosion and protect livestock and crops. For increased effectiveness, it is recommended that shelter belts compose of a mixture of tree species and many rows.
- Trees can also be grown on lands with marginal productive value but prone to erosion in protection forestry.

Soil Improvement

Where soil improvement is the main objective, several techniques can be used for planting trees, the techniques used for planting are mostly agroforestry techniques such as alley cropping. The concept of agroforestry is very simple that trees and crops are planted and managed on the same piece of land. The role of trees, in this case, is limited to adding organic matter to the soil and nutrient cycling to improve soil structure and function. The only difference between the planting techniques used is the configuration of trees and crops on the land.

Traditional uses of trees in farming: Traditional farming practices in Ghana and most of Africa had and still conserve trees on the landscape. Most of these trees are those with economic benefits and hence the concept of integrating trees with food crops is not new to farmers in Ghana. The configuration of trees was random, as these trees were not planted on purpose but were the result of natural regeneration. The few exceptions to this are economically valuable trees such as mango, shea and dawadawa that were planted to meet household needs and to generate additional incomes for households. The traditional practices did not cut back the branches of trees to reduce shading over food crops, as is done in modern agroforestry practices. However, shade tolerant food crops such as peppers and sweet potatoes were often planted in the shade as a way to utilise the farm space. In traditional systems, trees are often scattered and limited in number on the landscape, thereby reducing the benefits of incorporating trees with food crops. For example, *Acacia albida* was one popular tree species on farmlands that had multiple benefits. The tree attracts animals to its shade during the hot, dry season and these animals often leave their manure under the tree to fertilise it. The tree fixed nitrogen and its falling leaves added organic matter to the soils. There was little competition between the tree and crops because *A. albida* sheds its leaves in the rainy season, therefore not shading the crops and draws nutrients from levels that the roots of food crops do not reach.

Line planting: In this practice, trees are planted in rows and food crops are then grown between the lines of trees.

Band cropping: trees are planted in a band or "mini-forests" and food crops are planted in-between these band of trees. The outer band of trees is cut back before the planting season to avoid shading the food crops.

Alley cropping: A much more complex agroforestry technique is alley cropping, as developed by the International Institute for Tropical Agriculture (IITA). This involves the growing of crops between closely-spaced lines of trees or shrubs, usually legumes. To prevent competition with crops for light and water, the trees are regularly cut back and are kept pruned during the growing season. Larger cuttings are available for firewood. Leaves and twigs are spread as mulch among the crops. This mulch replenishes the soil with humus and nutrients, as well as providing a barrier against erosion (Pétry 1991).

10.3.4. Management

Management practices to control erosion and improve soil conditions should focus on the correct choice of tree species, appropriate planting and tending operations and integrated management practices to protect the site during the establishment phase of the trees. In erosion control, maintaining adequate vegetative cover is the main goal. As a result, weeding may be inappropriate in most cases and if done at all, should be restricted to around the trees.

If trees are planted as part of soil improvement in an agroforestry or farm forestry system, the trees should be cut back at the beginning of the farming season to provide organic matter to the soil and also to reduce shading to the planted food crops. Most agroforestry species have uses as fodder for animals, so care has to be taken to protect the trees from animals. Protection of planted trees from browsing and physical damage by animals is one of the biggest challenges of growing trees in and around human settlements, and appropriate measures must be taken to ensure trees survive.

After trees are harvested, the roots of the remaining stumps continue to hold onto the surrounding soil and provide protection against erosion. However, this protection only lasts as long as the roots have not decomposed. As a result, it is important to replant the site immediately to avoid erosion setting in. Artificial regeneration using the best practice for the species that allows the immediate establishment of the trees should be used. Thinning should be done only in exceptional cases to ensure continuous tree cover for the soil.

10.4. Wind Control

10.4.1. Role of Trees in Wind Control

In areas of high winds and low vegetation cover, trees can be grown and used as windbreaks around farms, households, livestock, crops and public buildings. Throughout the African continent, farmers have and continue to use windbreaks to protect crops, water sources, soils, and settlements on plains and gently rolling farmlands (Nair 1993). Windbreaks are narrow strips of trees, shrubs and/or grasses planted to protect fields, homes, canals, and other areas from the wind and blowing sand. A shelterbelt is a type of windbreak, usually with long, multiple rows of trees and shrubs.

Properly planted windbreaks or shelterbelts confer the following benefits: reducing the wind; acting as visual screens; controlling blowing dust/sand, protecting buildings and decreasing energy consumption; capturing atmospheric carbon; helping control soil erosion; providing habitat for wildlife; beautifying protected areas and increasing property values. Trees and shrubs in combination can be used as windbreaks and act as natural barriers that help reduce wind speed and/or redirect the wind to reduce damage to lives and property. The most important feature of trees that make them effective as windbreaks is the thickness of their crowns that do not allow the wind to go through but rather deflects the wind over and above the crowns. As a result, the wind speed is reduced and deflected at an angle to the target being protected.

The amount of wind speed reduction and the area affected depend on the height, density, width, and shape of the windbreak, although it is widely recognised that the height of the windbreak is the most important determinant of its effectiveness (Iowa State University 1997). The continuity of the windbreak is important, and holes or gaps in the windbreak may result in increased wind speed and reduced protection. The most effective protection afforded by a windbreak extends to a distance of 10 times the height of the windbreak's tallest trees though this area can be 15 to 20 times the height of trees. As an example, if a windbreak is 20 metres tall (H), the effective area protected (where wind speed is reduced by 50%) will be 200 metres (i.e., 10xH) from the windbreak.

Apart from reducing wind speed and acting as wildlife habitats, windbreaks also moderate local temperatures and reduce cooling and heating costs. If well designed, they add to the aesthetics and add values to homes. They also play useful roles in reducing sound, visual screens and odour reduction from neighbourhoods and protect surroundings from animals as they can act as fences. Finally, windbreaks prevent soil erosion by providing vegetative cover around the buildings they were intended to protect, and they can produce fodder for farm animals in and around homes or farms.

Small living fences and hedgerows can also act as windbreaks for small sites such as home-gardens and nurseries. However, windbreaks are distinguished from boundary plantings and living fences by their orientation, which must face the wind, and by their multi-story, semipermeable design. Windbreaks can improve the microclimate in a given protected area by decreasing water evaporation from the soil and plants (Nair 1993).

10.4.2. Choice of Species

Choice of species for windbreak should depend to some extent on soil characteristics and location of planting, as well as the personal preferences of the landowner. There are many indigenous and exotic species that are suitable for use as windbreaks. However, environmental hazards such as insect pests (especially termites), wild and domestic animals, poor soil, water usage, rooting system and drought, will narrow the choice as well as reduce the tree's growth rate. It is advisable to not use a single species for the windbreak, but rather a diverse mix of species as the site and conditions will allow. In fact, it is more effective to use as many different species as there are rows in the windbreak, which can be achieved by block planting different species. If there is a secondary intention to improve wildlife habitat, greater species diversity will be essential to achieving this objective and minimise the impacts of pests and diseases.

Tree species selected for windbreaks should have the following characteristics:

- Easy to establish and maintain in poor soils and not browsed by animals;
- Should ideally be evergreen trees, with foliage all year-round;
- Must have dense foliage that will reduce the amount of wind penetrating the crown;
- Trees that have lower limbs that do not self-prune are more effective; and
- Have aesthetic values.

The best tree species for use as windbreaks in Ghana are listed in Table 2 of this Chapter.

10.4.3. Planting Methods

As with most forest plantation establishment in Ghana, planting of trees for windbreaks should occur early in the rainy season. Good site preparation is always essential to ensure successful establishment of seedlings. Several choices exist for planting stock. Bare root seedlings or transplants are appropriate for shrubs and trees such as teak. The use of transplants, potted, and a container-grown stock will more likely result in higher tree survival and better the establishment rates. Larger stock tends to speed up the establishment and helps the windbreak to become functional in a shorter period. Mulching, watering and other best practices should be used as necessary.

To ensure maximum effectiveness, trees and shrubs planted for use as windbreaks should have rows that are perpendicular to the prevailing winds. Windbreaks should have at least three rows to achieve good wind protection – the greater the windbreak's density, the greater the reduction in wind velocity. It is recommended that shrubs comprise the outer rows while trees are in the inner rows to provide a semipermeable barrier to wind over their full height, from the ground to the crowns of the tallest trees. The inner rows could consist of species such as *Eucalyptus* spp., *Casuarina equisetifolia*, or *Azadirchta indica*, and two rows each of shorter spreading species such as *Cassia siamea*, *Prosopis chilensis* or *Leucaena leucocephala* in the outer rows.

Although species diversification is encouraged, they should not be mixed within rows, they should rather be alternate rows for each tree or shrub species. A properly planted windbreak should have staggered rows to create a barrier to the wind to increase its effectiveness. Windbreaks with rows that are not staggered can result in a wind tunnel effect, increasing wind velocity and scouring the soil and reduce the efficiency of the windbreak. On the other hand, care should be taken to avoid very dense windbreaks, which may do more harm than good by creating strong turbulence that will scour the soil on the windward side and damage property on the leeward side.

10.4.4. Management

The overall objective of managing windbreaks is to maintain and improve the vigour of individual trees, maintain and improve the structure and function, and promote the longevity of the windbreak. For windbreaks to remain effective, it is important to not leave open gaps in the design and so to replant open gaps as trees die or are destroyed by animals. Early filling in of gaps during establishment is also essential to ensure that the windbreak establishes in line with the original intent of the landowner. The effective life for windbreaks will vary greatly depending on the species and health of the trees.

Pruning should be used to maintain a single main stem per tree for as long as possible. When multiple stems occur, they should be corrected immediately. Pruning should avoid removing the lower branches as the main goal is to maintain as much foliage as possible to improve the effectiveness of the windbreak.

Weed control should be undertaken to control competing vegetation to ensure the growth of the windbreak. Control measures include weeding with hoes/cutlasses and mulches. If windbreaks are established around homes, the use of chemicals to control weeds and/or pests should be avoided.

As with any trees grown around human settlements, interference from livestock and browsing may break limbs, remove the bark, or compact the soil. Hence, adequate precautions should be taken to protect windbreaks from physical damage.

It is usual practice in forest management that trees and shrubs that are initially planted at close spacing to become functional early may require some thinning later to maintain the effectiveness of the windbreak. Thinning should be used sparingly as large open gaps left in the windbreak will make it less functional. The landowner should continuously observe the trees and shrubs for early signs of damage or poor growth such as premature loss of foliage, the poor colour of leaves, reduced live crown, and incidences of disease and insects.

10.5. Water Conservation and Watershed Protection

10.5.1. Role of Trees in Water Conservation and Watershed Protection

Forests play an important role in water conservation and watershed management. This section will focus on two main aspects of plantation forestry and water conservation: watersheds and riparian buffers around dams and streams/rivers. Watersheds are land areas that drain surface water and ground water to a downstream water body or outlet, such as a river, lake, or estuary. Forests make excellent watersheds chiefly because their soils usually have a high infiltration capacity— they are capable of quickly absorbing large amounts of water (Cotrone 2008). Therefore, rainstorms or melting snow in woodlands produce relatively little surface runoff with the associated problems of *erosion* (detachment and movement of soil), *sedimentation* (the deposition of soil) and very low turbidity (Cotrone 2008).

Planting and maintaining woody vegetation along streams provide a wealth of benefits and stream health is dependent on the presence of woody vegetation along its banks. Riparian forest buffers filter sediment from streams during storm events; remove nitrogen and phosphorous leaching from adjacent land uses such as agriculture; provide stability to the bank (wood root systems); shade and modify stream temperatures, critical for habitat and pollution reduction; provide aquatic and wildlife habitat for many species; reduce stream velocity; and reduce downstream flooding (Cotrone 2008).

The ability of trees to intercept rainwater and store it makes them play a vital role in water conservation. Most surface soils are capable, to varying degrees, of absorbing water, however, the quantity and type of vegetation cover on these soils greatly affects the ability and capacity to absorb water. Planted trees act as a buffer, absorbing and retaining water and later releasing moisture through a combination of evaporation and transpiration.

In areas without vegetative cover, soil erosion washes off soils and sediments that are eventually deposited in rivers, streams and other water bodies, which leads to reduced water quality. Planted trees can improve water quality by stabilising soil with their extensive root systems, thereby reducing soil erosion and runoff. As a result, water draining from a forested area is generally cleaner than from other land use types such as grazing or agriculture.

Forests conserve and naturally filter water, recharging the ground water supply and preventing the transport of sediments and chemicals into streams. Shade provided by trees around water bodies reduce evaporation losses and helps to conserve water for human purposes. Finally, trees also store water on the canopy, stems, and plant tissues and, therefore, retain moisture above and below ground level.

Trees in watersheds have the ability to regulate stream flow and reduce flooding. Because trees trap water and release it slowly into the drainage systems, tree cover regulates the amount of water that flows into rivers and streams. The advantage is that it reduces flooding following heavy rainfall. However, this also reduces the amount of water available for use in areas of water scarcity. In this regard, there is a debate on whether trees should be planted close to water bodies or not. The answer may lie in the purpose of the tree planting exercise. It is clear that vegetation cover around water bodies reduces erosion and improves water quality. However, this comes at the expense of reductions in water quantity through water loss by transpiration.

Forested catchments significantly reduce water yield compared to non-forested ones. Mathur et al. (1976) observed up to 28% reduction in water yield following the establishment of eucalyptus. Transpiration from trees, together with the evaporation of water intercepted by the forest canopy, combine to reduce the amount of water available in a forested watershed compared to one that is not forested.

In an application of meta-analyses to studies comparing water flows in tropical watersheds, Locatelli and Vignola (2009) found that total water flow and base flow were higher in watersheds under planted forests than non-forested land uses. This can be explained by the high transpiration rates of exotic species, especially eucalyptus. Infiltration capacity may be higher under planted forests than non-forest land, but may not be enough to offset the higher loss of water by transpiration. Some authors have reported that in degraded soils, planted forests may increase infiltration more than transpiration, thus increasing base flow compared to non-forest land (Bruijnzeel 2004).

The following points should be noted when deciding whether to plant trees around water bodies:

- Reduced runoff from forested land reduces the frequency and volume of downstream flood events. As a result, removal of vegetation cover from catchment areas leads to more severe flooding (increased peak flows);
- Streamflow from forested catchment areas is prolonged in the dry season mainly because of better infiltration and less runoff; and
- Removal of trees from water catchment areas leads to a temporary increase in water yield

10.5.2. Choice of Species

The overall objectives of planting trees around water bodies and watersheds are to: reduce soil erosion and sediment deposition into the water; reduce winds blowing over the water to minimise evaporation; and provide shade to reduce evaporation from the water surface. All these objectives are to be achieved with vegetation that uses as little of the stored water in the dams or streams as possible. Therefore, the choice of tree species for riparian afforestation should be limited to species with less water requirements. Fast growing exotic species that use large quantities of water should be avoided. In general, tree species that grow well in dry conditions have low water requirements compared to those that cannot survive in dry weather. Examples of species that are suited for planting to conserve water are *Acacia mangium, Alnus acuminate, Prosopis*, *Leucaena*, *Melia Volkensii*, *Azadirachta Indica,* and bananas.

Secondly, species that are used to control soil erosion can be used as long as they are planted within a reasonable distance (15-30 m) from the water. Tree species that are suitable for planting as shelterbelts can also be used to plant around streams, dams or dugouts to reduce the prevailing winds and further reduce evaporation of the water.

10.5.3. Planting Methods

Any vegetation situated near water consumes some of that water and loses it through transpiration. The amount of water lost depends on the species planted. A fully grown tree may lose several hundred litres of water on a hot, dry day (or nothing on a cold, humid day) through transpiration and evaporation (evapotranspiration). The main factors affecting the actual quantity lost are the tree species, the size of the tree, relative humidity and the atmospheric temperature. Fast-growing tree species will demand more water than slow growing ones. This is the principle that underpins the use of some trees to dry up marshy areas. Planting around water bodies should be restricted to a reasonable distance (e.g., 20 m) from the water in order to minimise the quantity of water used by the trees and still offer vegetative cover to the surroundings to reduce soil erosion.

Tree planting around dams and dugouts is influenced by many considerations such as its size, location and orientation, the purpose for which the water has been impounded, and, if for irrigation purposes, how it is to be applied, i.e., by pump or flood irrigation. Planting trees to protect watersheds or as buffers around water bodies such as streams must start with a careful analysis of whether trees are the answer to the problem, and then analysing the vegetative cover of the area for planting. Secondly, develop forest cover goals for the area under consideration, and then, identify, conduct and prioritise existing reforestation/afforestation opportunities in the area. This approach allows the landowner to assess the needs of the watershed or buffer before starting to plant any trees. Afforestation or reforestation in and around water bodies and watersheds planting should as much as possible take advantage of existing vegetation and landscape characteristics.

When trees are planted around dams or streams that are intended to provide water for human or animal use, the trees should be arranged to permit ready accessibility to the water by trucks, animals or people. Trees should be planted from 15 to 30 metres from the edge of the dam or stream to prevent the trees from sucking up and losing the stored water through transpiration. The space between the edges of the water to the trees should be planted in contours with vegetative cover with low transpiration rates compared to the trees, such as grass. The distance from the edges to the water can be less than 15 metres if the trees chosen have low transpiration. The trees should be planted on all sides of the dam or stream, although the most benefit is derived by trees that planted towards the higher levels of the drainage area. The trees should be protected from physical damage by animals. As most dams are created by blocking a stream or river, it is equally important that such streams are protected from erosion and evaporation with trees appropriately planted on either side of the stream or river that supplies water into the dam.

10.5.4. Management

Management of planted trees around water bodies and in watersheds should be targeted at reducing runoff and soil erosion, and improving soil and water quality. Tree planting should be seen as part of a larger and integrated approach to managing water resources, rather than from a solely forestry perspective. If these watersheds or resources are located near human settlements, many more disciplines such as in community development, sociology, urban planning, etc. will need to be involved.

If the overall objective of the tree planting is to reduce erosion and sediment deposition into streams, rivers and dams, then the tree species chosen, the planting methods and management techniques will be identical to those described in Section 10.3 of this Chapter.

10.6. Rehabilitation of Degraded Mined Sites

10.6.1. Role of Trees in Land Rehabilitation

Mining, whether large or small scale, legal or illegal, often leaves behind sites with depleted vegetation cover, compared to the original site before the disturbance. The problems associated with mined sites include accelerated erosion, water pollution, the danger of slides and a barren landscape in and around the mine. To reduce the impacts of these problems, land reclamation and remediation efforts often include land-use planning, physical alteration of topography, re-vegetation and subsequent management of reclaimed land. The planting of trees on mined sites, therefore, should fit well into the overall reclamation plan. As discussed above, trees provide a suitable long-term means of minimising erosion and pollution that make them suitable to be used on sites that have been previously subject to mining activities.

If the disturbance is due to the activities of commercial mining companies, different countries have diverse requirements for the amount of financial assurance to be provided for mine closure. For example, many jurisdictions in Australia determine the financial assurance required on a case-by-case basis, India requires a fixed sum per hectare of mine site, and Suriname and Botswana require funding for closure as an ongoing expense (Miller 2012). The jurisdictions of Arkansas, New Brunswick, and Ontario require financial assurance covering the complete cost of mine clean-up while Ghana requires 5-10% of the estimated clean-up costs to be provided (Miller 2012). If environmental degradation is caused by small-scale illegal miners, popularly called *galamsey* in Ghana, then there is no company or person to hold responsible for the restoration of the site, and hence the State bears responsibility.

The extent of land destruction and the need for reclamation and remediation often depend on the type of mining undertaken. Large-scale commercial mining may involve underground mining, which results in very large excavations deep into the ground. On the other hand, surface mining is more extensive, destroying more vegetation but not as deep into the earth's crust. There are also small-scale surface mining activities that are a legal means of mineral exploitation in Ghana that have serious environmental consequences.

The terms reclamation, remediation, rehabilitation, and restoration are all used to describe mine closure activities that attempt to alter the biological and physical state of a site (Finger et al., 2012). Finger et al. (2012) and Dutta et al. (2005) provide the following definitions to clarify these terms:

- *Remediation*: The cleanup of the contaminated area to safe levels by removing or isolating contaminants. At mine sites, remediation often consists of isolating contaminated material in pre-existing tailings storage facilities, capping tailings and waste rock piles with clean topsoil, and collecting and treating any contaminated mine water if necessary.
- *Reclamation*: The physical stabilisation of the terrain (dams, waste rock piles), landscaping, restoring topsoil, and the return of the land to a useful purpose.
- *Restoration*: The process of rebuilding the ecosystem that existed at the mine site (where applicable) before it was disturbed. The science of mine reclamation has evolved from simple revegetation activities to a discipline that involves using native plants to mimic natural ecosystem development over an extended period (Errington 1992).
- *Rehabilitation*: The establishment of a stable and self-sustaining ecosystem, but not necessarily the one that existed before mining began. In many cases, complete restoration may be impossible, but successful remediation, reclamation, and rehabilitation can result in the timely establishment of a functional ecosystem.

From these definitions, it is obvious that the process of using plantation forestry on mined sites fall under rehabilitation, and to some extent, mine restoration. The initial remediation and reclamation activities are essential to a successful land restoration/rehabilitation with trees.

10.6.2. Choice of Species

The objective of land restoration is to return the site to the original ecological function or vegetation that was present before the degradation or mining operations took place. As a result, the first choice of tree species will be to use the species that existed on the land or adjacent to the degraded site. Actual species selection may be a little more complicated than this because the process and chemicals used in mining often change soil characteristics in ways that may no longer support the vegetation that had existed on the landscape. Species should be chosen based on their ability to survive and thrive on the new site. Species choice should be undertaken by a person who is knowledgeable of local tree species, mine site conditions, and landowner and reclamation goals.

Another feature of mined sites is that the soil and site conditions are often quite variable due to the differences in the disturbance on the landscape (hills, valleys, slopes of different steepness, etc.). It is, therefore, advisable to use multiple species to match the varying site conditions and to help a plant community withstand pest or disease attacks on one or more of its species.

In Ghana at the Tarkwa Mines, tree species that have been found suitable for planting on mined sites are *Acacia magium*, *Gliricidia sepium*, *Leucaena leucocephala* and *Cassia siamea,* which are often planted in a mixed stand (Tetteh 2010). These exotic species used for restoration are in contrast to the original species that were on the site before extraction. According to a baseline survey by the Environmental Advisory Unit contained in a document at the Environmental Department of the Company (Tarkwa Mines), within the primary forest, the emergent trees included *Piptadeniastrum africanum, Ceiba pentandra, Canarium sp., Tieghemella heckeli* and *Milicia excels* (Tetteh 2010).

10.6.3. Planting Methods

Tree planting for purposes of mined sites remediation is undertaken in the later parts of the reclamation process. Trees need mineral soil to grow and survive and on disturbed sites, the greatest challenge will be to use appropriate methods to prepare sites sufficiently to expose mineral soil of sufficient quality and depth. Before planting trees, the general steps involve creating a suitable rooting medium for good tree growth that is no less than four feet deep with a non-compacted growth medium followed by the use of groundcovers that are compatible with growing trees.

Therefore, the process of land remediation begins with site preparation and depending on the depth of excavation, restoration should involve importation and replacement of subsoil to a depth of 60 cm – 90 cm spread in 15 cm layers, and left to settle naturally for a period of 3 - 6 months (Aseidu 2013). Aseidu further describes the process of restoring vegetation on a typical mine in Ghana as follows. When the land is sufficiently settled, topsoil mixed with manure should be laid on the subsoil to a depth of 15cm minimum after settling. This is followed by sowing of nitrogen-fixing leguminous green manure cover crop by broadcasting to provide the first blanket of vegetative cover to protect the soil from the direct effect of the elements.

An example of a suitable species is *Crotalaria juncea* (Sunhemp), which is highly recommended because it is a nitrogen-fixing leguminous annual with high biomass production for highly degraded land. Other area-specific nitrogen fixing plants can be used. About 60 days after sowing of the green manure cover crop, when it is in bloom, it is smothered and worked into the soil. Seedlings of both local and exotic tree plants can then be established on the land. The reclaimed land should be managed by maintaining a balance between introduced exotic tree seedlings and native sprouted tree seedlings by weeding, staking, and occasional pruning for at least 5 years before any assessment for the success of the reclamation can be done.

At the AngloGold Ashanti mine in Tarkwa, Ghana, the processes identified in mine reclamation starts with earthwork/slope battering, spreading of oxide material, spreading of topsoil, construction of crest drains, raising of cover crops, tree planting, field maintenance, monitoring and measuring of success criteria (Tetteh 2010). Management of planted trees will depend on the objective of tree planting as described in previous sections of this Chapter.

10.7. Trees in Rural and Urban Environments

10.7.1. The Role of Trees in the Rural and Urban Environments

Trees play many important roles in the rural and urban environment (cities, towns and villages). That is why since time immemorial humans have incorporated trees into their living environment. Unfortunately, this practice is waning in Ghana, and many homes, private business surroundings, and public facilities such as schools, hospitals and health facilities lack tree cover. Below are some examples of the importance of trees in rural and urban communities:

- Due to their aesthetic appeal, trees increase the beauty and hence the property values of homes. When well-designed and planted, trees are one of the best ways to improve the beauty of homes and real estate properties (Friends of the Urban Forest 2015).
- Trees produce oxygen, clean the air and reduce global warming. Through the process of photosynthesis, trees release oxygen and absorb carbon dioxide, a greenhouse gas that contributes to global warming (McPherson et al. 2010). Trees capture airborne particles such as dirt, dust and soot (Bradley 1995). Researchers at Columbia University found that more trees in urban neighborhoods correlate with a lower incidence of asthma (Lovasi et al. 2008).
- Trees reduce flooding and water pollution. The role of trees in reducing runoff and flooding has already been discussed. Trees reduce flooding by capturing rain; thus reducing runoff and increasing infiltration. In urban environments with drainage systems, the presence of trees reduces the large quantities of water that can

potentially overload these systems during rainstorms and cause flooding (McPherson et al. 2010).

- Trees along streets promote exercise as people tend to enjoy walking in areas that are beautified by trees and also people judge walking distances to be less, and are therefore more likely to travel on foot, which has health benefits (Friends of the Urban Forest 2015).
- Street trees create a physical and mental barrier between the street and the sidewalk, keeping pedestrians, children and pets out of harm's way, provide a natural habitat for birds and insects, absorb traffic noise and increase privacy.
- When planted in play grounds and school yards, trees provide shade to both humans and animals during the dry season. Especially in areas of the country where temperatures can reach up to 40°C in the open sun, residents usually cool off under the shade provided by trees planted within their compounds and around their houses. The benefits of such trees planted initially to provide shade extend to cleaning the air, providing fruits for humans and fodder for domestic animals.
- Areas inhabited by humans tend to be much warmer than the surrounding open spaces outside of cities and towns. Planting trees can help keep urban surroundings cooler by providing shade and evaporating water from their leaves. The ability of trees to lower dry season time temperatures in towns and communities also helps to reduce energy use and hence carbon dioxide emissions resulting from energy generation. This makes a compelling case for planting trees strategically for energy conservation.
- Trees within urban and rural communities can provide food for the population. These foods may help meet their dietary requirements such as vegetables (baobab and kapok leaves), parkia condiment (dawadawa), tamarind, etc. or for medicinal purposes. The sale of tree products from urban trees is a source of revenue to households who often live on the margins of societies, and hence urban greening provide socio-economic benefits to urban and rural dwellers. This is even more so if the land is scarce and it is impractical to dedicate large tracts of land for tree planting.
- In many rural and to some extent urban communities in Ghana that still rely on the dwindling natural forests and savannah woodlands to provide their woodfuel needs for cooking and heating, woodlots and trees incorporated into their environment can help alleviate fuelwood shortages.

For these reasons, it is important that vegetation be included in rural and urban planning processes.

10.7.2. Choice of Species

The choice of tree species for greening the urban environment depends to a large extent on the purpose. For example, if the purpose is for shade, then evergreen tree species that do not shed their leaves in the dry season will be suitable. In most cases, however, tree species that provide multiple benefits, such as shade, fruits, fodder are preferred as they can satisfy more than one need at a time. Trees selected for street planting should have deep rooting systems to ensure that they do not easily blow down in a storm and cause damage to the road and /or vehicles. Mahogany and neem have been used extensively in Ghana as street trees for many decades, and these are suitable for the purpose.

All tree species that are suitable for fodder, shade, fruits, building poles, firewood and fencing can all be used for planting in urban greening initiatives. Local species include kapok, mahogany, neem, mango, shea tree, ebony, tamarind and dawadawa. Exotic tree species suitable for use in communities include cassia, moringa, albezzia, gmelina, teak and luecaena.

10.7.3. Planting Methods

It is important that tree planting in urban areas be based on a well-thought out plan. Some urban communities have developed tree canopy cover targets they plan to meet over a certain period of time. First, one needs to assess the tree canopy cover of the area that needs planting. Tree canopy cover refers to the proportion of land area covered by tree crowns, as viewed from the air. This assessment can easily be done using aerial imagery to provide a good picture of the urban forest. In rural areas, if an assessment of tree cover needs to be done for a public space such as a primary school, it would be cheaper for a trained person to simply count the number of trees on the property and estimate the average crown cover to determine the tree canopy cover. Canopy cover depends a lot on the age of trees, the size of crowns and tree species. Another important factor to consider is the distribution of the trees. It is possible that overall the area assessed has a high tree canopy cover, but majority of the trees are concentrated in a few places and the remaining areas need to be planted.

Tree planting in urban or rural communities needs to take into account the root systems of the species. Species with extensive root systems, when planted very close to buildings, can cause damage to the foundations of those buildings and risk the safety of the building and human lives. The design of trees will depend on existing vegetation and other structures/infrastructure already in place. In general, though, trees can be incorporated as windbreaks, shelterbelts, row planting or random planting within the landscape.

10.7.4. Methods of Estimating Tree Canopy Cover

The most accurate and quickest way to obtain estimates of tree canopy cover is to use aerial photography or other remote sensing techniques. This is especially true for large urban areas where ground sampling is not practical, such as the City of Accra. However, cost and the lack of technical expertise may be a concern for small towns, where less accurate estimates are required. We present a simple though less accurate, method to estimate tree canopy cover for small areas, such as a school yard or a local health centre. A trained forestry technical officer will be able to estimate this using the following procedure:

1) Estimate the total area covered by the facility (e.g., 2 ha of land)
2) If there are very few trees, count and record the number and tree species on the land. If the area is large with many trees, use appropriate sampling techniques to inventory the trees.

3) Estimate the size of the crown of each species using existing information or by measurement
4) Calculate the area covered by the crown of the average tree (you can approximate the shape of the crown to a circle for simplicity. Refer to Section 5.3.6)
5) Multiply the average crown area for each species by the total number of trees on the space to get the total tree canopy cover
6) Divide this area by the total area of space (e.g., 2 ha) and multiply by 100 to get the percentage canopy cover

For example, neem has a crown spread of about 5-10 m. Based on this we can calculate the area covered by a typical mature neem tree as about 20-80m^2 (assuming the crown is a perfect circle). Assuming an average of 50m^2 per neem tree, if there are ten neem trees on the 2 ha land, the canopy cover is 500m^2 divided by 20,000 m^2, multiplied by 100. This gives a tree canopy cover of 2.5%. If the target canopy cover for this space is 30%, it means we need a canopy cover of 6000 m^2, but given that the canopy cover of an average tree is 50m^2, we will need to plant 6000/50 trees, which translates to a total of 120 trees. However, we already have ten trees on the site, so 110 trees will be needed to meet this target. To generalise this formula, the number of trees (N) needed to meet a target is given as:

$$N = \frac{PCxTA}{CC_A x100}$$

Where:

PC = target percent cover

TA = total area covered by the facility in m^2

CC_A = canopy cover of average tree in m^2

It should be noted that this method is limited in application to small areas of land and /or where there is a labour supply that allows for complete enumeration or intensive sampling of trees. The approximation of the canopy cover of the tree can be based on the literature, estimated by the technician or based on the use of equations that relate diameter at breast height of the particular species to the crown cover. There is no universally accepted tree canopy cover for urban areas. Different urban areas have come up their targets based on their needs. However, the common values for cities in North America are between 30-40%. In the case of Ghana and the example of neem, a 30% target looks reasonable, as this will result in 60 trees per ha. Considering that these urban spaces usually have other infrastructure development such as buildings, roads, electric and water supply systems and in the big cities are located in tight spaces, it will be difficult in many cases to go higher than a 30% canopy cover. The number of trees can be changed to reflect different species and ages of the trees planted. Another word of caution in using targets is that it may take several years after planting to reach these canopy cover targets, as small trees have small crowns, and so while the number of trees to meet the target would have been planted, the actual canopy cover desired may be 5-10 years into the future. One way to mitigate this is to plant more trees than are required to meet the target and then thin them out as the trees mature, so as to maintain a relatively high crown cover in the initial stages of tree growth.

10.7.5. Management

Management of planted trees in rural and urban environments will depend on the purpose of planting the trees. As a result, only management practices that are consistent with the objectives of the planting should be undertaken. For example, pruning should be avoided for trees planted for shade, while if the purpose is for fodder for animals, pruning would be a main component of the management plan. It is difficult to provide management prescriptions here to cover all cases and trained forestry staff should be consulted when in doubt. Table 2 summarises examples of tree species that can be used for various wood and environmental forestry purposes.

Table 2. Summary of examples of tree species for planting for non-timber functions

Purpose	Example Species		Suitable Plantation types
Fuelwood	*Leucaena leucocephala,* *Cassia siamea,* *Eucalyptus spp.* *Azadirachta indica* *Gmelia arborea* *Albizia lebbeck* *Anogeissus latifolia* *Terminalia catappa*	*Gliricidia sepium,* *Acacia spp.* *Tectona grandis* *Tamarindus indica* *Khaya senegalensis* *Eucalyptus spp.* *Anogeissus* *leiocarpus*	Woodlots, farms, line planting, multipurpose plantations, agroforestry
Soil improvement and Erosion control	*Acacia spp.* *Eucalyptus spp.* *Anacardium occidentale* *(Cashew)* *Albizia lebbek* *Casuarina equisetifolia* *(false pine)* *Moringa oleifera* *Mimosa pigra*	*Pinus spp.* *Pithecolobium sp.* *Prosopis chilensis,* *Parkinsonia* *aculeate,* *Leucaena* *leucocephala* *Acacia Senegal* *Cassia siamea* *Prosopis spicigera* *Parkinsonia* *aculeate*	Spaced planting, pair planting, stream bank, riparian plantings, hills and slopes, line planting, agroforestry
Rural Development	Depends on purpose of plantation, but would include fruit trees, and species that produce timber, electric poles or have economic values. Examples are:		Dedicated forest plantations, farmlands, home gardens, fruit orchards
	Mangifera indica *Tectona grandis* *Khaya spp.*	*Anacardium* *occidentale* *Vitellaria paradoxa* *Parkia biglobosa*	
Fencing	*Parkinsonia aculeata* *Pithecellobium dulce*	*Leucaena* *leucocephala* *Gliricidia sepium*	Around homes, public buildings (schools, clinics, etc.), gardens, compounds, nurseries
Shade	*Tamarindus indica,* *Gliricidia sepium* *Cassia siamea* *Khaya spp.* *Albizia lebbek*	*Pithecellobium* *dulce* *Blighia sapida (Akee* *apple)* *Ficus spp.* *Mangifera indica* *Azadirachta indica*	Planting as individual trees or in lines, on sidewalks, parks and playgrounds, planting around homes and schools

Table 2. (Continued)

Purpose	Example Species		Suitable Plantation types
Building / construction poles	*Diospyros mespiliformis* *Anogeissus latifolia* *Azadirachta indica* *Vitellaria paradoxa* *Acacia spp.* *Cassia siamea* *Khaya spp.*	*Tamarindus indica* *Gliricidia sepium* *Eucalyptus spp.* *Ptericarpus erinaceous* *Gmelina arborea* *Tectona grandis*	Woodlots, dedicated forest plantations, individual trees, farms and agroforestry, line planting, windbreaks
Fruits	*Blighia sapida* *Anacardium occidentale* *Balanites aegyptiaca* *Psidium guava* *Moringa oleifera* *Ficus spp* *Carica papaya*	*Parkia biglobosa* *Magifera indica* *Citrus spp.* *Vitellaria paradoxa* *Tamarindus indica* *Diospyros mespiliformis*	Individual trees, line planting, plantings on farms, fruit orchards
Windbreaks	*Cassia siamea* *Eucalyptus spp.* *Leucaena leucocephala* *Casuarina equisetifolia* *Cedrela odorata*	*Azadirachta indica* *Gliricidia sepium* *Mangifera indica* *Anacardium occidentale*	Line planting around buildings of interest, woodlots, farms
Watershed protection and buffers	*Acacia mangium, Alnus acuminate, Prosopis, Leucaena*	*Melia Volkensii, Azadirachta Indica*	Water catchment areas, watersheds, around dams and dugouts, along river/stream banks
Mine rehabilitation	*Acacia magium, Gliricidia sepium,*	*Leucaena leucocephala Cassia siamea,*	Mining sites and other disturbed/excavated sites
Rural and Urban spaces	*Khaya sp.,* *Azadirachta indica* *Mangifera indica* *Tamarindus indica* *Gmelina arborea*	*Cassia siamea* *Tectoca grandis* *Albizia lebbeck* *Leucaena leucocephala* *Ficus spp.*	Planting around schools, clinics, parks, sidewalks, compounds and other public facilities and spaces

10.8. Conclusion

Though the main purpose of growing trees in most cases is to produce wood to satisfy human needs, there are increasing examples of tree planting to meet other environmental needs of society. In some cases, these environmental needs will be considered the main benefits of tree planting. In Ghana, the need for fuelwood and planting trees to improve soils and control erosion is critical in many parts of the country. This Chapter has provided information on the role that trees play in these environmental situations and how plantation forestry can be used to achieve these objectives.

PART IV. FOREST PLANTATION ECONOMICS AND INTERNATIONAL CLIMATE CHANGE POLICY

Chapter 11

FOREST PLANTATION ECONOMICS

11.1. IMPORTANCE OF ECONOMIC ANALYSES IN FORESTRY

Economics is the study of how scarce resources are allocated among competing uses and how the wealth that is generated is distributed amongst users to further social objectives. The main reason economics exists as a field of study is because there are not enough resources for everybody to have all he or she wants, i.e., resources are scarce. Scarcity is a result of resources having alternative uses. The application of economic principles to managing natural resources is not new, although there have been considerable shifts over the years from sustained yield management paradigms towards the concept of sustainable development. If current resources are managed sustainably, it will ensure that future generations also have access to some form of the resources to meet their needs.

Economic considerations are important to any successful plantation development exercise, irrespective of whether the plantation is established by an individual, community or a private firm. Forest economics applies economic principles to the management of forests. Forest economics has three main unique characteristics that distinguish it from other branches of economics: 1) a growing forest crop is both capital and an end product at the same time; 2) trees grow on long rotations; and 3) forests produce marketable and non-marketable products simultaneously. Economic analyses are applied in forestry to ensure that the economic implications of forestry practices are understood to enable forests to be grown optimally. Since resources have alternative uses (opportunity costs), it is important that if these resources are used to grow trees, the choice can be justified in economic terms. Below, we elaborate on the applications of economic principles to forest management. The list is not exhaustive and the tools used are not mutually exclusive since there may be some overlap in their application.

First, economic tools can be applied to determine *ex-ante* whether undertaking particular activities associated with plantation development and investments make economic sense. Such decisions, which may include taxation, trade, investment, conservation, efficient use and optimal policy development, rely on economic principles for analysis and insights. Economic principles can be applied to choose among competing investments and between consumption and investments. For example, the decision of whether a piece of land should be used to plant trees or cultivate food crops can be made through economic analyses that examine all costs and benefits of each option.

The application of benefit/cost analysis and other efficiency criteria are very important in making investment choices among forestry projects and between forestry and non-forestry investments.

Secondly, forests produce timber and non-timber products. The challenge for forest managers is to understand the nature and degree of tradeoffs that may be associated with the provision of multiple benefits. To understand these tradeoffs, we need to know the nature of the relationships between the various forest uses. Determining this relationship is a task of considerable complexity, although these relationships are known to fall into three main categories: independent, complementary, or competitive (Teeguarden 1982). When the relationships between forest uses are independent or complementary, there are no tradeoffs. However, most forest uses are more likely competitive, especially at higher levels of use, making tradeoffs inevitable. In this case, economic tools can be used to guide efforts to assess these tradeoffs and to optimise the benefits from the production of multiple forest resources.

The free market system has been a cornerstone of economic theory for many decades. The perfect competitive market system is efficient in allocating resources only when the strict assumptions of perfect competition are met. However, markets can fail as a result of violations of some or all of these assumptions. When markets fail, inefficient outcomes that lead to the overuse of resources or pollution of the natural environment may result. Market failures abound in forestry due to the joint production of goods and services, some of which are not traded in markets. Under these circumstances, economic analyses can provide guidance on the best approaches for governments to intervene to correct such market failures and ensure efficient use and management of forest resources.

One interesting application of economic principles is to predict human behaviour. The assumption of utility maximisation in microeconomic theory, which applies to firms and consumers, provides a powerful foundation to predict human behaviour. Such theories and models can be used to predict the impact of policy measures on the forest industry as well as how people will react to the changes in policy. For example, the ability to predict how people would react to changes in legislation, taxation, subsidies, etc. is invaluable to governments in the design and implementation of forest policies.

Fifth, economic principles can be applied to choose instruments for forest policy. Once a forest policy decision has been made, there may be more than one approach to achieving the policy goal. Economics provides powerful insights into how best to structure policy measures to achieve the goals in the most effective and least costly way. In general, governments can apply command-and-control techniques or use economic instruments. Economists tend to favour the latter approach unless it can be proven that they will be ineffective. An example of economic instruments used in Ghana is the provision of incentives such as subsidies to plantation developers.

On the other hand, legislation that mandates TUC holders to reforest harvested areas in their contract areas is a command-and-control measure to plantation development.

The subsequent sections discuss the basic concepts and tools needed to undertake economic analyses of plantation forestry investments.

11.2. Basics of Discounting and Compounding

11.2.1. Interest and Discount Rates

Interest is a fee for borrowing money, which is paid to the lender over and above the amount borrowed (principal). The amount of interest paid by the borrower is determined by the *interest rate*. Most people who borrow money from a bank or other financial institutions are familiar with this concept. Investments in forest plantations are long-term in nature and require huge capital outlays in the first few years while benefits accrue after many years. Many investors may not be able to fund these projects from their resources hence there may be a need to take out a loan to finance the project. The loan will continue to attract interest until it is fully paid back to the lender.

The interest rates charged by the banks or other financial institutions are made up of three main components. The first component is the real rate (or pure) interest rate. The real interest rate is the return on money borrowed that would prevail if borrowers and lenders expect repayment to be made in money having constant purchasing power (i. e., the interest rate in an inflation-free situation). The real interest rate is approximately the rate of interest charged on Treasury Bills in the commercial banks. Added to this pure rate is an expected rate of inflation during the period of the loan. The third component is the risk rate, which the lender determines based on a risk assessment of the investment itself and /or the risk of the borrower. As an example, if the pure interest rate is 20%, the expected level of inflation is 10%, and the risk rate is 5%, the total interest rate would be about 39%[1].

The *discount rate* is the interest rate used to determine the present value of future cash flows. This means that the discount rate is not *necessarily* the same as the interest rate. In general, the discount rate is the interest rate the Bank of Ghana charges on loans it makes to banks and other financial institutions. The banks in turn use the discount rate as a benchmark for the interest they charge on loans to consumers. Therefore, most borrowers who go to the bank get to deal only with the interest rate, rather than the discount rate.

The choice of an appropriate discount rate to use in forestry investment analyses is controversial. This is because the discount rate plays a very crucial role in determining whether a given project or investment is economically feasible or not, especially the long-term investments encountered in forestry. The idea behind discounting is to reflect time preference, and the discount rate is, therefore, the key to comparing values accrued over time. A high discount rate assigns less weight to future benefits (and costs) compared to the present, and vice versa. Assigning less weight to benefits in the future is a way of compensating for future consumption for the inconvenience of forgoing present consumption. A second reason for discounting is that capital has an opportunity cost. Therefore, when capital is tied up in a particular investment, the loss of use of that capital in alternative investments must be taken into account. The main controversy in natural resource analysis is not whether future costs and benefits should be discounted, but what rate to use. Using a lower discount rate has a potential to make many projects seem profitable, and so could result in more environmental damage as more projects are undertaken and vice versa. A compromise solution may be to use the social rate of time preference, which reflects the

[1] Note that the correct way to calculate the total interest is not a simple sum of the three rates, rather it is the multiplication of them: (1.2)(1.1)(1.05) = 1.386 or 39%.

weighted average of individual preferences of present versus future consumption. A major problem in Ghana is the habitual high bank interest rates (ranging from 20 – 40%). Since discount rates show some correspondence with interest rates, this will tend to make most long-term resource investments unprofitable from a purely economic standpoint. In conducting economic analyses involving discount rates, sensitivity analyses should be performed to assess the impacts of various discount rate scenarios on the results.

11.2.2. Present and Future Values

One important characteristic of forestry investments is the long separation in time between the time the investments are made (costs incurred) and the benefits (revenues) accrue. This means that many costs and revenues occur over time, and hence there has to be a way to compare these costs and benefits. As noted above, these comparisons are made using the discount rate. The comparison can be made either at the maturity of the investment (final harvesting of trees) or the beginning of the investment (the initial year of planting). In the latter case, discounting is used to bring all future costs and benefits to the present (base year), while in the former case, compounding is used to move all future costs and benefits to the maturity date of the investment.

The following equations are used to calculate future values (compounding) and present values (discounting) respectively:

Future value equation:

$$V_T = V_0(1+d)(1+d)........(1+d) = V_T = V_0(1+d)^T \tag{1}$$

Present value equation:

$$V_p = \frac{V_T}{(1+d)^T} \tag{2}$$

where V_T = value of an amount T periods into the future

V_0 = value of the initial investment

Vp = present value of future investments

T = number of periods over which the investment is evaluated

d = periodic discount rate, stated as a decimal

In forest plantation investments, V_0 would represent the costs of site preparation, seed acquisition, planting and tending operations, and T would represent the rotation age. If the initial cost of planting is Gh¢1500/ha, the rotation length is 50 years, and the discount rate is 10%, then the future value of the investment is calculated as:

$$V_T = 1500(1+0.1)(1+0.1)........(1+0.1) = V_T = 1500(1.1)^{50} = 176,086.28 \tag{3}$$

If, on the other hand, we want to compare investments in the present year (year 0), then we will use the present value formula. If the future value of the investment is known (assume Gh¢200,000), then using the same discount rate and timeframe, the present value is given by:

$$V_p = \frac{200{,}000}{(1+0.1)^{50}} = 1{,}703.71 \quad (4)$$

Costs and benefits that occur between year 0 and T can easily be incorporated into this framework, by discounting to the present or compounding to the future maturity date of the investment.

11.3. THE CONCEPTS OF VALUE AND VALUATION

The calculation of present values requires knowledge of the values of the forest resource either in the present or future years. The value of a good or service is an expression of its worth to an individual or group at a given time and context. Value expresses how much someone is willing to give up for that good or service, and hence it is often measured by the willingness to pay (WTP). The more something is worth to an individual, the more the person is willing to pay for it. In general, therefore, when someone pays GH¢5 (five Ghana cedi) for a good, it is worth more to the person than another good for which he pays GH¢2. The *process* of assigning a value to a natural resource product or service is referred to as valuation.

Because value is a measure of worth, it is no surprise that there are different types of values; including social, economic, cultural and spiritual. Not all values are, or can be, expressed in monetary terms. Even then, it is still possible to compare the worth of various things in relative terms. For example, barter trade practised before the advent of paper and coin money relied on the valuation of different goods based on values of other goods. Also, how long a person is willing to travel to acquire fuelwood relative to other substitute products nearby would be an indication of its worth or value to the person. Therefore, the fact that the value of a good or service cannot be expressed in monetary terms does not mean it has less value than one that can. This is why it is critical that forest resource valuation should take into consideration all values produced by the ecosystem, and not be restricted to only those for which monetary values can be assigned.

Another important dimension of value is the question of who is assigning the values to whom. In other words, whose values? For example, if we are discussing the value of banana fruits in a forest, is it the value of the fruits to a monkey who badly needs them for lunch, or the value to human beings when sold in a market? In most valuation, it is the value of the product or service to humans as assigned by humans, not other living things that may also have uses for that product or service. Hence, natural resource valuation is essentially anthropogenic (human-centered).

11.4. Information Requirements for Economic Analyses

Benefit /cost analysis (BCA) is the process of estimating the economic values of the benefits and costs of a particular course of action to decide if it is worthwhile to implement. It is the main tool used in the evaluation of new technology and policies in natural resources. Costs and benefits are determined based on the economic principles of consumer choice and the role of markets in expressing social preferences as reflected in prices. BCA for policies or projects enhances transparency and accountability, adds rigour and discipline to the analysis, contributes to more informed decisions by assessing the full impacts of decisions on society, and provides common measures of comparing options.

Economic analyses of forest resources rely on information from various biophysical, social and economic spheres. Biophysical data include the growth and yield patterns of the trees and other resources within the forest, such as animals, birds, other vegetation, etc. Social information includes preferences and needs of the landowner while economic variables include prices of goods and services produced from the forest, the costs of production, opportunity costs, and interest and discount rates.

Chapter 6 discussed growth and yield models and how they can be constructed and used. The following sections describe how benefits and costs are estimated to provide the essential data to undertake economic analyses of forestry investments.

11.5. Economic Valuation of Forest Resources

Estimating the prices (or value/benefits) of forest resources involves many different techniques. These techniques can be divided into two main types: market-based and nonmarket-based valuation. Market-based valuation can further be divided into two: those based on direct market prices, and those that rely on indirect market prices.

11.5.1. Market-Based Approaches

Market-based approaches are applied to value resources that are traded in the market place. In most cases, these prices are relatively easy to obtain. In a free market, the price of a commodity is established based on the interaction of demand and supply. The assumptions of perfect competition are pretty strong and are the foundation of price theory in the neoclassical economic theory of product markets. These assumptions include profit maximisation, many buyers and sellers (so no one has market power), homogenous products, no market entry/exit barriers and perfect information. When market prices are used to value resources, it is often assumed that these assumptions are satisfied, and hence the prices reflect the economic scarcity of the resource.

A. Direct Market Prices

To estimate values of resources traded in markets, the forest resource goods and services to be valued have to be clearly identified. Once these are known, data can be collected through surveys or the use of secondary data. For example, to determine the value per kilogramme of fuelwood or charcoal, a survey in the market of interest can easily reveal this. The prices of non-timber forest products that are sold in markets can also be obtained by surveying local markets. In the forestry sector, prices of wood products exported can be derived implicitly by using volumes and values exported published by the Food and Agriculture Organisation (FAO) of the United Nations. The other two approaches for obtaining prices of marketed products are direct observation of the prices in the specific market or conducting experiments (usually called experimental markets).

B. Indirect Market Prices

When the assumptions of perfect competition outlined above are violated, market prices of resources are distorted and hence indirect prices are the appropriate alternative measures of value for resources traded in markets. The following are examples of indirect methods used to derive market prices from the prices of other goods and services under these circumstances.

Residual Value Method

This method subtracts the costs of production and profits from the prices of final goods and services to arrive at an estimate of residual value. In forest valuation, it is often the case that there is a need to estimate the value of the trees in a forest. The value of the trees as they stand in the forest (on their stumps) is called the stumpage value. The residual value method is commonly used to determine stumpage values in forestry using the formula:

$$\text{Stumpage value} = \text{log price} - \text{harvesting costs} - \text{transport costs} - \text{processing costs} - \text{operators profit} - \text{risk allowance} \quad (5)$$

Note that the operator's profit and risk allowance have to be subtracted from the final price because it is necessary to pay entrepreneurship its return too, as this is a cost of production. The main advantage of this method is that it uses what is known about wood products values traded in a market to logically deduce stumpage values. However, figuring out all the costs is complex as the costs vary by forest location. The complexity increases given that many forest products companies are vertically integrated. Therefore, one may need to go through several stages of processing (and estimate processing costs at each stage) to obtain residual prices of trees. Another downside of this approach is that it can be difficult to find an upstream wood product market that is competitive.

Surrogate Prices

A surrogate price is the price of a close substitute for the product being valued. In this method, the price for a close substitute for the product is used as a proxy for the unpriced good or service. In the example of valuing fuelwood or charcoal above, if it is believed that the market is not competitive, one way to indirectly deduce the price of fuelwood would be to use the value of a close substitute in that location such as kerosene and adjust for calorific values of the two fuels. For example, see Kengen (1997) for descriptions of how the value of fuelwood can be derived from that of kerosene.

Resource Replacement Cost Method (RCM)

Resource replacement cost methods (RCM) are used where damage has occurred to a component of the forest ecosystem. This method can be used to value improved water quality by measuring the cost of controlling effluent emissions or valuing erosion protection services of a forest or wetland by measuring the cost of removing sediment load from downstream areas (King and Mazzotta 2000). The approach involves assessing the cost of restoring, rehabilitating or replacing a damaged or depleted forest asset to its pre-damage condition. This cost is then used as an estimate of the benefit of restoration. The simplest way to apply this method if the damage is restored is to use the actual cost of the restoration. If the damage is not restored, the cost could still be estimated if the resource has close substitutes. In this case, sample values of the substitutes can be obtained from primary or secondary data and used to estimate the cost to replace the resource. For resources that possess unique characteristics, application of this method would be quite challenging. In practice, there are several other implementation challenges. First, unless the damage attracts widespread national attention, it may never be repaired, in which case it may be difficult to estimate the benefits. Second, even when the damage is repaired, the resource may appear identical to the original state though the ecological functions of the damaged ecosystem may not be completely restored.

Appraisal Method

Another valuation method for assessing forest resources that have been damaged is the *appraisal method*. In the case of a forest that has been burnt, or a forest land that has been damaged, a knowledgeable appraiser may be able to identify a fair market value of the damaged property using comparable undamaged land. This value should reflect, as closely as possible, the price at which the resource would sell in the market place at the time of the damage (Ulibarri and Wellman 1997).

11.5.2. Nonmarket-Based Approaches

There are some forest resources that are not traded in markets, or the appropriate markets do not exist for trading in them. As a result, prices for goods and services provided by these kinds of resources cannot be determined using market prices (the interaction of market demand and supply). These resources include the contribution of forests to clean air, value of forest watersheds, or erosion control. These benefits are difficult to assess using market-based valuation techniques. The various techniques that have been developed to estimate such values rely on observed and unobserved (hypothetical) behaviour to deduce the values of these resources to individuals and society. Below are brief descriptions of the major methods.

A. Travel Cost Method (TCM)

Every year, millions of people around the world travel to many natural resource destinations (parks, forests, lakes, streams, etc.) for recreational activities such as hunting, camping and fishing. These natural systems, therefore, provide valuable services to people. Since the benefits of such natural resources are not traded in markets, economists have developed methods to capture the values that people derive from these services. The values of these natural ecosystems and the benefits they provide can be used to provide useful information to support multiple forest resource management decisions.

The travel cost method has become popular in the natural resource valuation literature for estimating the value of recreational activities such as changes in access costs for a recreational site, elimination of an existing recreational site, addition of a new recreational site, or changes in environmental quality at a recreational site. All the models used in estimating the value of recreation sites rely on the fact that a visit to a recreation site involves transactions costs (travel cost, entrance fee, opportunity costs of time, etc.). These costs are incurred solely to gain access to, and enjoy the benefits or services provided by, the site. The basis for the estimation technique is quite simple. For a given site, different individuals will face different travel costs while one individual will face different costs to different sites. Individuals will respond to the variation in costs (implicit prices) and this will impact whether and how often they visit the site. For instance, if we were interested in estimating the value of the Mole National Park using the travel cost method, we would note that people coming from different parts of Ghana and abroad will face different travel costs. The cost of travelling from Bunkpurugu to Mole will differ from that from Najong No. 1 or Accra to Mole. This observed variation in travel costs are used to estimate a travel cost function and to evaluate the willingness to pay (WTP) to visit the site. An individual demand function for a site is generally of the form:

$$D = f(T_r, M, Q) \tag{6}$$

where:

D = the number of visits to the recreation site

T_r = the full cost of a given visit

M = individual's income

Q = environmental quality of site

The value of the flow of services from the site is then aggregated over all those who visited the site. The data to implement the TCM can easily be obtained by phone, onsite visits, mail surveys, or information from site registration data. Using this method to estimate the value of changes to a recreational site is premised on the knowledge that the environmental attributes of sites influence how often people visit the sites. Therefore, changes in visitation rates may reflect changes in the quality of the site.

B. Random Utility Model (RUM)

If we are interested in estimating the value of a forest site based on its characteristics, we would not be able to use the standard TCM described in the preceding section. A discrete choice, or random utility model, explains the choice among sites as a function of the characteristics of that site, and hence is more suitable for this purpose.

The following description of the RUM follows McFadden (1981).The basis for the RUM is the concept of indirect utility functions in microeconomic theory. Indirect utility functions characterise the maximum utility that can be achieved given prices and income. Discrete choice theory follows similar reasoning except consumption can only be in specific quantities and so allows for choices of zero or "corner solutions" in consumption. When modelling recreation demand, discrete, rather than continuous, choices are more realistic since recreationists make decisions based on going to a site or not and also because sites cannot be sub-divided infinitely or continuously. Recreational demand models typically have a finite set of alternative sites or are discrete. The choice of alternative sites is dependent on the utility, U; recreationists derive from various attributes, Q, of the site:

$$U_{in} = f(M, Q_{in}, S) \tag{7}$$

where the utility is a function of income (M), the quality of the site or attributes describing site i as perceived by recreationist n (Q_{in}), and other socio-economic variables (S). The choice set is defined as C_n, and n is the number of alternative sites (or a subset of sites). If site i is chosen, we assume that the utility associated with visiting site i is higher than for any other site j, i. e.,

$$U_{in} > U(Q_{jn}) \qquad \forall i \neq j; \quad i, j \in C_n \tag{8}$$

Utility in this framework is treated as a random variable since researchers do not have perfect behavioural information (McFadden 1981; Smith 1989). More formally, utility is modelled to include a systematic/observable component and a random or unobservable component:

$$U_{in} = V_{in} + e_{in} \tag{9}$$

where V_{in} is the systematic component of utility and e_{in} is a random element. The random element captures any unexplained factors that are not directly modelled or observed by the researcher.

C. Hedonic Price Method (HPM)

The RUM described above models the demand for a site based on its characteristics. The hedonic price method (or hedonic travel cost method if applied to recreation) goes further by attempting to value the characteristics of the site, using the information on site demand. The method can be used to estimate economic benefits or costs associated with environmental quality, including air or water pollution, and benefits of environmental amenities such as aesthetic views or proximity to recreational sites (King and Mazzotta 2000).

The HPM is premised on the assumption that people choose goods or services because of their characteristics. It was developed for and has been applied extensively in the real estate sector for valuing real property. When applied to real estate, for example, the characteristics of a house that determine its value include location, the number of bedrooms, the size of the kitchen, etc. It is possible that by building two identical houses and changing only the number of bedrooms, the value of the additional bedroom(s) can be estimated as the difference in price between the two houses, all other things held constant.

To apply the HPM method to natural resource valuation, we determine market value differences for two similar goods or services that differ only in terms of one characteristic as a measure of the value of that characteristic (Kengen 1997). For example, if there are two National Parks located in the same geographical area, and these parks are identical in every aspect except that one has a good road network within the park and the other does not, then the difference in value between these two parks can be attributed to the good road network.

The main advantage of this method is that it can be adapted to consider several possible interactions between market goods and environmental quality. However, it is a relatively complex valuation method to implement and interpret, requiring a high degree of statistical expertise (King and Mazzotta 2000). Also, large amounts of data must be gathered and manipulated, and time and expense to carry out an application depends on the availability and accessibility of data (King and Mazzotta 2000).

D. Contingent Valuation Method

The contingent valuation method (CVM) uses structured surveys to elicit information from people about their preferences for a natural resource. This approach is suitable for the valuation of nonmarket goods and services where other approaches are deemed inappropriate. The value obtained by this method is contingent on the existence of a market for the good or service as described in the questions put to respondents (hence the name contingent valuation). Contingent valuation is, therefore, a hypothetical, direct (survey-based) method that does not rely on observed behaviour to elicit information, and hence is the only method suited for estimating non-use values of resources.

Contingent valuation surveys may be conducted as face-to-face interviews, telephone interviews, or mail surveys based on a randomly selected sample or stratified sample of individuals. Face-to-face interviews are the most expensive survey administration format, but they are generally considered the best, especially if visual material needs to be presented (Ulibarri and Wellman 1997). The central goal of the survey is to generate data on respondents' WTP (or willingness to accept compensation) for some programme or plan that will impact the well-being of the respondents (Ulibarri and Wellman 1997). The most commonly used hypothetical questions simply ask people what value they place on a specified change in environmental amenity or the maximum amount they would be willing to pay to have it occur. The responses, if truthful, are direct expressions of value and would be interpreted as measures of compensating surplus (Freeman 1993).

The CVM has been subjected to various criticisms for biases arising from instrument choice, implementation and data analysis. Some of these biases include strategic bias, scenario misspecification bias, aggregation bias and starting-point bias. The method requires a lot of expertise not only to implement, but to detect and avoid, or at least minimise, these biases. In situations where people are not used to purchasing a particular natural resource product or service, they find it difficult to attribute a monetary value to it (Kengen 1997). For example, if we ask people living around the Gambaga East Forest Reserve how much they are willing to pay to ensure that the soil is preserved from erosion, it would be difficult to get a reasonable value simply because it is not a concept they usually encounter in their daily lives.

Monetary value can be meaningless for many subsistence resource users. Thus, modifications are required to ask, for example, about relative preferences that can be easier to express than monetary valuations; such modifications require specific expertise (Kengen 1997). According to Brown et al. (1995), CVM is more effective when the respondents are familiar with the environmental good or service and have adequate information on which to base their preferences, and hence is likely to be far less reliable when the object of the valuation exercise is a more abstract aspect, e.g., existence value. Critiques argue that the CVM is somewhat implausible for serious decision-making because it does not relate to real market conditions (Winpenny 1992). A further problem with CVM is that because it involves a survey, it is time-consuming and costly to design and implement. Despite its shortcomings, CVM is considered the most widely accepted method for estimating total economic value, including non-use values (existence and bequest) and use values such as option values.

Asuming-Brempong (2003) applied the CVM to assess the important factors influencing the maximum WTP for agroforestry attributes and Sudan Savannah farming systems such as improved scenery, increased soil fertility, improved water retention, soil erosion prevention, gender and age of the farmer, household income, and supply of fuelwood.

E. Productivity or Factor Income Method

So far, we have looked at methods that deal with the outputs from forest resources that provide benefits to people. In this section, we turn to the question of whether forest resources used as inputs into the production of other goods and services have value to people, and if so how such values can be measured. The productivity method is one way to estimate such values. While the method of factor income is not as well defined or widely referenced as the hedonic price or travel cost methods, it is recognised by some analysts and organisations such as the U.S. Department of Interior's natural resource damage assessment regulations (Ulibarri and Wellman 1997). The method is applied in cases where the products or services of an ecosystem are used, along with other inputs, to produce a marketed good.

This method relies on the economic costs of production as an important source of information. Several types of resources for which the factor income approach is potentially well-suited have been identified by Ulibarri and Wellman (1997) to include: surface water and groundwater resources, forests and commercial fisheries. Surface and groundwater resources may be inputs to irrigated agriculture, to manufacturing, or to privately owned municipal water systems. The products in these cases (agricultural crops, saw logs, manufactured goods, and municipal water) may all have market prices. Similarly, commercial fishery resources (fish stocks) are inputs to the production of a catch of saleable fish (Ulibarri and Wellman 1997). To apply the productivity method, data must be collected regarding how changes in the quantity or quality of the natural resource affect costs of production for the final good, supply and demand for the final good, and supply and demand for other factors of production (King and Mazzotta 2000).

King and Mazzotta, (2000) further highlight the following disadvantages of the productivity approach: 1) when valuing an ecosystem, not all services will be related to the production of marketed goods. Thus, the inferred value of that ecosystem may understate its true value to society; 2) information is needed on the scientific relationships between actions to improve quality or quantity of the resource, and the actual outcomes of those actions. In some cases, these relationships may not be well known or understood; and 3) if the changes in the natural resource affect the market price of the final good or the prices of any other production inputs, the method becomes much more complicated and difficult to apply.

11.5.3. Mixed Market and Nonmarket-Based Approaches

Natural resource valuation professionals have developed techniques that combine elements from market-based methods with pre-existing estimates of natural resource values based on either direct or indirect nonmarket valuation techniques. The major motivation for using these mixed approaches is the relative simplicity of using a pre-existing study based on an accepted method, as well as the cost considerations in undertaking a fresh natural resource valuation study (Ulibarri and Wellman 1997). The two main approaches that have been applied are the benefits transfer and unit day value methods.

A. Benefits Transfer Method

Benefits or value transfer involves taking economic values for addressing a particular natural resource valuation question and applying it to another context (Pearce et al., 2006). The idea of benefit transfer emerged in the early 1980s when the U. S. Environmental Protection Agency, proposed the use of desk studies as the basis for cost-benefit analysis in environmental impact assessment because of capital and time constraints (Yaping 1999). The method uses a data set developed for addressing a particular natural resource valuation question to another context. For example, fuelwood values in savannah woodlands in Ghana may be estimated by applying measures of woodland values from a study conducted in Zimbabwe.

Studies to value natural resources may be expensive, especially in Ghana and other developing countries where research budgets are limited. Benefits transfer, therefore, offers a cheaper alternative to performing a full-scale study of any particular issue. The benefits estimated from expert opinion, observed behaviour or CVM can be transferred to another context. Benefits transfer should only be done under certain conditions to ensure that the information is useful. The analysts needs to know the purpose of the original value estimates, the user group(s) that were considered in generating the initial estimate, and other factors that could have influenced the results of the study being considered for benefits transfer (Ulibarri and Wellman 1997).

The accuracy of the estimated benefits from the benefits transfer method is somewhat suspect, as it depends on the accuracy of the original estimates and how they were derived. This is more so if the benefits transfer technique extrapolates beyond the range of characteristics of the initial study. Moreover, it may be difficult to access and assess existing studies if they are not published or easily available to the general public.

B. Unit Day Value Method

The *unit day value method* is similar to the benefits transfer method. However, instead of relying on the benefits derived from a single estimate, the unit day value method derives an average value based on multiple value estimates from existing studies. Consequently, the unit day value of the underlying resource reflects a resource that has average preference-related attributes, amenities, or qualities (Ulibarri and Wellman 1997). The application of the unit day value method may also involve groups of experts attempting to interpret from the existing set of estimates (regardless of method used in the original study) the best estimate for each of a set of generic types of environmental resources or activities. The unit day value approach then combines and converts these estimates into a standardised unit of measure that reflects the average value of one unit of the resource on a per-day basis (Ulibarri and Wellman 1997).

11.6. Measuring Economic Costs

The previous section discussed the benefits side of benefit/costs analysis. This section will focus on the measurement of costs to be included in economic analyses. Costs are easier to ignore or miss and more difficult to estimate than benefits in a BCA framework. The results of BCA can be severely biased if costs are not accurately and completely estimated and included. There are three main types of costs: operations and maintenance costs, costs of regulations and social costs.

11.6.1. Operations and Maintenance Costs

Costs of production in forestry plantations include several types, such as the costs of labour to prepare the site, weed, prune, thin, harvest, etc. Operations and maintenance (O&M) costs also include the costs of purchasing inputs: seedlings, working gear, transportation, seedling production, and water and electricity costs. In addition, there may be other overhead costs of operating and maintaining facilities such as road networks, which may extend over the life of the plantation project. Many of these costs are relatively easy to obtain and include in BCA, since they are usually out of pocket expenses and managers keep track of these for financial and tax purposes. It is also possible for engineers and other technical persons involved in the design and implementation of the project to provide fairly accurate estimates of costs for the project. In fact, most investors would order an ex-ante economic feasibility analyses before undertaking a project, and hence all costs would be evaluated at this stage. Table 1 shows a template for calculating the costs of production in a typical plantation forest project. Other costs can be added or deleted from this template as required to arrive at the total cost of a specific project.

Table 1. Template for calculation of typical costs in a plantation project

Operation	Unit costs (Gh¢/ha/yr)		Total (Gh¢/ha/yr)	Overhead (%)	Overhead plus unit costs (Gh¢/ha/yr)
	Materials*	Labour			
Site preparation (machinery or manual)					
Transportation of planting materials					
Seedlings (nursery or purchase)					
Planting					
Infrastructure maintenance (water, electricity, etc.)					
Road building and upgrading					
Fire protection					
Road maintenance					
Weed control					
Pruning					
Thinning					
Final harvest					
Total					

* Note: Material costs include vehicle running and maintenance costs.

11.6.2. Costs of Regulations

Whenever governments introduce regulations, there is a potential for the observance of the regulations to impose costs on both producers and consumers. In developing forest plantations, there are many legal and institutional requirements that impose direct and indirect costs on the plantation enterprise. If, for example, a plantation owner requires a timber utilisation permit to harvest timber on his plantation, this would be a direct cost to the plantation enterprise. Direct costs are usually known with certainty, and can be added to the costs of operations described in Table 1. However, regulations that do not require direct payment of fees but which by being observed lead to changes in the cost of production of the output would be considered indirect regulatory costs. A typical example of such a regulation would be the requirement to have large-scale plantations subjected to environmental impact assessments. The cost of the assessment would be borne by the investor and would influence the cost of production and probably the price of the output. As regulations change, these costs will change, and hence they are not only difficult to measure with certainty, they are also difficult to predict.

From an economic perspective, the information needed to measure the costs of regulations include: the extent to which the marginal costs/supply function is shifted up by the regulation; and the extent of any output adjustments that the firms will make as a result of the cost changes associated with the regulations (Field 2001). To estimate the costs of regulations, we must know the costs and the demand function facing the industry (Fields 2001). Fields (2001) recommends that one way to obtain such costs is to use cost surveys of the industry, using a sufficiently detailed questionnaire to examine the cost structures of the firms and how these might be affected by regulations. Of course, there are problems with this technique, as firms will have an incentive to inflate the costs to argue for reduced government regulations (Fields 2001). Therefore, estimating the costs of regulation will continue to be challenging, although efforts should be made to identify and include these in any economic analyses of forestry plantations.

11.6.3. Social Costs

When individuals or firms measure costs and benefits to include in their analyses of their projects, it is often the case that they consider only the costs they bear when making decisions, not the costs that may be borne by others. But in reality, individual decisions often have implications for society as a whole. The costs that are borne by society as a result of private decisions are called *social costs*. Social costs are incurred through two main ways: opportunity costs and external costs. When resources are used to produce forest products in plantation forestry, the same resources could have been used to produce other goods and services that are of benefit to society. The opportunity cost of using resources (inputs) in a particular way is the highest-valued alternative use to which the resource could have been put (Fields 2001). The consideration of opportunity costs is one of the key differences between the concepts of economic cost and accounting cost because the latter does not consider opportunity costs of using particular inputs.

If the use of the resources in plantation forestry is not the best use of those resources, then society is losing out, as their utility is not being maximised. When forest plantations are established on land, the same land has alternative uses, such as for rearing animals or producing food crops. Also, the inputs such as labour and materials could have been employed for other uses as well. If an input has no alternative use, then its opportunity cost is zero; otherwise it has a positive opportunity cost, even if it is difficult to estimate. In a well-functioning and competitive market, the market price of an input is a good reflection of its opportunity cost. To use a concrete example, in undertaking economic analyses of a forest plantation, the opportunity cost of the land can be calculated by considering all other alternative uses of the land and the benefits that could have been obtained if the land was put to those uses. Assuming the same land would have yielded a net benefit (NPV) of Gh¢500/ha in food crops, Gh¢300/ha if used to rear animals, and Gh¢200/ha if left as fallow, then the best alternative use in this example would be food crops with an NPV of Gh¢500/ha. If the forestry plantation investment earns Gh¢450/ha it means that the plantation owner has lost Gh¢50/ha by not investing in the best alternative use of the land, that is, in food crops. The Gh¢50/ha will constitute the opportunity cost of the land.

The second type of social cost is external costs or costs due to externalities. Externalities are side effects of the actions of economic agents. They can be negative - when the actions of one party impose costs on another party, or positive - when the actions of one party benefit another party. Let us consider a forest plantation that produces timber, but also contains wild animals of benefit to society in terms of meat production and game viewing. If the production of the timber negatively affects the number of animals (for example in a situation where some animals die as a result of the timber production activities), then those who depend on the animals for their food are negatively impacted due to the actions of the forest manager for which they have no control. This is an example of a negative externality. The death of the animals by the timber production process imposes a cost on society that is not represented by the forest manager's production costs. In a perfectly competitive market situation, the forest manager chooses to produce where the marginal private cost equals marginal revenue.

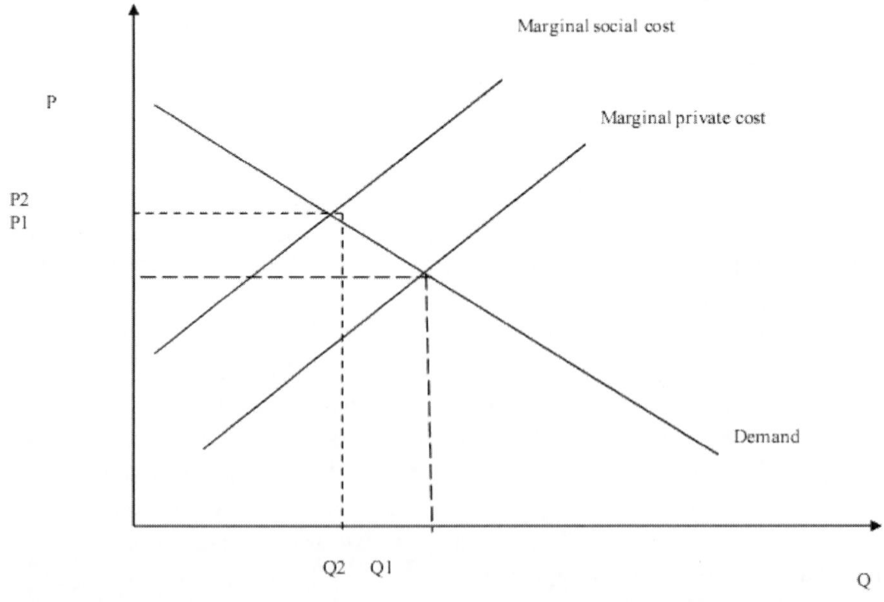

Figure 1. Production, market marginal cost and private cost of producing timber and wildlife.

Figure 1 shows the plantation timber market, with total production taking place where market marginal private cost equals market demand. This is at a quantity of Q1, and a price of P1. Now consider a marginal social cost curve that takes into account production costs plus the costs to society. The marginal social cost curve lies above the forester's marginal cost curve because the death of some of the animals is a cost to society. The intersection of the marginal social cost curve with the demand curve shows that the socially optimal output of timber is Q2, at a price of P2. Society is better off because, at this output, fewer animals are dying due to the reduction of timber produced.

However, in competitive markets, in general, this optimal output will never be reached because firms (in this case the forest manager) do not consider the environmental concerns of the consumer. In the case of negative externalities, competitive markets will provide too much of the good at too low of a price and consequently too much externality. Competitive markets will fail to provide the efficient output. The external cost to society from Figure 1 is P2-P1, which a government agency can impose (e.g., as a tax) on the forest manager. With an appropriate tax, it will be unprofitable for the manager to produce anything but the socially optimal output (Q2). This way, an efficient allocation of resources would occur. The external cost of P2-P1 can now be used as a cost in the BCA analysis. In this example, because the marginal social cost curve is above the marginal private cost curve, Figure 1 illustrates the case of a negative externality. If the marginal social cost curve were below the marginal private cost curve, it would be a positive externality and social optimality would require a greater output than Q1 rather than a reduction of output to Q2.

11.7. Economic Measures of Plantation Profitability

The descriptions of the various criteria that are used to conduct BCA - referred to as discounted cash flow (DCF) techniques- are given below.

11.7.1. Net Present Value (NPV)

The net present value (NPV) is the sum of the difference between the discounted benefits and discounted costs of investment. The feasibility of the investment is determined based on the NPV. The computations and application of this investment criterion are very simple. If the NPV is greater than zero, it implies the benefits are larger than the costs, and so the project is economically feasible. On the other hand, if the NPV is negative, then it is an unprofitable venture. Since the benefits and costs usually occur over time, these are discounted to a chosen base year using an appropriate discount rate. The NPV is calculated using discrete time as:

$$NPV = \frac{p_1V_1 - C_1}{(1+r)^1} + \frac{p_2V_2 - C_2}{(1+r)^2} + ... + \frac{p_tV_t - C_t}{(1+r)^t} = \sum_1^t \left(\frac{p_tV_t - C_t}{(1+r)^t} \right) \quad (10)$$

where p_t is the price per unit of output, V_t is the volume of output resulting from the investment, and C_t is the cost of producing the output; all three variables evaluated at time t. The discount rate is given by r, and t is the time span over which the investment is being evaluated. Usually, t is taken as intervals of one year. The time subscript is eliminated in subsequent formulae to simplify the notation, except where it is required to avoid confusion. In some circumstances, shorter discounting periods may be required, such as half a year or one month. Using the McLaurin expansion of $e^r = 1 + r + \frac{r^2}{2!} + \frac{r^3}{3!} + ...$ and taking r to be small with *continuous* discounting, e^{rt} is approximately equal to $(1+r)^t$. The NPV formula from Equation (10), written in continuous discounting form is therefore given as:

$$NPV = \sum_{1}^{t}\left(\frac{pV - C}{e^{rt}}\right) \tag{11}$$

The continuous discounting version is used in the rest of this Chapter. Note that when calculating the NPV for a particular time t, (i.e., when not summing over more than one time period) the summation operator would not apply, instead, Equation (11) would become:

$$NPV = \frac{pV - C}{e^{rt}} \tag{12}$$

In analysing forestry investments where regeneration usually occurs in the first year (taken as t = 0) without any income, the regeneration costs are not discounted (or would remain as C if discounted in year zero).

11.7.2. Soil Expectation Value (SEV)

The soil or land expectation value is the standard tool used in forestry for investment analyses. The NPV criterion is appropriate if we are evaluating an investment for a single rotation of crop trees. If however, we are interested in determining the sustainability of the investment (for multiple rotations), then we will need to include all costs and benefits after the first rotation. Faustmann (1849) was the first to realise the implicit value of the site as part of the incremental cost of time. The value of the bare land, also known as the Soil Expectation Value (SEV) of a forest stand is the NPV of an infinite series of rotations. Faustmann (1849) showed that if conditions remain constant for an infinite period, and if at the end of each rotation the same species is planted and managed in the same manner as previous rotations, then the soil expectation value at time t is given by:

$$SEV = -C + [pV - C]\left[\frac{1}{(1+r)^t} + \frac{1}{(1+r)^{2t}} + ...\right] = \frac{pV - C}{(1+r)^t - 1} - C \tag{13}$$

where all the variables are as defined above. With continuous discounting, this becomes:

$$SEV = \frac{pV - Ce^{rt}}{(e^{rt} - 1)} \tag{14}$$

When intermediate income and expenses such as thinning, fertilisation or insect control are incurred between initial investment and final harvest, the Faustmann (1849) model can be generalised (Chang 1984). It can be shown that if $A(s)$ is the annual net expense incurred on year s ($0 < s < t$), then the SEV is:

$$SEV = \frac{pV - Ce^{rt} - \int_0^t A(s)e^{r(t-s)}ds}{(e^{rt} - 1)} \tag{15}$$

If $A(s)$ is a net benefit, such as income from a thinning operation, then the arithmetic operator before the integral sign in Equation (15) is positive, rather than negative.

11.7.3. Internal Rate of Return (IRR)

The internal rate of return (IRR) is the interest that is actually earned by the money put into an investment. IRR is also commonly defined as the discount rate at which the NPV = 0. Many economists, including Boulding (1955) argue that the rate of return should be maximised instead of the NPV because a rational manager should strive for a maximum rate of asset growth. In this model, the interest rate earned on the investment in the resource i, is maximised with respect to t. Here, r remains the lending and borrowing bank rate while i is the *actual* interest earned on the investment. The IRR principle uses the basic Faustmann or NPV models except that instead of maximising the NPV or SEV, the interest is maximised. To be able to do this, the initial value of the asset (NPV) must be known. Using the NPV formula,

$$NPV = \frac{pV - C}{e^{it}} = 0 \tag{16}$$

where i = IRR.

The solution to the maximisation problem for IRR usually leads to problems (Fortson 1972). The procedure used, therefore, is to select the IRR nearest zero where the discounted costs and returns are equal. This is equivalent to solving the equation:

$$i^* = \frac{\ln(pV - C)}{t} \tag{17}$$

If $i^* < r$, then no investment will be made as a negative NPV arises. The principal advantage of the IRR is that it does not require *a priori* specification of a discount rate. However, a potential problem with the IRR criterion is that maximising the IRR may maximise something else besides the owner's wealth. Therefore, it is possible that if investments are ranked based on the IRR alone, the largest IRR may select a project that does not maximise the NPV. Furthermore, the IRR maximisation assumes that the amount of land available for forestry is infinite and that access to all capital markets is closed. Thus, the owner will continue to invest all the returns from the original investment and never consumes from the earnings (Newman 1988). Another problem with the IRR criterion is that the decisions relating project scale and timing may be incorrect and that changes in prices or the market interest rate do not affect the optimal relation or the level of investment. If the planting costs are equal to zero, then the IRR is undefined, and if intermediate costs and benefits are received, there could be multiple IRR's that solve the maximisation problem (Schofield 1987). In forestry applications, if the planting costs occur in year zero and an NPV criterion is applied, then from Equation (17):

$$i^* = \frac{\ln\left(\frac{pV}{C}\right)}{t} \tag{18}$$

11.7.4. Benefit/Cost Ratio (B/C)

Another version of the NPV that has commonly been used is the benefit/cost ratio. Unlike the NPV, the B/C ratio is calculated as the ratio of discounted benefits to discounted costs, rather than the difference between them. A ratio greater than one means the project is feasible, and vice versa. The B/C ratio is calculated as:

$$BC\ ratio = \frac{pV / e^{rt}}{C / e^{rt}} \tag{19}$$

Not all projects that show a positive NPV or IRR greater than the discount rate can be carried out. There are budget constraints to be satisfied, and so limited resources require that projects or investments be prioritised. It is common for analysts to rank projects based on the highest B/C ratio or IRR. However, it is possible for the investment with the highest B/C ratio to fail to maximise the NPV (Buongiorno and Gilless 1987).

Given the problems with the B/C ratio and IRR criteria, Buongiorno and Gilless (1987) suggest that the best way to choose among projects is to select the one with the highest NPV. However, the B/C ratio and IRR values provide information that complements the NPV. Furthermore, the concept of rates of return is easily understood by managers and other non-technical staff.

11.7.5. Cost-Effectiveness Analysis

A cost-effectiveness analysis (CEA) is a form of economic analysis that compares the relative costs and outcomes (effects) of two or more activities. The United States Department of Defence developed cost-effectiveness as a method for adjudicating among the demands of the various branches of the armed services for increasingly costly weapons systems with different levels of performance and overlapping missions (Hitch and McKean 1960). Since then, it has become widely used as a tool for analysing the efficiency of alternative activities and programmes outside of the military (Hitch and McKean 1960). In short, CEA is a tool for comparing two or more alternative resource allocation options in identical units.

CEA is often used where there is general agreement on what to do, and the outstanding question is how to achieve the objective agreed upon efficiently, such as how best to comply with a legal requirement. For instance, if a community decides that they need to increase their fuelwood supply; several options could be evaluated to determine which approach would achieve the goal in the most cost-effective way. In this example, unless an estimate of the exact return on the investment or a quantitative value of the benefits is required, a full benefit/cost analysis (BCA) does not have to be conducted.

The cost-effectiveness ratio is the criterion used to calculate and compare the cost-effectiveness of various options. This ratio is calculated as the cost per unit of outcome effectiveness. It is defined by Boardman et al. (2001) as the ratio of the cost of implementing alternative *i* (*Ci*) to the benefit (*Bi*) (or effectiveness) of that alternative:

$$CE_i \ ratio = \frac{C_i}{B_i} \tag{20}$$

The interpretation of this ratio is straightforward. The alternative with the lowest ratio is the most cost effective. An example of an analysis that would apply the above method is in preventing soil erosion. It may be difficult to monetise the value of soil erosion control due to a given programme or policy. In this case, the cost per kilogramme of soil saved will be an appropriate effectiveness measure of the various alternatives.

The literature also shows examples of situations where a decision has been reached to implement a policy (e.g., implement a government regulation) and there are more ways than one to achieve that goal. In such a case, an effectiveness-cost ratio (Boardman et al. 2001) can be calculated, which is essentially the reciprocal of the above formula:

$$EC_i \ ratio = \frac{B_i}{C_i} \tag{21}$$

This ratio measures the average benefits per unit cost. Some authors refer to this as the cost-effectiveness ratio. The interpretation of this ratio is that the policy alternative with the largest ratio is the most cost-effective one. If a community decides to plant 20 ha of trees to increase their wood supply, there could be several ways to achieve this goal, and the method that gives the lowest cost for the same benefit would be the preferred method. The appropriate measure of effectiveness to use is Equation (21).

To avoid confusion, analysts need to keep in mind the goal of the analysis and clearly define what the measure of cost effectiveness is for the particular study. It is important to note that in reality, CEA measures only technical efficiency, not allocative efficiency (Boardman et al. 2001). CEA is good at ranking various alternative policies, but cannot tell us whether any alternative makes economic sense to implement (i.e., has a positive net present value). Also, note that in resource allocation decisions where the benefits and costs accrue over long periods of time, the benefits and costs will have to be discounted using an appropriate discount rate. Given the drawbacks of CEA discussed above, it is usual in economic analysis, to broaden the analysis by conducting full benefit/cost analyses.

11.7.6. Real Options Theory

The traditional DCF techniques discussed above are based on the assumption that future cash flows follow a constant pattern that can be accurately predicted from regeneration up to the rotation age. The profitability or otherwise of forestry projects are then determined based on these cash flow assumptions. The uncertainty that is inherent in an afforestation or reforestation project and the management's reactions to changes in the assumptions are only dealt with superficially (Morck et al., 1989). Furthermore, analyses of profitability of forestry investments have generally focused on the forest stand level, with little attention to the fact that most forest industry firms are vertically integrated, and so investment decisions at the stand level are not determined only by the profitability of the forest stand, but the overall profitability of the firm.

Therefore, in analysing investment decisions even at the stand level it is important that the forestry firm be considered as a corporate entity in which managers make decisions that commit the firm's resources, across business lines and over time. The disjunction between stand-level analysis and firm-level management decisions has often led to unfavourable conclusions regarding forestry investments. Another problem with the DCF forest investment analyses methods is that the DCF analysis is linear and static in nature and assumes that either the investment opportunity is reversible or if irreversible it is a now-or-never opportunity (Dixit and Pindyck 1994). Consequently, the DCF techniques fail to address adequately the business valuation of growth opportunities or strategic alternatives arising from investments in large-scale commercial projects. These limitations render the conclusions of DCF valuation of somewhat suspect.

The field of forestry entails significant amounts of uncertainties, which make strategic managerial decision-making paramount. Particularly, investments in plantations are characterised by relatively large sunk costs, as well as significant amounts of risks and uncertainties in production and prices, both input and output. Due to the irretrievable nature of most forestry investments, a greater focus must be placed upon investment valuation. Previous efforts at incorporating risk and uncertainties in future prices of forestry products have focused on the determination of the optimal rotation age when prices and costs are stochastic. One of the first attempts to model changing prices and costs was McConnell et al. (1983), who developed an optimal control model of timber production, assuming that the production technology is a function of timber age, while costs and prices are functions of calendar time. An extension of this study to model the impact of evolving prices and costs on rotation length was conducted by Newman and Yin (1995). Also, Plantinga (1998) examined the role of option values in influencing the optimal timing of harvests under uncertainty and non-stationary prices. The valuation of a timber lease when prices follow a Geometric Brownian Motion (GBM) and inventory, a GBM with constant drift term was studied by Morck et al. (1989), using contingent claims analysis. Reed and Haight (1996) have estimated the mean and variance of the present value of plantations when prices and yield follow a GBM. While these studies have incorporated uncertainties, the benefits of managerial flexibility that can be obtained by the application of the real options theory have received little attention.

Only a few studies have addressed forestry investment analysis from a real options perspective. These studies include Thorson (1999), Abildtrup (1999), and Thomson (1992). When real options are present, the traditional DCF methodology may fail to provide an adequate decision-making framework because it does not properly value management's ability to revise the initial operating strategy if future events turn out to be different from originally predicted, or to account for future investments or disinvestments (Trigeorgis 1993).

Theoretical Forestry Investment Real Options Model[2]

The analysis presented in this section makes use of standard tools of option pricing and investment in discrete time, using the multiplicative binomial approach (Cox et al., 1979). The binomial process was chosen because of the advantages discussed below; although there are several other types of models that can be used (for example see Trigeorgis 1999). The binomial approach is the simplest of the option pricing formulas (Elton and Gruber 1995). With the binomial approach, it is possible to price options other than European options, like American options, which in contrast to European options can be exercised at any time up to maturity. This is because the binomial model makes it possible to check at every point in an option's life (at every step of the binomial tree) the possibility of early exercise. Also the binomial model does not depend on the probability of certain outcomes. This means that the model is independent of investors that have different subjective probabilities about an upward and/or downward movement in the underlying asset. The binomial model requires relatively less mathematical background and skill to develop and use, compared to the Black-Scholes model[3].

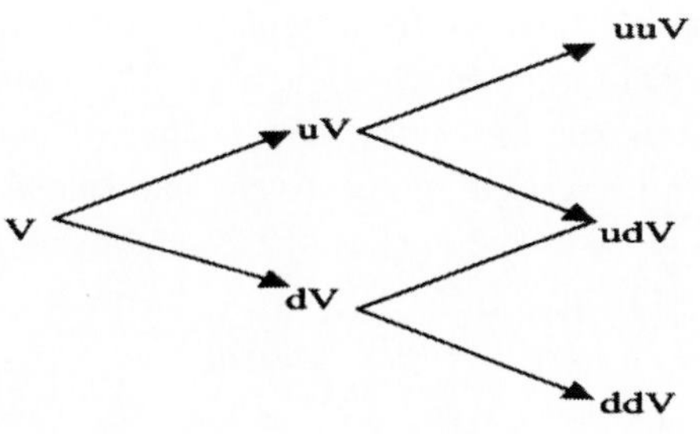

Figure 2. A Two-step binomial tree.

[2] This model was first developed and published in Duku-Kaakyire and Nanang (2004).

[3] The Black-Scholes model (Black and Scholes, 1973) is a continuous time option-pricing model that is suitable for valuing European options.

The binomial model breaks down the time to expiration of the option into a potentially very large number of time intervals or steps. A tree of the gross present value of the underlying asset is initially produced working forward from the present to expiration. At each step, it is assumed that the value of the asset will move up or down by an amount calculated using volatility and time to expiration. The tree represents all the possible paths that the asset value could take during the life of the option. At the end of the tree, that is, at the expiration of the option, all the terminal values for each of the final possible asset values are known, as they simply equal their intrinsic values. The option values at each step of the tree are then calculated working back from expiration to the present. The value of the option at each step is used to derive the option value at the next step of the tree using risk neutral valuation based on the probabilities of the project value moving up or down. Figure 2 presents an illustration of a two-step binomial tree for the paths of an asset value, *V*[4]. The binomial model assumes that the future values of the underlying asset follow a multiplicative binomial distribution over discrete periods. Also, the model assumes that the volatility of the underlying asset (σ) is constant and known and uses risk-neutral probabilities (p and $1-p$) rather than subjective probabilities for valuation.

In Figure 2, the gross project value of the underlying asset at the beginning of a given period, *V*, may increase (by a multiplicative factor *u*) with probability *p* to *uV* or decrease with complementary probability *(1-p)* to *dV* at the end of the period. From these, the risk neutral probability is given as:

$$p = \frac{\exp(r\Delta t) - d}{u - d} \tag{22}$$

$$u = \exp\{\sigma(\Delta t)^{1/2}\} \tag{23}$$

$$d = 1/u \tag{24}$$

where Δt is the time interval or step size and *r* is the risk-free interest rate. It is important to note that as $\Delta t \to 0$ the parameters of the multiplicative binomial process converge to the Geometric Brownian Motion (GBM) given in Equation (25). We use this continuous approximation to determine the parameters in Equations 23-24 after which we apply the discrete binomial process to estimate the option values. The continuous GBM assumes that the value of the project, *V*, is determined competitively and follows the exogenously given continuous stochastic process given by the stochastic differential equation:

$$dV = \mu V dt + \sigma V d\omega \tag{25}$$

where μ is the deterministic drift (instantaneous expected return of the project), $d\omega$ is the increment of a standard Wiener process and σ is the volatility, defined as the instantaneous standard deviation of *V*. The discrete version of this model is called a random walk with a drift.

[4] Within the real options section of this Chapter, V will be used as asset value, while Q will refer to volume of timber. This should not be confused with the use of V for volume in other sections or chapters of this book.

If timber prices and management costs are constant over time, then we can compute the soil expectation value (SEV) of a regenerated stand at any point in time t using a continuous time version of the Faustmann (1849) formula as:

$$SEV[P,Q_t] = \left[\frac{PQ_t - Ce^{rt}}{(e^{rt}-1)}\right] \tag{26}$$

where Q_t is the volume of timber harvested at time t, P is the price per /m^3 of timber, and C is the cost of regeneration. This is the classical Faustmann formula for calculating the present value of an infinite series of rotations, also known as the soil expectation value (SEV). The gross project value V, for an infinite series of rotations is calculated from Equation 27 as:

$$V_t[P,Q_t] = \left[\frac{PQ_t}{(e^{rt}-1)}\right] \tag{27}$$

If on the other hand, timber prices change over time, and follow a binomial process, then the expected price at time t will be used instead of the constant price, and the expected SEV is given as:

$$E(SEV) = \left[\frac{pSEV[uP_t,Q_t] + (1-p)SEV[dP_t,Q_t] - Ce^{rt}}{(e^{rt}-1)}\right] \tag{28}$$

Where $E(SEV)$ is the expected soil expectation value and $SEV[uP_t,Q_t]$ and $SEV[dP_t,Q_t]$ are the values of the stand in period t if the up and down price states have occurred, respectively. These two values of the stand are calculated using the SEV formula (Equation 26) with the up and down values respectively.

11.8. Application of Discounted Cash Flow Methods

11.8.1. Neem Plantations

Neem plantations are grown by individuals and communities in the savannah zones of Ghana to produce small timber to meet local building needs and for fuelwood. This example analyses the economic profitability of such investments, using realistic costs and variables. It compares the various criteria for assessing plantation profitability and sensitivity analyses by changing the initial assumptions to examine their impact on the results. A discount rate of 5% is assumed and volumes of wood output per year were determined from the yield equation for a Site Class I neem plantation from Chapter 6. The value of wood is GH¢10/m^3 and the cost of planting is assumed to be GH¢500/ha and occurs in year zero (this means it is not discounted). It is further assumed that there are no additional maintenance costs.

Table 2. Benefit/cost analysis of neem plantations in Northern Ghana at 5% discount rate

Plantation Age (yrs)	Volume (m^3/ha)	NPV (GH¢/ha)	SEV (GH¢/ha)	B/C Ratio	IRR (%)
1	1.03	-490.20	-10051.10	0.02	-388
2	13.67	-376.33	-3954.64	0.25	-65
3	32.35	-221.55	-1590.52	0.56	-15
4	49.77	-92.48	-510.19	0.82	0.0
5	64.46	1.99	9.01	1.00	5.0
6	76.58	67.32	259.73	1.13	7.0
7	86.61	110.34	373.63	1.22	8.0
8	94.99	136.72	414.72	1.27	8.0
9	102.06	150.76	416.04	1.30	8.0

The results in Table 1 show that all four criteria consistently conclude that the plantations only begin to make a profit in year 5. The value of the NPV is less than that of the SEV as expected because the SEV includes the value of future rotations as well. In analysing profitability using the IRR, an investor will compare the actual interest rates earned by the investment in Table 2 with the interest she or he would have earned if the money were deposited in the bank (i.e., interest rates). Hence, unlike the NPV and SEV, IRR is a relative measure of profitability. Given the high-interest rates experienced in Ghana, most economic valuation of forestry investments using the IRR criterion will indicate non-economic viability.

Sensitivity analyses were applied to assess the impacts of changes in the variables on the NPV, SEV and B/C ratio (Table 3). A 30% discount rate was used to assess the impacts of high discount rates on the profitability of the plantations. The results showed that for all criteria, the plantation investment was not profitable at this discount rate, as the benefits were less than the costs, and the B/C ratio was less than one in all years. An increase in the price of wood from GH¢10 to GH¢20 per m^3 showed that the plantation was now profitable in year 3 and with larger NPVs and SEVs than when the price was GH¢10/m^3. A final change to the initial assumptions was to increase the planting costs from GH¢500/ha to GH¢800/ha. This increase resulted in negative NPVs and SEVs, and B/C ratios less than one for all ages.

Table 3. Sensitivity of NPV, SEV and B/C ratio, to changes in the discount rate, price of wood and regeneration costs

Plantation age (yrs)	*r=30%*			*p= GH¢20/m³*			*C= GH¢800/ha*		
	NPV	SEV	B/C	NPV	SEV	B/C	NPV	SEV	B/C
1	-492.37	-1899.69	0.02	-480.40	-9850.12	0.04	-790.20	-16202.35	0.01
2	-424.99	-941.94	0.15	-252.67	-2655.12	0.49	-676.33	-7107.14	0.15
3	-368.47	-620.91	0.26	56.91	408.55	1.11	-521.55	-3744.27	0.35
4	-350.08	-500.97	0.30	315.04	1737.95	1.63	-392.48	-2165.19	0.51
5	-356.18	-458.48	0.29	503.98	2278.42	2.01	-298.01	-1347.24	0.63
6	-373.41	-447.36	0.25	634.63	2448.61	2.27	-232.68	-897.76	0.71
7	-393.94	-448.91	0.21	720.68	2440.39	2.44	-189.66	-642.24	0.76
8	-413.83	-455.12	0.17	773.45	2346.05	2.55	-163.28	-495.26	0.80
9	-431.41	-462.49	0.14	801.52	2211.87	2.60	-149.24	-411.84	0.81

Table 4. Benefit/cost analysis of teak plantations in the Guinea savannah zone at 15% discount rate

Plantation Age (yrs)	Volume (m^3/ha)	SEV (GH¢/ha)	NPV (GH¢/ha)	B/C Ratio	IRR (%)
5	22.70	-3378.69	-1782.71	0.64	6.18
10	97.55	1969.49	1530.03	1.31	17.67
15	186.09	988.39	884.22	1.18	16.09
20	273.49	-963.11	-915.16	0.82	13.99
25	355.67	-2550.63	-2490.64	0.50	12.24
30	431.79	-3600.96	-3560.96	0.29	10.85
35	502.04	-4231.86	-4209.66	0.16	9.73
40	566.91	-4589.81	-4578.43	0.08	8.82
45	626.93	-4785.38	-4779.78	0.04	8.06
50	682.65	-4889.44	-4886.73	0.02	7.43
55	734.53	-4943.72	-4942.43	0.01	6.88
60	782.99	-4971.62	-4971.01	0.01	6.42
70	871.06	-4992.94	-4992.80	0.00	5.65
80	949.21	-4998.28	-4998.25	0.00	5.05

The main conclusions from these sensitivity analyses are that increasing the discount rate and the planting costs decrease the profitability of the investment while increasing the price of output increases the economic attractiveness. It is not surprising that discount rates and costs have identical impacts on profitability because a high discount rate increases the cost of borrowing and the opportunity cost of capital tied up in the plantation venture.

11.8.2. Teak Plantations

Teak plantations are grown by individuals, communities and industry for the provision of fuelwood, small timber to meet local building needs, timber for export, and for electric and telephone transmission poles. In this example, the economic profitability of teak is analysed using realistic costs and price variables for growing teak in Ghana. Volumes of wood output per year were determined from Site Class I teak plantations in the Guinea savannah and semi-deciduous vegetation zones of Ghana from Nunifu (1997). A discount rate of 15% is assumed in the analysis, with a price of teak wood of GH¢300/m^3 and the present value of the costs of planting and maintenance is assumed to be GH¢5000/ha.

Table 5 presents the results of the economic analyses for Site Class I teak in the High forest zone, with site index 26 m at 20 years.

In Tables 4 and 5, the SEV, NPV and BCR criteria reach the same conclusions about the profitability of the plantations. The two plantations in the savannah and HFZ only make marginal profits in year 10 and 15 in the savannah zone and year 10 only in the HFZ. The profitability results mainly from the high price paid for teak wood around the world that translates into higher local prices as well. The IRR criterion shows a return on investments of up to 17% in both ecological zones. From an economic investment perspective, if the interest rate at the bank or the interest that would be earned on alternative investments is actually less than what is earned on the plantations, the latter might still be a better investment than the alternatives.

These results show that the plantation in the savannah zone was more profitable than that in the semi-deciduous zone. Although, in general, the soils in the semi-deciduous zone are more productive than those in the savannah zone, the lower profitability of the teak plantations in the HFZ examined in this example is due to their lower stocking levels compared to the savannah plantations (see Nunifu 2010). Of course, if the price of teak is less than what is used in this example, or if there is an increase in the cost of production or an increase in the discount rates, these would all result in reductions in the profitability, as shown in the example on neem above. As an extreme example, if the discount rate is 15%, the present value of the costs of planting and tending is GH¢10,000/ha, and the price of teak wood is GH¢400/m^3, then these two plantations would not make an economic profit.

11.9. Applying Real Options Theory in Plantation Investments

The purpose of this example is to show how firm-level management decisions that allow for managerial flexibility affect the value of the forest stand compared with the traditional Faustmann criterion. Consequently, I have not included the value of a wood processing plant itself, but recognise that the stand level decisions are taken in the overall interest of the processing mill. In this example, the interest is in examining the values of various options that the firm can exercise. These options include the option to delay the plantation project by up to 10 years, the option to double the processing capacity (and hence the area planted) in 20 years, and the option to abandon the plantation by 50 years if market conditions turn out to be worse than originally predicted, or the processing plant is taken over due to a corporate merger. To achieve these objectives, I describe simple models and show how the values of these options can be estimated.

Table 5. Benefit/cost analysis of teak plantations in the semi-deciduous zone at 15% discount rate

Plantation Age (yrs)	Volume (m^3/ha)	SEV (GH¢/ha)	NPV (GH¢/ha)	B/C Ratio	IRR (%)
5	28.52	-1,817.65	-959.05	0.81	10.74
10	92.27	1,514.58	1,176.63	1.24	17.11
15	155.24	-102.09	-91.33	0.98	14.88
20	211.68	-1,934.58	-1,838.26	0.63	12.71
25	261.58	-3,230.46	-3,154.48	0.37	11.01
30	305.81	-4,025.56	-3,980.84	0.20	9.70
35	345.29	-4,479.94	-4,456.43	0.11	8.66
40	380.79	-4,728.56	-4,716.84	0.06	7.82
45	412.94	-4,860.64	-4,854.95	0.03	7.13
50	442.26	-4,929.34	-4,926.62	0.01	6.56
55	469.14	-4,964.53	-4,963.23	0.01	6.07

Plantation Age (yrs)	Volume (m^3/ha)	SEV (GH¢/ha)	NPV (GH¢/ha)	B/C Ratio	IRR (%)
60	493.91	-4,982.33	-4,981.71	0.00	5.65
70	538.19	-4,995.69	-4,995.55	0.00	4.96
80	576.76	-4,998.97	-4,998.94	0.00	4.43

It is assumed that the present cost of regenerating a hectare of land is *C*, which could include the present values of stand enhancement investments such as thinning and fertilisation that occur in the future. The real cost of capital is given as *r* and is assumed known and constant throughout the planning horizon. The prices of timber, on the other hand, are assumed to follow a GBM similar to that given in Equation [25]. We can compute the gross present value of a regenerated stand at any point in time *t* using Equation [27]. The volatility of timber prices, σ was estimated as 0.10 using historical prices of teak wood on the international market and the drift term, μ. The drift term and volatility are the mean and standard deviation respectively of the time series $\ln(P_{t+1}/P_t)$. The series contained drift terms of 0.04 (i. e., $\mu \approx 0.04$). This example uses teak grown in the semi-deciduous vegetation zone in Ghana, and a rotation age of 60 years.

The current price/m^3 of timber used in the analysis was Gh¢500. A step size of 1 ($\Delta t = 1$) was used in this analysis. Smaller step sizes of Δt = 0.5 years and Δt = 0.25 years did not result in significant differences from the Δt =1 case, therefore, the results are reported for only Δt =1. To keep the analysis simple and in perspective, the example is restricted to a per hectare basis, which is a common form of stand level analysis in forestry. The various options were determined based on the backwards recursion solution technique described previously using Microsoft Excel. Table 6 summarises the information used in the binomial tree analysis. The options evaluated are briefly described below.

A. Option to Delay Plantation Development by 10 Years

This option gives management the flexibility to defer forest regeneration and therefore allows management to benefit from favourable random movements in the project value, whilst at the same time they cannot be hurt by unfavourable market circumstances, because they do not have the symmetric obligation to invest. Therefore, management will wait and make the investment within 10 years if the option value within that period turns out to be favourable. In other words, the option to delay reforestation is seen as a call option on the gross project value V_t, with an exercise price equal to the required outlay in 10 years' time, C_{10}. This gives the investor the option to choose the maximum of the project value minus the required investment, or zero since management will simply not exercise the option if project value turns not to cover the necessary costs. This can be expressed as:

$$V_D = \max[V_t - C_{10}, 0] \quad (29)$$

Table 6. Summary of input variables used in the binomial tree analysis

Variables	Values
Investment (regeneration) Cost (C) (Gh¢/ha)	1500.00
Present Value of Cash Flows (V) (Gh¢)	1452.16
Annual Volatility (%)	10.00
Risk-Free interest Rate (%)	8.00
Δt	1.00
Starting price of timber (Gh¢/m^3)	500.00
Additional Investment Cost to Expand (C') (Gh¢)	321.82
Abandonment (salvage) value (Gh¢)	3198.18

The subscripts "t" on the right hand sides of Equations (29) to (31) refer to the option values at different points in time (at different steps of the binomial tree), whilst the left-hand sides refer to final single values of the options (hence no reference to time).

B. Option to Expand Wood Processing Plant

Suppose that in this example, management has the option to expand their production (or sawmill) by doubling the size and scale. This means that instead of one hectare, the firm now regenerates an additional hectare by 20 years after the mill starts operations (or 20 years after afforesting the first hectare of land). Then by Year 20, management has the flexibility either to maintain the same scale of operation (one hectare) and receive project value V_t, at no extra cost, or double the scale and receive twice the project value by paying the additional cost, whichever is higher. That is:

$$V_E = \max\left[V_t, 2V_t - C'\right] \tag{30}$$

where C' is the present value of the additional cost of reforestation and V_E is the option value of the project when exercised.

C. Option to Abandon

We incorporate the option for management to abandon the sawmill (and consequently the reforestation project). This could happen if the sawmill is abandoned in exchange for the salvage value of the mill and forest if market conditions turn out to be unfavourable or there are corporate mergers or takeover. The option to abandon mill operations for any number of reasons before the rotation age (50 years) will result in a sale of the forest stand. The option to abandon the project early in exchange for the salvage value translates into the flexibility to choose the maximum of the project's present value in its present use, V_t, or its value in its best alternative use, A. We should notice that if the trees have no alternative use, then A = 0. However, if the processing plant is being sold to another firm, then the trees are also sold and, therefore, A can have a positive value. Since it is difficult to tell what A will be in reality, we set A to zero and so the expression for the option to abandon the mill is given as:

$$V_A = \max\left[V_t - C,\ A\right] \tag{31}$$

Computing the Option Values

The details of calculating the various option values using the binomial method require a large number of steps that result in extremely large binomial trees. It is therefore not possible to include all details of calculations here. The backwards computations use the following equation:

$$V_{ot} = Max\left[\frac{puV_{o(t+1)} + (1-p)dV_{o(t+1)}}{(1+r)},\ \ V_t - C\right] \tag{32}$$

where V_t, refers to the stand values, V_{ot} is the value of option at time t, $uV_{o(t+1)}$ is the *up* value of the option in the previous year, and $dV_{o(t+1)}$ is the *down* value of the option in the previous year.

Table 7 shows the descriptions of the different types of models estimated, the NPVs and the option values for the real option models. The calculated SEV using the Faustmann formula for the reforestation project is negative, which suggests that the investment is not profitable to undertake. Valuing the project with its inherent options presented more favourable results. Using the binomial tree approach discussed earlier, the option to defer afforestation/reforestation by 10 years alone is valued as an American call option on the project, with an exercise price equal to the necessary investment outlays. As shown in Table 5, this option increases the project's value to \$406.52/ha, in contrast to the no flexibility base case SEV of -Gh¢60.29/ha. Alternatively, the value of this option is \$466.81/ha (project value = SEV +option value). This result indicates that waiting for the 10 years prior to exercising the option might improve timber prices, but might not necessarily resolve all the risks and uncertainties associated with the plantation investment. However, this is an option that could be valuable for this investment if exercised.

Table 7. Summary descriptions, project values, and the option values of the models examined

Model #	Model description	Project value (Gh¢/ha)	Option Value* (Gh¢/ha)
1	Faustmann, with no price volatility	-60.29	0
2	Option to delay regeneration by 10 years	406.52	466.81
3	Option to double mill size by 20 years	1247.26	1307.55
4	Option to shut down mill by 50 years	1685.74	1746.03
5	Multiple options: Options 2, 3, and 4 combined	1685.74	1746.03

* Note: the option values are the differences between the project values of the other models and Model #1.

Similarly, the option to expand the operation scale is worth Gh¢1307.55/ha. This option is valued analogous to American call option to expand by paying an extra outlay as the exercise price. The present value of the additional investment required for the expansion was calculated to be approximately Gh¢321.82/ha. Including this option increases the project's expanded value to Gh¢1247.26. Should the market show improvement above management's initial expectations, this is an option that management might want to consider seriously exercising.

The option to permanently abandon operations for salvage value is also valued as an American put option with an exercise price equal to the salvage value. The salvage value was realised from the sale of the standing timber at age 50 years and was estimated as Gh¢3198.18. The option to abandon was Gh¢1746.03, which is worth a lot to this project due to the high salvage value. It is an option that should be exercised for this particular project if market conditions deteriorate severely, or if for other reasons the wood processing plant has to go out of business.

As pointed out previously, the value of the option in the presence of others may differ from its value in isolation. The presence of subsequent options increases the effective underlying asset for prior options (Trigeorgis 1993). Moreover, exercise of a prior option may alter the underlying asset and the value of subsequent options on it. Thus, combining all the options and valuing them together could lead to a result that is different from the sum of their isolated values. It could be higher when there are positive interactions among the options or lower when interactions are negative. With this knowledge, all the three option were combined and valued together as multiple options. When all the three options are combined, their value is estimated to be Gh¢1746.03, which is equal to the option to abandon. Although the value is less than the sum of the isolated values, the combined flexibility that they afford management may be economically significant. The multiple options happened to have the same value as the abandon option because the years between the options are large and the valuation is done such that the maximum value is picked among the lot.

The results show that the inclusion of managerial flexibility can affect project value drastically and lead to different conclusions regarding the profitability of forestry investments. Stand values calculated when prices follow a diffusion process are higher than the Faustmann soil expectation value as the former method values the flexibility of forest management and investment decisions. The option theory approach holds great promise for use in forestry because static financial analyses often lead to conservative conclusions regarding the viability of investing money in forestry projects. Most forestry projects, when compared with other land use options are often at best of marginal profitability. To remove this bias against forestry projects, the values of price uncertainties and managerial flexibility need to be incorporated into the analytical framework.

Although this analysis was restricted to three options, there are many other managerial options that can be analysed in forestry investments. For example, it should be possible to examine the option to switch the species planted after the first rotation (or after several rotations) if market conditions turn out to be unfavourable. In the traditional Faustmann framework, it is often assumed that the same species is planted on the same piece of land *ad infinitum*. Furthermore, silvicultural investments in fertilisation and thinning may also be evaluated as options for forest industry firms.

Despite the usefulness of the option theory approach, its applicability in certain circumstances requires that some difficulties be overcome. For example, how to incorporate effectively some risks, especially risks associated with security of forest tenures remains a challenge. Whilst in theory we can include risk by estimating a risk premium and adding it to the chosen risk-free interest rate, estimating the risk premium for qualitative variables such as tenure security and other legal operating environment variables is complicated at best. It should also be emphasised that these results are sensitive to the risk-free interest rate and volatility of the timber prices. Future research could focus on quantifying the interactions among the different options and determine where the option interactions are small and, therefore, simple option additivity could be a good approximation, and where high interactions will seriously invalidate options additivity.

11.10. CONCLUSION

Plantation forestry is fundamentally an economic venture and the ability to make a profit from forestry plantations may be the single most important determining factor as to whether individuals and companies will undertake plantation ventures. Ex-ante economic analyses are indispensable in evaluating the economic viability of the plantation enterprise. In cases where it is not possible to make profit, government intervention in the form of subsidies or grants may make the venture viable, but as we have seen in Chapter 3, these kinds of external support may not be sustainable, and hence in the long run, may be counterproductive. Economic sustainability depends on species, site, economic conditions, markets, and management practices. The long-run sustainability of plantation forestry requires that these factors be seriously considered in the establishment and management of plantation forests to ensure that they are by themselves independently sustainable.

Chapter 12

THE FOREST ROTATION DECISION

In this Chapter, the age-old question of the optimal rotation age in forestry is discussed, beginning with the economic foundations for determining the optimal rotation of a stand of trees. Following this, the various models in the literature for deciding the optimal rotation are described. Data from teak and neem plantations are used to present two case studies of optimal rotation ages under various combinations of prices, costs, discount rates, and risks of a forest fire.

12.1. BEHAVIOUR OF ECONOMIC AGENTS

The optimal forest rotation question has been debated by foresters and economists for many decades. The optimal rotation problem can be defined as finding the optimal output and input mixes of forest management to maximise the goods and services produced by the forest. There are two main areas of economics that are relevant to solving this optimisation problem: capital theory and the theory of the firm.

Standard economic theory assumes that individuals choose actions that optimise their expected utility, which is the amount of satisfaction derived from consuming goods and services. Capital theory analyses the links between the theories of production, growth, value and distribution to explain why capital produces a return that keeps capital intact yet yields interest or profit that is permanent (Jorgenson 1963). According to the neoclassical theory of capital, a production plan for the firm is chosen so as to maximise utility over time. Under certain well-known conditions, this leads to maximisation of the net worth of the enterprise as the criterion for optimal capital accumulation (Jorgenson 1963).

Capital theory, as exemplified by the theory of the firm, is important in understanding how resources are allocated to produce goods and services in any forest enterprise. There is clearly a link between the decisions of economic agents and the forest rotation problem. The neoclassical economic theory of the firm is the basic starting point for understanding the kinds of economic decisions that are made by firms and individuals in forest plantation management. Even if the plantation is not managed for industrial purposes by a firm, the concept of profit (or utility/benefit) maximisation is still central to the type of decisions that are made by those who invest in forest plantations.

12.1.1. The Neoclassical Theory of the Firm

In neoclassical economic theory, the firm is an entity, which maximises profit (minimises cost) subject to constraints imposed by its technological capabilities. In a static framework, the firm may be modelled with either profit maximisation (where profit is defined as revenue minus cost) or cost minimisation. The assumption of cost minimisation is consistent with profit maximisation because if a firm operates to maximise profits, it will, after selecting an output level, select inputs to minimise costs. It is, however, a weaker assumption because cost minimisation is concerned only with the choice of inputs, while profit maximisation deals with the choice of both inputs and outputs. This basic model is based on several assumptions including perfect competition, the firm being a price taker in input and output markets, perfect information, a large number of buyers and sellers, etc. Given these assumptions, the firm solves the following basic problem:

$$Max \quad \pi = R - C \tag{1}$$

Where π is the profit, R is revenue received from the sale of outputs, and C represents cost the firm incurred in the production process. According to this theory, firms maximise profits (Equation 1) by operating at the point where marginal revenue equals marginal cost. In a perfectly competitive market, marginal revenue equals the price of the commodity. In a monopolistic situation, however, the firm maximises profits by setting price greater than marginal revenue. Without accounting for opportunity costs of the factors of production used in Equation 1, we measure the accounting profits, rather than economic profits.

In this static framework, factors of production are classified into fixed and variable factors, that is, the shorter the run, the fewer the number of factors that can be varied. One criticism of this static model is that they focus on conditions that prevail when full static equilibrium is achieved (long-run) to the exclusion of the adjustment process (Fernandez-Cornejo et al., 1992). To overcome this problem, restricted profit and cost functions were introduced, where one or more inputs are fixed. However, in a forestry context where costs and benefits occur at different points in time, and inter-temporal decisions are made, the static model is inadequate in modelling such a process and will, therefore, need a dynamic framework.

In the dynamic approach, a firm considers a multi-period horizon and inter-temporal allocation of resources to be an integral part of the analysis (Fernandez-Cornejo et al., 1992). According to Treadway (1971, 1974), dynamic models of the firm should incorporate both dynamic optimisation and the idea of adjustment costs. Brechling (1975) showed that multi-period profit maximisation (or cost minimisation) models that do not include adjustment costs result in static decision rules, even if the firm has multi-period objectives, whilst a model with adjustment costs lead to multi-period decision rules. Adjustment costs may be due to institutional requirements (e. g., government regulation of the forest industry), market imperfections, or internal causes. Brechling (1975) further indicates that internal adjustment causes may be associated with integrating new capital equipment into on-going operations, or reorganising production lines. In general, therefore, the long-run solution derived from dynamic models within the cost of adjustment framework is different from the long-run solution using a static optimisation (Treadway 1971; 1974). Using this dynamic theory, we can state the problem of the firm as trying to:

$$Max \quad \beta^{t-1}\left[\sum_{t=1}^{T}(R_t)-\sum_{t=1}^{T}c_t\right] \tag{2}$$

whereas before, R_t is revenue and c_t represents the cost in period t. These are all discounted using a discount factor, β.

12.1.2. Extensions to the Basic Model of the Firm

The basic assumption of the forest industry firm is the same as that of any other firm; that it seeks that harvest rate through time subject to the constraints that it faces that maximises the present value of profits. Extensions of the basic models (static and dynamic) of the firm to a forest industry firm require that the institutional environment in which the firms operate be taken into consideration. When this is done, the model will be modified appropriately to allow for a correct analysis of the firm. For the purposes of this discussion, a forestry firm will be considered as a vertically integrated[1] firm that is involved in the production of forest crops, which are used as inputs in the manufacture of wood products.

There are certain characteristics that differentiate a forest industry firm from other manufacturing firms. Nautiyal (1988) discusses some of the following important decisions that face forestry firms. First, most of the production decisions of the forestry firm will bear results after a relatively long period. Comparisons of different costs and benefits over time mean that the notion of profit maximisation in the classical theory of the firm has to be replaced with maximisation of net present value (or net present worth). Secondly, the decision- making process is dynamic, that is, decisions taken today can have effects several years later. For example, stand management decisions and investments in silviculture (e. g., thinning) affect future growth and yield of forest crops. However, most of these effects cannot easily be determined at the time of making the decision, which leads to a considerable amount of uncertainty for the firm. The major characteristics of the forest industry firm that distinguish it from other firms and how these factors modify the classical theory of the firm are now discussed.

Joint Production

Joint production refers to a situation where a single production process results in more than one output. The process of growing forest trees, for example, may also lead to the production of wildlife, recreation opportunities, water conservation, etc. Even if one use seems dominant (usually timber production), there are often other uses, even if these are not explicitly recognised. Even for timber production alone, the idea of joint production occurs where different species mixes are produced Nautiyal (1988). These different species and the end products they can be used to manufacture can be considered as different products. Because of the inevitability of joint production in forestry, the forestry firm must attempt to

[1] Vertical integration can be differentiated from horizontal integration. In the later, firms produce outputs at the same time from the same production process e. g., lumber and sawdust or timber and recreation.

produce an optimal combination of outputs that maximises profits or minimises costs, and so it is possible to modify the theory of the firm to handle multiple outputs.

The determination of the optimal product mix requires that all products be quantifiable and measurable. The discussion on the quantification and measurement of non-timber benefits was provided in Chapter 11. We can represent the trade-off between different outputs using a production possibility frontier (PPF), given in Figure 1. The PPF shows the various combinations of the outputs that maximise the returns to the investment. For example, it shows the optimal level of timber and non-timber outputs to be X and Y respectively. The required modification to the theory of the firm will, therefore, be to maximise net present value overall outputs. That is,

$$Max \quad \beta^{t-1}\left[\sum_{p=1}^{P}\sum_{t=1}^{T}(R_{pt})-\sum_{p=1}^{P}\sum_{t=1}^{T}c_{pt}\right] \tag{3}$$

Where R_{pt} is revenue received from the sale of product p in period t, and c_{pt} is the cost of producing product p in period t. The revenues and costs are summed over all time periods ($t=1....T$) and outputs (p=1,...,P). The problem facing forestry firms is not only that of joint production but also, joint costs. Joint costs refer to the difficulty of attributing inputs to the production process unambiguously to one or the other output (Nautiyal 1988). In a forestry context, it could be the use of inputs to produce both timber and recreational opportunities, or in the production of lumber and chips (which is a by-product of the lumber production process).

An interesting question, however, is whether the vertically integrated forestry firm should maximise net present value from each of the two production processes separately, or to maximise returns to both processes together. Nautiyal (1988) argues that the forestry firm will make more profit by maximising the returns from each process than by combining the entire enterprise. Therefore, vertical integration is not desirable if there are good markets for logs, for example. However, given that mills have to process their logs or sell logs in local markets due to the ban on log exports, vertical integration for these firms is inevitable in many cases. Luckert and Haley (1991), also note that in most cases, investments in log production alone is not profitable, however, by combining with the manufacturing process, the overall firm can make some profits.

Externalities

One of the cornerstones of economic theory is that, under certain conditions, markets will produce Pareto efficient allocation of resources. A Pareto optimal allocation is one in which at least one person is made better off and no one is made worse off. When conditions prevail such that Pareto optimal solutions are not obtained a market failure occurs as a matter of definition. One of the causes of market failures in forestry is externalities. Externalities occur when an economic agent affects the production possibilities or the utilities of others in a way that is not reflected in the market. Externalities are common in forestry due to joint production of multiple outputs. In the case of growing trees, a forestry firm may at the same time improve habitats for wildlife (external economies) or destroy the habitat that results in reduced wildlife (external diseconomies). Also, in harvesting one tree species from a land base that contains other tree species, some of the species not being harvested may be damaged leading to negative externalities. Another example of an externality relating to an integrated forestry firm is that of effluent emissions from processing (especially pulp) mills. To reduce this externality means firms have to spend extra money on pollution control. Theoretically, it is possible to eliminate most externalities in forestry by changing property rights arrangements. For example, negative externalities against wildlife can be internalised by expanding the rights of tenure holders to include wildlife, rather than the rights to just timber alone.

Nonmarket Goods and Services

As discussed above, the joint production of outputs for the forest industry firm is the rule rather than the exception in many cases, and so we will have to deal with non-timber values. Associating monetary values to non-timber benefits has been a challenge to economists for some time. In terms of maximising net present value for a forestry firm, the only complication presented by the production of non-timber benefits is how to value these outputs. Once a monetary value is assigned to these benefits, the firm then simply maximises the NPV of the combined timber and non-timber benefits. That is, the firm then solves the following problem:

$$Max \quad \beta^{t-1}\left[\sum_{t=1}^{T}(RTB_t + RNTB_t) - \sum_{t=1}^{T} c_t\right] \tag{4}$$

where:

RTB_t = *revenues* received by firm in period t from *harvesting and processing* timber;

$RNTB_t$ = *revenues* received by firm in period t from non-timber resources;

c_t = joint costs of harvesting and processing timber and providing non-timber benefits in period t.

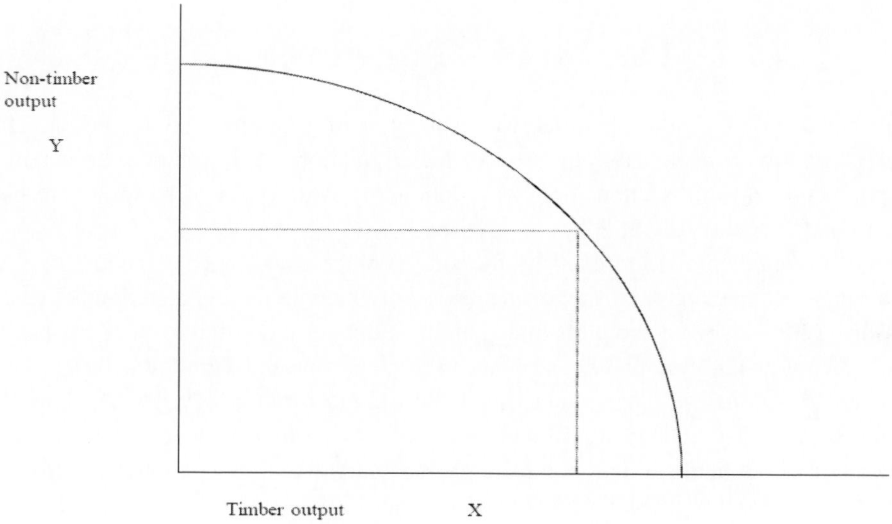

Figure 1. A production possibilities frontier curve for timber and non-timber outputs.

12.2. The Concept of a Forest Rotation

The decision by a landowner as to when to harvest a forest plantation affects the flow of raw material to the firm, revenues from the sale of the raw material as well as future regeneration decisions. A rotation in forestry refers to the number of years between the time a stand regenerates and the time of final felling (or final harvesting). The rotation age is then the age of the plantation when final felling occurs. The rotation length is an important tool for controlling tree size in forestry: the longer the rotation, the larger a tree can be grown. However, the rotation length also markedly influences yield, profitability, and regeneration methods (Evans and Turnbull 2004). The correct rotation age is the age when the utility from the forest is maximised (capital theory).

The question of the correct optimum rotation of a forest stand is as old as forestry itself and has not been without controversy. Historically, foresters have used the age at which the mean annual increment (MAI) peaks as the basis for deciding the optimal rotation age of a stand. Though the practice is still in use today, economists have argued that this rotation is too long to be efficient and that it ignores the opportunity costs of labour and capital (Anderson 1992). Many different economic models have been proposed for the determination of the optimum rotation age varying in their basis and degree of sophistication (e.g., Faustmann 1849; Fisher 1930; Anderson 1976). The large number of approaches can be taken as an indication that no one criterion has received more than partial recognition. One reason for the difficulty of the rotation question stems from the fact that investments and benefits in forestry are spread over the growth cycle of the forest. The uncertainty involved in predicting accurately inputs and outputs in monetary and physical terms (prices, costs, yields, etc.), interest rates, and other risks well into the future (sometimes about 100 years or more) are factors which compound the difficulty of the optimal rotation question. Furthermore, forests are capable of providing both timber and non-timber products. These non-timber values in most circumstances are difficult to estimate and include in the optimal decision framework.

12.3. Types of Rotations

Evans and Turnbull (2004) identified four categories of rotations used in forestry. These are physical rotations, technical rotations, financial rotations, and rotation of maximum volume production. In this section, brief descriptions of these types of rotations are given (after Evans and Turnbull 2004).

Physical Rotations are determined by the site or other environmental factors that may prevent a stand reaching a certain size or maturity. For example natural disturbances such as cyclones or bushfires, severe droughts and shallow soils and the build-up of lethal pests or pathogens can physically limit the size a stand can reach (Evans and Turnbull 2004).

Technical Rotations are those planned to yield the most output of a specified size and type of forest product to satisfy a particular end-use, which may include lower and upper limits (Evans and Turnbull 2004). For example on individual and community plantations grown for fuelwood and small-scale construction timber, short rotations are preferred because the products can easily be harvested and handled with locally available implements. Moreover, these products are harvested to meet the local demand for small wood. In other tropical areas where fast grown exotic species are grown for pulpwood, the rotations are determined based on the technical qualities of the wood produced.

Rotations based on economic criteria are determined by the rotation age that yields the highest financial return under a particular set of circumstances. Economic analyses may be done in several ways using different kinds of criteria such as the soil expectation value (or the net present value), the internal rate of return or the benefit/cost ratio. However, these criteria produce different results regarding the profitability of the forest enterprise, depending on the assumptions.

The maximum sustained yield rotation (also called the biologically optimum rotation or rotation of maximum MAI) is the rotation that yields the greatest average annual production of timber. In terms of wood yield, it realises the full growth potential of the site (Evans and Turnbull 2004). This rotation is reached when the annual growth increment of the stand falls to the level of the overall mean annual increment. In addition to the above, there are also a few other *minor* factors determining rotation age such as silvicultural reasons, opening of new markets, failure of anticipated markets, inaccessibility of stands, uneven age-class distribution, etc. (Evans and Turnbull 2004).

12.4. OPTIMUM ROTATION MODELS

A review of the literature reveals a wide variety of models used in determining the optimal forest rotation. For example, a major review by Newman (1988) showed at least six different approaches. However, only a handful of these are frequently used, and so only these will be discussed in this section. Unless otherwise stated, the following are some of the basic assumptions used in the analysis of the various rotation models presented in this review. These assumptions are made in part to ease the mathematical manipulations but more importantly because the world of 19th century Germany, where the models were first presented, fit many of these static assumptions (after Newman 1988).

- *Even-aged management:* Starting from bare land, a single stand of trees is grown and all trees are cut at the same time (rotation age). The land has unchanging growth potential, and the technology available for growing trees does not change. The land is capable of being regenerated at the same fixed costs instantaneously after harvest. A regulated forest was also shown by Faustmann (1849) to give the same results as a single stand. Chang (1983) and Hall (1983), among others, have shown that the optimal conditions will change under uneven-aged stand management.
- *Perfect certainty of future growth and yields:* There must be perfect certainty regarding the stand's growth function, future market prices, interest rates, and costs. All are assumed known and constant, or at least predicted with certainty.
- *Access to capital markets*: It is assumed that there is unlimited access to capital markets and money can be borrowed or lent at the same interest rate, r.
- *Constant net stumpage price*: Net stumpage price is not a function of tree quality but rather is constant per unit of volume produced. This assumption is needed because multiple optima can arise when management objectives favour different products or

when discontinuous jumps in prices create non-concave sections in the total revenue function (Newman 1988).

It is possible to vary these assumptions through sensitivity analyses to assess the impacts of changes in interest rates, costs, prices, and changing technology on the optimal forest rotation.

12.4.1. Maximum Sustained Yield or Maximum Mean Annual Increment

In forestry, a popular rule for determining the optimum rotation of a stand is the age at which the current annual increment (CAI) intercepts the mean annual increment (MAI). The annual allowable cut is often set based on the maximum sustained yield (MSY) concept. This is the point that maximises the MAI (shown by point t^* in Figure 2). The MAI is given by

$$MAI = \frac{V(t)}{t}$$

and the CAI by

$$CAI = V'(t) = \frac{dV(t)}{dt}$$

The optimum rotation according to the MSY is to cut the trees when the current annual increment equals the mean annual increment, CAI = MAI. That is, $\frac{V(t)}{t} = V'(t)$ where V(t) is the volume production at time t (plantation age in years) in both cases. The optimal rotation age under this framework is based only on biological considerations and is not affected by changes in market conditions and hence is relatively easy to apply.

12.4.2. The Net Present Value (NPV) or Single Rotation (Fisher) Model

The maximum sustained yield criterion does not include any economic considerations. When regeneration and other silvicultural costs and interests on the capital tied up in the forest are considered with a finite number of rotations, the decision rule for optimal rotation changes. The NPV model calls for the maximisation of the net present value (NPV) of one rotation.

$$\max NPV = \frac{pV(t) - C}{e^{rt}} \tag{5}$$

Using calculus, at the optimal point, the first derivative of the function equals zero. Hence, this requires that at the optimum rotation age,

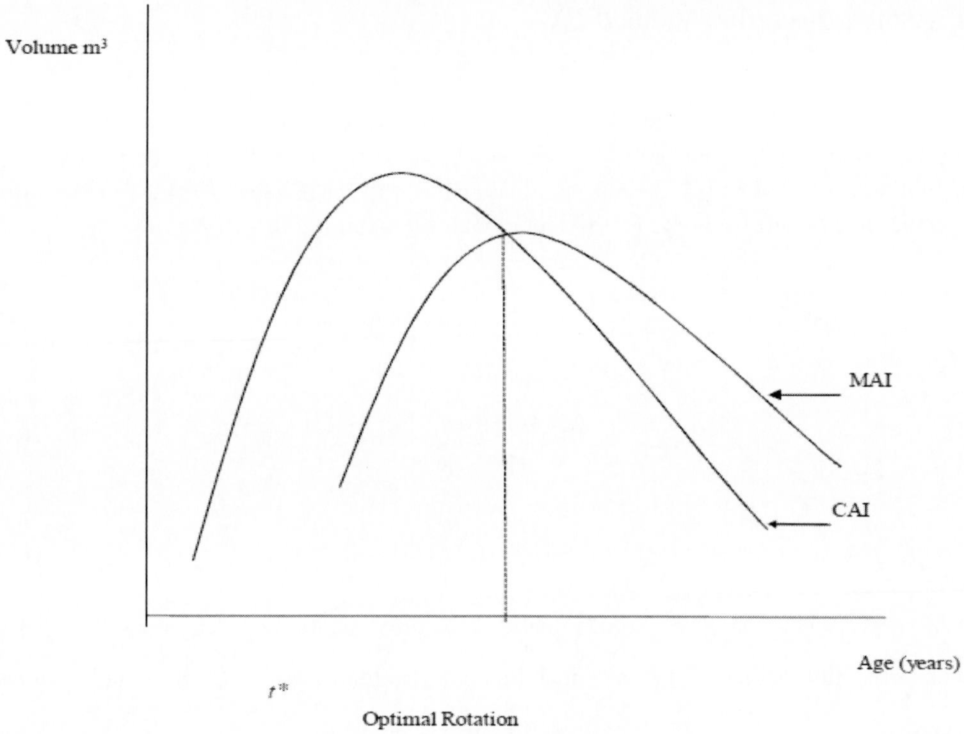

Figure 2. An illustration of the optimal rotation decision in the MSY model.

$$\frac{dNPV}{dt} = 0 \tag{6}$$

$$\Rightarrow \frac{dNPV}{dt} = Y'(t)e^{-rt} - re^{-rt}Y(t) = 0 \tag{7}$$

$$\frac{Y'(t)}{Y(t)} = r \text{ or } Y'(t) = Y(t)r \tag{8}$$

where:

Y(t) = V(t)P(t), the stumpage value of a stand t years old, P(t) is today's stumpage price for trees of various ages; V(t) is the timber volume of the stand at age t, Y'(t) = dY(t)/dt

r = discount rate

t = rotation age

This model states that the forest should be harvested when the rate of value increment equals the interest rate.

12.4.3. Soil Expectation Value (SEV)

The NPV criterion is appropriate if we are evaluating the optimal rotation age for a single rotation of crop trees, while the SEV or Faustmann model is used when evaluating the optimal rotation age for an infinite series of rotations. The following derivation and interpretations of the optimisation and marginal analysis follow Chang (1984). To determine the optimal rotation age using the Faustmann model, we maximise the SEV:

$$\max SEV = \frac{pV(t) - C}{e^{rt} - 1} - C \text{, which requires that } \frac{dSEV}{dt} = 0 : \tag{9}$$

$$\frac{dSEV}{dt} = \frac{Y'(t)(e^{rt} - 1) - re^{rt}[Y(t) - C]}{(e^{rt} - 1)^2} = 0 \tag{10}$$

$$\Rightarrow Y'(t)(e^{rt} - 1) - re^{rt}[Y(t) - C] = 0 \tag{11}$$

Where $Y(t) = p_t V_t$ (total revenue). Equation (11) provides the necessary condition and a sufficient condition is that: $\frac{d^2 SEV}{dt} < 0$ (since this is a maximum point). Equation (11) can be re-written as:

$$Y'(t) = \frac{re^{rt}[Y(t) - C]}{(e^{rt} - 1)} \tag{12}$$

From (11), $Y'(t)$ which is equal to $\frac{dY(t)}{dt}$, represents the marginal revenue product (MRP) of waiting out the rotation. The right-hand side represents the marginal input cost (MIC). The marginal analysis interpretation of the optimum rotation criterion is that if the MRP of letting the stand grow one more year is greater than the MIC of doing so, one should allow the stand to grow another year. On the other hand, if the MRP of letting the stand grow one more year is less than the MIC of doing so, one should harvest the trees. The above marginal rules can be written as:

If $Y'(t) > \frac{re^{rt}[Y(t) - C]}{(e^{rt} - 1)}$, let the stand grow another year;

However, if $Y'(t) < \frac{re^{rt}[Y(t) - C]}{(e^{rt} - 1)}$, then harvest the trees.

The MIC includes the cost of holding the growing stock and the cost of holding the land for future rotations.

$MIC = \frac{re^{rt}[Y(t)-C]}{(e^{rt}-1)}$ can be separated into these two costs as follows:

$$MIC = rY(t) + r\left(\frac{[Y(t)-C]}{[e^{rt}-1]} - C\right) \tag{13}$$

where $rY(t)$ is the cost of holding the growing stock - the amount of interest payment the landowner would get if he sold the stand at year t and invested the money at an interest rate r for one year, and the last term on the right-hand side is the cost of holding the land - the soil rent the landowner should charge himself for waiting one more year (Note that this is equal to $rSEV$).

Chang (1984) notes that past attempts at explaining the optimum rotation age determination in terms of marginal analysis were unsuccessful because $Y'(t)$ has traditionally been *mislabelled* as marginal revenue (MR). Since MR means change in total revenue per unit change in total output, while MRP is defined as the change in total revenue per unit change in input, and time is an input to, rather than an output of, timber production, Chang (1984) contends that $Y'(t)$ should be called MRP rather than MR. Secondly, the use of the discrete discounting version of the SEV formula does not enable one to separate the MIC into the cost of holding the trees and the cost of holding the land (Chang 1984).

12.4.4. Relationship between the Models

Relationship of the MSY Model to the Faustmann Model

When *p(t),* the stumpage price is a constant price (p) for all age classes, the cost of regeneration C is zero, and the interest rate is zero, maximising MAI is the same as maximising SEV.

$$\max SEV = \frac{pV(t)-Ce^{rt}}{(t-1)} = \max \frac{V(t)}{t} \tag{14}$$

In general, the optimal rotation determined from the MSY approach results in a rotation age longer than the Faustmann rotation. However, Binkley (1987) showed that if $r \leq 1/t^*_{MSY}$, where r is the discount rate and t^*_{MSY} is the optimal rotation determined from the MSY model, then, the MSY model gives a shorter rotation than the economic rotation models.

Relationship of the Single Rotation Model to the Faustmann Model

We can re-write the optimum condition of the Faustmann principle as:

$$Y'(t) = rY(t) + rSEV \tag{15}$$

We recall that the optimum condition for the NPV model is given by as:

$$Y'(t) = rY(t) \qquad (16)$$

This implies that when SEV = 0, the NPV and Faustmann models are the same. The NPV model disregards all the income that could be generated by future rotations and hence is always smaller than the SEV in absolute terms. Also, the NPV model does not include the opportunity cost of holding the land, leading to a longer rotation than the Faustmann model. Finally, this model shows that the regeneration cost has no effect on the determination of the optimal rotation age. The two formulae for computing the SEV and the NPV are reproduced below.

$$SEV = \frac{pV(t) - Ce^{rt}}{(e^{rt} - 1)} \text{ and } NPV = \frac{pV(t) - C}{e^{rt}} \qquad (17)$$

Further comparisons of the two formulae show that when t is small, the difference between (e^{rt} -1) and e^{rt} is large in proportion and the difference between the rotation lengths determined by the Faustmann and NPV models become very pronounced. However, for large values of t, the difference may be negligible.

12.5. Effect of Changing Prices and Costs on the Optimal Rotation

Comparative statics in the Faustmann model shows that if timber prices rise, then the harvest rotation lengths shorten and regeneration inputs used in production increase. Conversely, if input costs rise, then rotations lengthen and regeneration inputs decline (Jackson 1980; Hyde 1980; Chang 1984). However, if prices and costs change simultaneously over time, general insights cannot be gained from the Faustmann model because of its inability to incorporate comparative dynamics (Yin and Newman 1995).

If a is the rate at which prices and costs change, where a is a constant not greater than 1, then it can be shown that the rotation lengths do not change over time, as there will be no shift in the production function. The only effect of this equiproportionate change is a reduction in the discount rate from r to $r_1 = r(1 - a)$. This gives the Faustmann solution with a reduced discount rate. Yin and Newman (1995) deduce the following from the above analysis, based on the assumption that the changes in prices and costs are less than the discount rate itself: (1) if prices and costs change at the same rate, then input use and rotation length do not change; (2) if prices and costs increase proportionately over time, the effective discount rate is reduced by the same proportion; (3) different proportionate changes in prices and costs correspond to different effective discount rates and, therefore, different rotation lengths.

12.6. Incorporating Non-Timber Values

The above analysis of the optimum rotation age determination assumed that the forests are only capable of producing timber. However, there is increasing interest in incorporating the non-timber benefits that forests provide during their growth cycles into optimum timber rotation models. Harvest timing models have, therefore, been developed that incorporate both timber and non-timber values into the rotation decision.

Calish et al. (1978) incorporated non-timber values by assuming that the standing timber volume Y(t), provides a flow of benefits each period given by Z[Y(t)]. When this is done, there are now two advantages to holding a stand of timber during the current period: the increase in stumpage value caused by the growth of forest volume and the flow of non-timber benefits. The cost of holding the stand is, therefore, the carrying cost of the timber and occupied land. The optimal harvesting point occurs when

$$Y'(t) + Z[Y(t)] = rY(t) + rSEV \tag{18}$$

In addition, the value of the bare land (SEV) on the right-hand side of Equation (18) must now include both the value of future timber harvests from bare land and the value of the future non-timber benefits that will be provided by the future sequence of forests (Anderson 1992).

If $Z^* = Z[Y(t)]$ and the value of bare land due to non-timber benefits is A" = Z*/r, then (18) can be written as

$$Y'(t) + Z^* = r[Y(t) + SEV + Z^*/r] \tag{19}$$

The terms involving Z* cancel out so that the optimal rotation is unaffected by adding non-timber benefits to the timber optimisation framework. The only time that addition of non-timber values will lengthen the rotation is when the flow of non-timber benefits depends positively on the stage of the forest growth cycle such that forests containing larger timber volumes provide a larger stream of benefits to users (Newman 1988). In this case, the benefits exceed the carrying cost, provided the timber growth rate is positive (Anderson 1992).

Hartman (1976) derives an optimal harvesting rule in the situation where the amenity streams never diminish with age. If the landowner can capture non-timber and timber benefits, the optimal rotation is found by maximising:

$$B(t) = \frac{\int_0^t \gamma(t)e^{-rt}dt + pV(T)e^{rT} - C}{(1-e^{-rt})} \tag{20}$$

Since γ(t) is the flow of amenity values accruing to the standing forest, the integral of γ(t) over the time between rotations sums up and discounts the value of future benefits due to delaying harvesting. Using first derivatives, the first order conditions for the optimal rotation result in the following decision rule at the optimal rotation age:

$$\frac{pV'(T) + \gamma(T)}{pV(T) - C} = r\left[\frac{1}{1-e^{-rT}} + \frac{\int_0^t \gamma(t)e^{-rt}dt}{[pV(T)-C](1-e^{-rT})}\right] \tag{21}$$

The optimal interpretation of this condition is that the trees should be harvested when the rate of value increment of both timber and non-timber benefits equal the interest rate. Hartman (1976) has shown that if the amenity values are small, then the Faustmann rotation approaches the Hartman model. If however, the non-timber benefits are large the optimum rotation would be longer or shorter than the Faustman model depending on whether the non-timber values increase or decrease with forest age. In a situation where the non-timber benefits are very large, the optimal decision may be to not harvest the trees (optimal rotation age of infinity).

Plantations established in Ghana provide a wide range of timber and non-timber benefits that can be incorporated into the optimal decision framework. With regards to timber benefits, the obvious ones are wood for timber, construction, and fuelwood and charcoal processing. Non-timber benefits include fodder for animals, fruits, gums, honey, control of soil erosion, windbreaks, provision of shade, improvement in soil fertility, carbon sequestration, etc. If these values can be monetised, they can be included in Equation 21 to determine the optimal rotation age that optimises total benefits.

12.7. Effect of the Risk of a Fire on the Optimal Rotation

Wildfires are an annual occurrence in most parts of Ghana, especially in the savannah vegetation zones. In analysing the optimal forest rotation problem, the risk of forest fires is an important aspect since it affects the decisions by landowners. We can rewrite the optimal condition for the Faustmann model as:

$$\frac{pV'(t)}{pV(t)-C} = \frac{r}{e^{rt}-1} \tag{22}$$

Conrad and Clark (1987) derive the optimal rotation age decision rule when there is a constant risk of fire equal to λ per unit time. It is assumed that the probability of fire follows a Poisson distribution with a parameter λ. Using these assumptions, the optimal rotation condition is given by (Conrad and Clark 1987) as:

$$\frac{pV'(t)}{pV(t)-C} = \frac{r+\lambda}{e^{t(\lambda+r)}-1} \tag{23}$$

This optimal rotation rule effectively adds the probability of fire to the discount rate. If the probability is zero, we get the Faustmann rotation rule. If λ is very large, the rotation length will be very short. What this means is that landowners will harvest the trees early to avoid them being destroyed by fire, and in the extreme case where $\lambda = \infty$, the rotations will be so short that the optimal decision will be to not establish plantations at all. That means that in a situation where one is almost guaranteed that the planted trees will be destroyed by fire, it makes complete sense to not plant them at all.

Table 1. Determination of the optimum rotation age for neem plantations in Northern Ghana using the maximum sustained yield criterion

Plantation Age (yrs)	Volume (V_t) (m^3/ha)	CAI (m^3/ha/yr)	MAI (m^3/ha/yr)
1	1.03	1.03	1.03
2	13.67	12.68	6.86
3	32.35	19.30	11.00
4	49.77	16.91	12.48
5	64.46	14.72	**12.93**
6	76.58	12.03	12.78
7	86.61	10.21	12.41
8	94.99	8.48	11.92
9	102.06	7.58	11.44

12.8. Case Study 1: Optimal Rotation Age Determination for Neem Plantations

12.8.1. Base Case Analysis

Using growth and yield data of neem in the Guinea Savannah zone of Ghana discussed in Chapter 6 and given in Nanang (1996), the optimal rotation ages are estimated using the three criteria described above. Table 1 gives a summary of the growth and yield data for Site Class I neem plantations in the Tamale Forest District. Although non-timber benefits are recognised as being jointly produced from the forest, these are not taken into account in these simple examples. In Table 1, the MAI and CAI intersect at approximately plantation age 5, and hence this is the optimal rotation age based on this criterion. Beyond age 5, the MAI is now greater than the CAI.

Table 2 presents the results of the optimal age determination when the single rotation (or NPV) criterion is applied. The following assumptions are used in the analysis for the NPV and SEV criteria: r =5%, p= GH¢20/m^3 and planting costs of GH¢1000/ha. It would be recalled that the optimal rotation in the NPV model occurs where $Y'(t) = Y(t)r$; i.e., if the annual increment in value of the trees is larger than the interest that the money gained by harvesting the plantation the previous year would have accrued, then do not harvest the trees in that year. In Table 2, the optimal rotation age is at year 10. Holding the plantation to year 11 will make the cost MIC [rY(t)] greater than the annual value increment MRP [Y'(t)].

When the SEV criterion was applied, the optimum rotation occurred at 9 years. From Table 3, at plantation age 8 years, the MRP is GH¢167.60 and MIC is GH¢136.46. Holding the trees for another year reduces the MRP to GH¢141.40 and increases the MIC to GH¢143.66. As discussed previously in the theoretical section, the optimum rotation occurs where the SEV is at a maximum. The last column of Table 3 shows the SEV at each age, with a maximum value occurring at age 9. In this case holding the trees for another year makes the MIC > MRP. Therefore, the trees should be harvested at age 9 years.

Table 2. Determination of the optimum rotation age for neem plantations in Northern Ghana using the NPV criterion

Plantation Age (yrs)	Volume (V_t) (m^3/ha)	Y(t) (GH¢/ha)	MRP [Y'(t)] (GH¢/ha)	MIC[rY(t)] (GH¢/ha)	NPV
1	1.03	20.60	20.60	1.03	-980.41
2	13.67	273.34	252.74	13.67	-752.67
3	32.35	647.03	373.69	32.35	-443.09
4	49.77	995.49	348.45	49.77	-184.96
5	64.46	1289.14	293.65	64.46	3.99
6	76.58	1531.60	242.45	76.58	134.63
7	86.61	1732.22	200.63	86.61	220.68
8	94.99	1899.76	167.54	94.99	273.45
9	102.06	2041.19	141.43	102.06	301.52
10	108.09	2161.88	120.69	108.09	**311.25**
11	113.30	2265.91	104.03	113.30	307.32

Table 3. Determination of the optimum rotation age for neem plantations in Northern Ghana using the SEV rotation criterion

Plantation Age (yrs)	Volume (V_t) (m^3/ha)	Y(t) (GH¢/ha)	MRP [Y'(t)] (GH¢/ha)	rY(t) (GH¢/ha)	SEV (GH¢/ha)	rSEV (GH¢/ha)	MIC[rY(t)] (GH¢/ha)
1	1.03	20.60	20.60	1.03	-20102.4	-1005.12	-1004.09
2	13.67	273.34	252.74	13.67	-7909.28	-395.46	-381.80
3	32.35	647.03	373.69	32.35	-3181.03	-159.05	-126.70
4	49.77	995.49	348.45	49.77	-1020.38	-51.02	-1.24
5	64.46	1289.14	293.65	64.46	18.02	0.90	65.36
6	76.58	1531.60	242.45	76.58	519.46	25.97	102.55
7	86.61	1732.22	200.63	86.61	747.27	37.36	123.97
8	94.99	1899.76	167.54	94.99	829.43	41.47	136.46
9	102.06	2041.19	141.43	102.06	**832.07**	41.60	143.66
10	108.09	2161.88	120.69	108.09	791.03	39.55	147.65
11	113.30	2265.91	104.03	113.30	726.43	36.32	149.62

The difference in rotation age between the NPV and Faustmann models is mainly in the fact that the Faustmann model includes the opportunity costs of the land holding the trees. If this opportunity cost is high, the rotation will be shorter than that predicted by the NPV model.

The optimal rotations determined by the three approaches are 5, 10 and 9 years for the MSY, NPV and SEV models respectively. From the theoretical analysis, it was observed that the MSY usually results in a longer rotation than the remaining two models. However, it was also noted from studies by Binkley (1987) that when the reciprocal of the rotation age determined by the MSY is greater than or equal to the discount rate, then the MSY produces a shorter rotation length than the SEV or NPV models. In this example, it can be verified that this condition holds. From a practical point of view, unless fuelwood and very small dimension timber are the ultimate objective of management, most landowners will not harvest their plantations at age five. As shown, this rotation age does not optimise the economic return on the investment. The economic return is maximised at age 9 years under the SEV model.

12.8.2. Sensitivity Analyses

From Table 3, the optimal rotation age for neem plantations using r =5%, p= GH¢20/m^3 and planting costs of GH¢1000/ha was 9 years. However, this optimal rotation age would change as the assumptions in the variables used in the analyses change. Table 4 reports the results of sensitivity analyses to examine the impacts of changes in the assumed discount rates, prices, costs and the risk of fire on the optimal rotation age.

The results are consistent with the theoretical analyses. The risk of fire reduces the optimal rotation ages for all scenarios examined. Except when the price of timber is 40/m^3, all other scenarios produce negative SEVs when the risk of fire is greater than 0%. This result is consistent with the reasoning that the probability of the risk of fire is equivalent to an increase in the discount rate. The main conclusions of these analyses are: Increasing the discount rate and the risk of fire each reduces the optimal rotation age while increasing the cost of regeneration increases the rotation length for the plantation. Increases in the price paid per m^3 of wood not only decreases the rotation age but also increases the profitability of the plantation, holding all other variables constant. Under the assumptions, the only scenarios that produce a net benefit for the plantation are when the discount rate is 5%, p = 20, and c= GH¢1000/ha, with no risk of fire. When the risk of fire is 10%, the plantation is still profitable only if the price of timber is GH¢40/m^3, with the discount rate at 5% and c= GH¢1000/ha.

12.9. Case Study 2: Optimal Rotation Age Determination for Teak Plantations

12.9.1. Basic Analysis

Table 5 gives a summary of the growth and yield data for Site Class I teak plantations in the Guinea savannah zone of Ghana from Nunifu (1997). The MAI and CAI intersect at approximately plantation age 30, and hence this is the optimal rotation age based on this criterion. Beyond age 30, the MAI is now greater than the CAI. If the decision to harvest the plantation is based solely on the age that yields maximum volume growth, then it will be at 30 years.

To apply the NPV and SEV criteria, the following assumptions are used: r =10%, p= GH¢100/m^3 and planting costs total GH¢1500/ha. With these assumptions, the NPV criterion shows that the optimal rotation age for these plantations occurs at age 23 years. From Table 6, the fourth column shows the marginal increase in the value of the plantation in each 5-year period. The last column shows the opportunity costs of holding on to the trees for each additional 5-year period. This column calculates the interest earned on the proceeds if the plantation were harvested in the previous period and the money deposited in a bank at an interest rate of 10% per annum. In deciding when to harvest the plantation, the landowner compares the benefits and costs of holding the trees for an additional period. As soon as the costs outweigh the benefits, the plantation should be harvested.

Table 4. Sensitivity of the optimal rotation age and SEV to changes in the discount rate, price of wood, regeneration costs and risk of fire

Risk of fire (λ%)	*r*=5%, *p*= 20, c=1000		*r*=5%, *p*=40, c=1000		*r*=5%, *p*=20, *c*= 2000		*r*=10%, *p*=20, *c*= 1000	
	Rotation	SEV	Rotation	SEV	Rotation	SEV	Rotation	SEV
0	9	832.07	7	4897.21	16	-1490.51	8	-265.83
10	7	-605.83	6	413.53	13	-1927.71	7	-760.34
20	7	-846.00	5	-366.24	12	-1981.33	6	-894.73

Note: *p* is measured in GH¢/m^3, *c* is in GH¢/ha, and rotation age is measured in years.

Table 5. Determination of the optimum rotation age for teak plantations in the Guinea savannah zone using the MSY criterion

Age	Volume (V_t) (m^3/ha)	CAI (m^3/ha/yr)	MAI (m^3/ha/yr)
5	22.76	11.33	4.55
10	97.8	17.21	9.78
15	186.56	17.87	12.44
20	274.17	17.06	13.71
25	356.54	15.87	14.26
30	432.85	**14.66**	**14.43**
35	503.27	13.52	14.38
40	568.28	12.5	14.21
45	628.45	11.58	13.97
50	684.3	10.77	13.69

Table 6. Determination of the optimum rotation age for teak plantations in Northern Ghana using the NPV criterion

Plantation Age (yrs)	Volume(V_t) (m^3/ha)	Y(t) (GH¢/ha)	Y'(t) (GH¢/ha)	rY(t) (GH¢/ha)
5	22.70	4540.67	2017.38	227.03
10	97.55	19510.39	3382.30	975.52
15	186.09	37218.59	3570.90	1860.93
16	203.87	40774.71	3556.12	2038.74
17	221.53	44306.91	3532.19	2215.35
18	239.04	47808.14	3501.23	2390.41
20	273.49	54697.52	3424.49	2734.88
22	307.07	61414.02	3335.44	3070.70
23	323.51	64702.29	**3288.28**	**3235.11**
24	339.71	67942.40	**3240.11**	**3397.12**
25	355.67	71133.76	3191.36	3556.69

Table 7. Determination of the optimum rotation age for teak plantations in Northern Ghana using the SEV criterion

Plantation Age (yrs)	Volume(V_t) (m^3/ha)	Y(t) (GH¢/ha)	Y'(t) [MRP] (GH¢/ha)	rY(t) (GH¢/ha)	rSEV (GH¢/ha)	MIC (GH¢/ha)
5	22.70	4540.67	2017.38	227.03	460.28	687.32
10	97.55	19510.39	3382.30	975.52	1313.15	2288.66
15	186.09	37218.59	3570.90	1860.93	1523.86	3384.79
16	203.87	40774.71	**3556.12**	2038.74	**1527.34**	**3566.08**
17	221.53	44306.91	**3532.19**	2215.35	1522.69	**3738.04**
18	239.04	47808.14	3501.23	2390.41	1511.33	3901.73
20	273.49	54697.52	3424.49	2734.88	1472.99	4207.86
22	307.07	61414.02	3335.44	3070.70	1419.74	4490.44
23	323.51	64702.29	3288.28	3235.11	1389.24	4624.36
24	339.71	67942.40	3240.11	3397.12	1356.88	4754.00
25	355.67	71133.76	3191.36	3556.69	1323.076	4879.764

Tables 6 and 7 show the optimal rotation ages for teak in the guinea savannah zones using the NPV and SEV models. Again, the SEV model results in a shorter rotation (16 years) than the NPV model (23 years) because of the difference in the opportunity costs of holding the land. The MSY model resulted in the longest rotation of 30 years. This is because the MSY framework does not account for the interest on capital investments or the opportunity costs of holding the land and trees.

12.9.2. Sensitivity Analyses

The optimal rotation age for teak plantations in the savannah vegetation zone was used to examine the impacts of changing prices, costs, discount rates and the risk of fire on the optimal rotation and profitability of the plantation project. Table 8 reports the results of sensitivity analyses.

The risk of fire reduces the optimal rotation ages for all scenarios examined. Even with a 20% risk of fire, the plantation still returns a profit as long as the price of timber is GH¢400/m^3.

Table 8. Sensitivity of the optimal rotation age and SEV to changes in the discount rate, price of wood, regeneration costs and risk of fire

Risk of fire (λ%)	*r=5%, p=* 200, c=2000		*r=10%, p=*200, c=2000		*r=5%, p=400, c=* 2000		*r=5%, p=200, c=* 4000	
	Rotation	SEV	Rotation	SEV	Rotation	SEV	Rotation	SEV
0	16	29638.86	12	8565.10	16	62,909.64	17	26,087.71
10	10	3029.28	9	797.84	10	8632.99	11	512.19
20	8	-297.42	8	-914.73	8	1718.20	9	-257.10

Note: *p* is measured in GH¢/m^3, *c* is in GH¢/ha, and rotation age is measured in years.

The difference between the effect of the risk of fire in the neem plantations and teak is because of the higher value of teak wood. The higher value of teak compensates for the increased costs (risks) of fire. What this implies is that, in general, a higher value for wood allows the investor additional latitude to take more risk. In these analyses, increasing the discount rate and the risk of fire all reduce the optimal rotation age while increasing the cost of regeneration decreases the rotation length for the plantation. Increases in the price paid per m^3 of wood, however, decreases the rotation age and increases the profitability of the plantation, holding all other variables constant. Due to the higher value of teak wood, the plantation is profitable under a wider range of prices, discount rates and costs, and risks of fire.

12.10. CONCLUSION

The determination of the optimal rotation age is all about optimising the benefits that society derives from the resource. If the trees are harvested too early before the optimal rotation age is reached, or too late after the optimal rotation age has passed, resources are wasted. Although there are several methods to determine the optimal rotation age, the economic model proposed by Faustmann (1849) and its subsequent modifications provide the correct approach. In Ghana, the most important factor that impacts on the optimal rotation age determination is the interest and discount rates. Using the data for teak and neem, this Chapter demonstrated that it is possible to grow plantations of these two species profitably even with timber benefits only. If other non-timber benefits are considered, the economic returns will improve even further.

Chapter 13

INTERNATIONAL CLIMATE CHANGE POLICIES AND FOREST PLANTATIONS

13.1. INTRODUCTION

This chapter examines the importance of international climate change agreements for plantation development and management. For example, the Kyoto Protocol and the Clean Development Mechanism (CDM) provided thereunder have provided a market for carbon sequestered in afforestation and reforestation projects as if carbon were a tangible commodity. The importance of the CDM to plantation forestry in Ghana is that it provides incentives to develop plantations to earn carbon credits in addition to other socio-cultural and environmental benefits. A good understanding of how the Kyoto Protocol and future international climate change policy could impact on plantation forestry is important to plantation forestry investors in Ghana.

These carbon credits and the potential income that can be generated could change the economics of the plantation enterprise in some circumstances from uneconomic to profitable. Carbon sequestration considerations could also impact on the type of management that would be applied to plantations. As we saw in Chapter 12, the inclusion of non-timber benefits could change the optimal rotation age of plantations and in some cases it will be optimal not to harvest the trees at all.

We begin this Chapter begins by looking at the larger context of the role forests play in the global climate change policy development and implementation. It then looks at the Kyoto Protocol and the flexible mechanisms in general. The economic basis for introducing the flexible mechanisms is examined along with the criticisms of these market-based approaches to climate change policy. The rules governing the CDM, the administrative structures in Ghana to manage CDM projects, and application of the CDM to forestry projects are discussed. An example application of how carbon credits could be awarded to a hypothetical teak plantation in Ghana concludes the Chapter.

13.2. THE ROLE OF FORESTS IN CLIMATE CHANGE MITIGATION

There is increasing scientific evidence that carbon dioxide (CO2) in the atmosphere contributes to the greenhouse effect. The greenhouse effect is a name given to the process whereby certain gases trap heat within the surface-troposphere system. These greenhouse gases (GHG) absorb and emit infrared radiation causing the heating at the surface of the planet. This heating of the planet is also believed to be leading to changes in global weather and climate patterns, a process known as climate change. The main source of CO2 from the earth is through the burning of fossil fuels and other organic materials, and through other natural processes.

Through photosynthesis, trees and other vegetation absorb CO2 from the atmosphere into their tissues and soil and store this as carbon. This ability to absorb and store carbon makes forests carbon sinks[1]. This carbon remains stored in the plant material until the material is either burnt or decays. Therefore, through growth, trees remove CO2 from the atmosphere, and the rate of removal is proportional to the rate of growth (or the rate of photosynthesis). The process of removing carbon from the atmosphere and storing it in plant tissues is known as carbon sequestration[2]. About 30% (4 billion ha) of the earth's surface is covered by forests (FAO 2005) excluding other vegetation types. The world's forests store an estimated 289 gigatonnes (Gt) of carbon in trees and vegetation, which is more than all the carbon in the atmosphere (FAO 2010). It is clear that forests can play an important role in reducing the amount of CO2 in the atmosphere and hence contribute to mitigating global climate change. Forests can also be sources of CO2 when they release the stored carbon through forest fires, decaying of forest biomass/products, pests and disease infestation and through deforestation activities.

13.3. THE KYOTO PROTOCOL

13.3.1. The Kyoto Protocol and the Flexible Mechanisms

The Kyoto Protocol (KP) is an international agreement that was negotiated on 11 December 1997 in Kyoto, Japan and commits industrialised countries (defined in the KP as Annex I countries) to reduce their (GHG) emissions by an average of 5.2% compared to 1990 levels, during the first commitment period from 2008 to 2012. The KP, which came into force on 16 February 2005, is linked to the United Nations Framework Convention on Climate Change (UNFCCC). Under the KP, each Annex I country has a binding target and countries are expected to meet these targets using domestic measures. However, three "flexible mechanisms" are included in the KP to help countries reach their targets in a cost-effective way. These mechanisms are: clean development mechanism (CDM); Joint implementation (JI); and emissions trading (ET). All three mechanisms were designed to help participating countries in meeting their domestic emission reduction targets at lower costs and to help developing countries and countries in transition in their sustainable development by

[1] In contrast, a carbon source emits GHGs into the atmosphere.

[2] Note that carbon sequestration is a general term referring to any process that increases a carbon pool other than in the atmosphere.

encouraging technology transfer. The CDM allows emission-reduction (or emission removal) projects in developing countries to earn certified emission reduction (CER) credits, each equivalent to one tonne of CO_2. These CERs can be traded and sold, and used by industrialised countries to meet a part of their emission reduction targets under the KP. These projects are also expected to result in sustainable development and emission reductions, as defined by the host country. Joint Implementation (JI) provides a means for countries or companies to invest in GHG reduction measures and sequestration projects in other industrialised countries or countries in transition (such as Russia), and gain certified credits. Emissions trading (ET), as set out in Article 17 of the KP, allows countries that have emission units to spare (those emissions permitted them but not used) to sell this excess capacity to countries that are over their targets. Therefore, ET provides a means for emitters to purchase emissions reduction credits through a special market that will be set up for this purpose. The only mechanism of interest in this Chapter is the CDM, and it will be the main focus of the remaining discussion.

13.3.2. The Economic Basis for the Flexible Mechanisms

The main objective of the introduction of flexible mechanisms under the KP was to reduce the overall cost of meeting binding targets by industrialised countries who are signatories to the KP. There are two main ways for countries to achieve their targets under the KP: a) take national measures through command-and-control approaches such as legislation, or direct and indirect taxation; or b) use economic instruments such as trading in GHG emissions and carbon sequestration projects. A legislative approach requires everyone to reduce their emissions irrespective of the cost of doing so. For example, if there are two firms with the same legislated limit on how much GHG they can emit, both will be expected to reach the same limit, even if it costs Firm A more money than it costs Firm B to meet those obligations.

The second approach to controlling emissions creates markets for trading in carbon emissions. There are two main schemes: the cap-and-trade system and carbon offset trading system. Under the former scheme, a cap on emissions is determined by the country and those industries that emit less than the cap can sell the excess "credits" to firms that cannot meet their target with existing technology. The national cap on emissions is set for a given time period, and this cap is divided into targets for the various industries within the country. Each country develops national regulations that determine which firms are covered by a cap on their emissions and those that are not. Firms are then given permits, and they cannot emit GHGs above what their permits allow them. Firms with caps must then decide the most cost-effective way to meet their targets, which could include cutting production, changing production processes through technology, or buying allowances from other firms (hence the term flexible mechanisms). What this means is that firms that have the technology and can reduce their emissions at a lower cost would do so; those with higher costs of reducing emissions would buy credits from others who have excess to spare. Theoretically, this will reduce the overall costs of meeting the national target. It is expected that as time passes, the availability of permits will be reduced, thus increasing the scarcity of the permits and hence the price and cost to pollute. The largest emissions trading scheme in the world today is the European Union Emissions Trading Scheme (EU-ETS) which was launched in 2005. As at

2008, the EU-ETS was worth about $63 billion (Gilbertson and Reyes 2009). There are other trading schemes in New Zealand, USA and Australia.

Under the carbon offset trading scheme, firms and individuals finance 'carbon offset projects' outside the industries with caps on their emissions to obtain emission credits that they can apply to offset their emissions. Recall that from above, the economy was divided into two main sectors for purposes of emissions trading: those industries with caps on their emissions; and those without caps. There are no emission caps on forests hence forestry activities can generate offset credits. The CDM and the JI are examples of the offset trading schemes, with the CDM being the largest offset trading scheme in the world.

Critiques of emissions and offset trading schemes point out that these schemes do not lead to emission reductions. In the case of the offset scheme, they argue that these actually increase global emissions rather than decrease them (e.g., Gilbertson and Reyes 2009). There are two main objections: leakage due to the project and emissions resulting from the buyer of the credits. Gilbertson and Reyes (2009) argue that even in cases where it can be verified by the seller that the emissions have been reduced due to the CDM project, there is increased emissions by the buyer which negates the benefits from the CDM project.

13.4. Plantation Forestry and Carbon Markets

The term *carbon market* refers to the buying and selling of emissions permits or credits that have either been distributed by a regulatory body or generated by GHG emission reduction projects. The largest market for carbon credits resulting from afforestation and reforestation (A/R) projects is within the CDM framework of the Kyoto Protocol. In this section, we examine the issues related to the CDM, important definitions related to A/R projects within the CDM and the rules governing the award and trading of credits. There may be markets for carbon credits within domestic offset trading schemes of Annex I countries. For example, The Government of New Zealand has already enacted an emissions trading scheme (ETS) under which owners of Kyoto Protocol compliant forests will receive units for increases in carbon stocks of their plantations (Manley and Maclaren 2010). There are several requirements, procedures and rules to be followed to be able to earn carbon credits through A/R projects under the CDM. These rules are not fully covered in this book, but interested readers can find information through the Environmental Protection Agency (EPA) in Ghana, or on several websites of the UNFCCC or the Intergovernmental Panel on Climate Change (IPCC).

13.4.1. General Provisions of the CDM

The Clean Development Mechanism (CDM), established under the Kyoto Protocol, is the primary international offset program in existence today, and while not perfect, it has helped to establish a global market for greenhouse gas (GHG) emission reductions. It generates offsets through investments in GHG reduction, avoidance, and sequestration projects in developing countries.

According to Article 12 of the KP, a CDM allows the countries with legally binding greenhouse gas (GHG) emissions reductions commitments to receive credits towards their obligations by investing in projects in developing countries. Article 12 stipulates that two criteria must be met for projects undertaken by Annex I countries to be eligible for crediting under the CDM:

- Projects must produce both "certified emissions reductions" that "are additional to any that would occur in the absence of the certified project activity" and contributions to the host country's sustainable development; and
- In addition, developing country parties will have access to resources and technology to assist in the development of their economies in a sustainable manner.

The KP identifies six GHGs and gas classes whose reductions qualify under the CDM: carbon dioxide, methane, nitrous oxide, hydrofluorocarbon, perfluorocarbon, and sulphur hexafluoride. Sectors that may qualify for CDM projects include energy, industrial processes, solvent and other product use, waste and land use, land use change and forestry.

The Marrakech Accords (UNFCCC 2002) recognises afforestation and reforestation as the only eligible land uses under the CDM. The KP recognises that forests, forest soils and forest products all play important roles in mitigating climate change. The incentives provided under the CDM is that tree growers can generate carbon credits that can be sold to entities subject to the KP to offset their emissions, in the form of Certified Emission Reductions (CERs). Since CDM projects can only be undertaken in developing countries, this offers interesting opportunities for the management of plantation forests for sequestering carbon in Ghana.

13.4.2. Important Definitions and Concepts Related to the CDM

Articles 3.3 and 3.4 of the KP and the Marrakech Accords (UNFCCC 2002) specified the following definitions for projects related to CDM initiatives:

Afforestation is the direct human-induced conversion of land that has not been forested for at least 50 years to forested land through planting, seeding and/or the human-induced promotion of natural seed sources.

Reforestation is the direct human-induced conversion of non-forested land to forested land through planting, seeding or the human-induced promotion of natural seed sources, on land that was forested, but that has been converted to non-forested land.

A *Forest* is a minimum area of land of 0.05–1.0 ha with tree crown cover of more than 10–30% of trees, with the potential to reach a minimum height of 2–5 m at maturity in situ. A forest may consist either of closed forest formations, where trees of various storeys and undergrowth cover a high proportion of the ground or open forest. Young natural stands and all plantations which have yet to reach a crown density of 10–30%or tree height of 2–5 metres are included under forest, as are areas normally forming part of the forest area which are temporarily not stocked as a result of human intervention such as harvesting or natural causes but which are expected to revert to forest. For purposes of implementing the CDM, Ghana has chosen a crown cover of 15%, a minimum area of 0.1 ha and a minimum height of 2.0 m to define a forest.

A *baseline* is a scenario that reasonably represents the anthropogenic emissions by sources and removals by sinks that would occur in the absence of a project activity. The development of project baselines is important to ensuring the project emission reduction/removal is real. The impact of a project typically is assessed relative to a baseline, with certified credits issued for above-baseline carbon sequestration or emission reductions. Methodologies for determining forest project baselines include modelling, control plots, and historical benchmarks. Whichever approach (es) is (are) chosen, it is expected that the onus will be on a project developer to support their baseline choice, and the Executive Board of the Clean Development Mechanism will review proposed project baselines to ensure standards of accuracy and consistency. Standardised baseline methodologies could also be developed and approved for use by projects. For specific projects within the CDM, baseline definitions must take into account the following elements, where applicable: natural and anthropogenic GHG emissions and removals, type of practice(s) and degree of implementation (or level), trends in practices and applicable technologies, and regulations and indirect climate change measures.

Additionality: For a CDM project, emissions reductions must be beyond - *or in addition to* – the reductions that would have occurred in the absence of the project. Additionality is assessed by comparing the carbon stocks and flows of the project activities with those that would have occurred without the project (i.e., its baseline). For example, the project may be proposing to afforest farmland with native tree species, increasing its stocks of carbon. By comparing the carbon stored in the 'project' plantations (high carbon) with the carbon that would have been stored in the 'baseline' abandoned farmland (low carbon) it is possible to calculate the net carbon benefit. Additionality is important to any cap-and-trade system that allows for projects from outside the cap as in the CDM. If reductions from projects are used to offset emissions elsewhere (i.e., reductions from GHG sources and sinks not covered by a cap, and are used to offset emissions under a capped system), then those reductions should not have happened anyway, or else there will be a net increase in atmospheric emissions by that amount.

Permanence is the length of time for which carbon will remain stored after having been fixed in vegetation. In other words, the ability to store carbon stocks generated by a project under the CDM until the crediting period expires. Any future emissions that might arise from these stocks need to be accounted for. Permanence is one of the main concerns related to the use of sinks as a greenhouse gas (GHG) mitigation option. In reality, the concern is about lack of permanence, or 'reversibility' of the benefits of storage, as a result of the possible loss of carbon stocks created or conserved by a project, whether on purpose or as a result of undesirable events (e.g., natural disasters). Natural disturbances or human activities can cause a reduction in net sequestration within a project boundary, and non-permanence occurs when this net sequestration rate (or sink size) falls below the amount for which offset credits have been issued. The two elements in a system to address non-permanence of offset projects are: the preparation of a plan to manage the risk of non-permanence events, reducing the likelihood of their occurrence, and a mechanism by which the validity of an issued offset credit is maintained when a non-permanence event occurs.

Project Boundary: The project boundary encompasses all anthropogenic emissions by sources and removals by sinks of greenhouse gasses under the control of the project proponents that are significant and reasonably attributable to the project activity. As it pertains to forest plantations, this favours a project boundary that encompasses not only the activity for which credits are being issued, but also other forestry activities under the direct control of the project proponent (that are significant and reasonably attributable to the project). Logically, a project proponent should not be issued credits for maintaining a plantation forest, for instance, if the resultant carbon benefits are effectively offset by a consequent increase in harvesting on other lands managed by the project proponent (the areas where this increased harvesting is occurring has to be in the leakage assessment area). Identifying afforestation and reforestation project boundaries and leakage is a two-tiered process. First, determining whether the project proponent *controls* a GHG source isolates potential leakage from potential inside-boundary emissions. Secondly, the source must be assessed in terms of its *significance, attribution,* and *cost-effectiveness,* and on this basis either dismissed or accounted for as leakage or inside-boundary emissions, as applicable.

Leakage is an increase in emissions or reduction in removals outside a project's boundary (the boundary defined for the purposes of estimating the project's net GHG impact) resulting from the project's activities. Leakage is associated with changes in reductions/removals that are significant and reasonably attributable to the project, but are not under the control of the proponent. The amount of potential leakage would depend mainly on the nature of the project and the type of activities. Afforestation/reforestation (AR) projects would likely have less potential for leakage especially if implemented on land that has few or no competing uses. For forest projects that change practices, these activities would generally not result in leakage because the underlying economic activity continues so that no activity shifting would be expected. However, leakage could occur if the changes in practices result in a reduction in the output of forest products. The potential for leakage implies that it may be necessary to: 1) identify if a project has any leakage potential; 2) try to mitigate the potential for leakage through project location and design; and 3) account for any leakage when estimating total project greenhouse gas (GHG) benefits. There is currently no standard method to identify or quantify leakage effects due to GHG mitigation projects in forestry.

13.4.3. Rules Governing CDM Forestry Projects

Below are the rules and conditions that will apply to CDM projects, especially to afforestation and reforestation projects, which provides the overall framework for approving projects and accounting for the carbon credits generated (UNFCCC 2002).

- All CDM projects must be approved by the CDM Executive Board.
- Only areas that were not forests on 31st December 1989 are likely to meet the CDM definitions of afforestation or reforestation.
- Projects must result in real, measurable and long-term emission reductions, as certified by a third-party agency.
- Emission reductions or sequestration must be additional to any that would occur without the project. They must result in a net storage of carbon and, therefore, a net removal of carbon dioxide from the atmosphere.
- Projects must be in line with sustainable development objectives, as defined by the government that is hosting them.
- Projects must contribute to biodiversity conservation and sustainable use of natural resources.

- Only projects starting from the year 2000 onwards will be eligible.
- Two percent of the carbon credits awarded to a CDM project will be allocated to a fund to help cover the costs of adaptation in countries severely affected by climate change. This adaptation fund may provide support for land use activities that are not presently eligible under the CDM, for example, conservation of existing forest resources.
- Some of the proceeds from carbon credit sales from all CDM projects will be used to cover administrative expenses of the CDM (a proportion still to be decided).
- Projects need to select a crediting period for activities, either a maximum of seven years that can be renewed at most two times or a maximum of ten years with no renewal option.
- The funding for CDM projects must not come from a diversion of official development assistance (ODA) funds.
- Each CDM project's management plan must address and account for potential leakage. Leakage is the unplanned, indirect emissions of CO_2, resulting from the project activities. For example, if the project involves the establishment of plantations on agricultural land, then leakage could occur if people who were farming on this land migrated to clear forest elsewhere.

13.4.4. Types of Carbon Credits Awarded for Plantations under the CDM

It is important that interested investors in afforestation and reforestation carbon sequestration projects under the CDM should also acquaint themselves with the types of carbon credits awarded under the CDM. Carbon credits are awarded for the CO_2 that is sequestered over and above the baseline, taking into consideration any leakage that might have occurred as a result of the project. This means that earning credits have to be based on the determination of a credible baseline for the particular project. The CDM Executive Board has published a guidebook on the various methodologies available for determining baselines for A/R projects (e.g., see UNEP 2005).

The main difference between forestry sequestration projects and emission avoidance projects (those that prevent carbon emissions) is that trees reduce CO_2 in the atmosphere by capturing and storing it in the biomass, whereas, many other projects prevent or reduce the release of CO_2 into the atmosphere. This means that CO_2 reduction from forestry projects is not permanent, as the CO_2 can be re-emitted into the atmosphere for example, through forest fires, pest attack related to dying forests, flooding of forest, etc. or human actions (for example, logging, deforestation or burning the wood) (UNEP 2005). For this reason, CO_2 reduction from forestry sequestration is considered temporary.

As a result of this potential of non-permanence of the sequestered CO_2, carbon credits issued for forestry A/R activities under the CDM are temporary Certified Emissions Reductions (tCERs) or long-term CERs (lCERs), which are different in characteristics from the CERs issued for projects that prevent the emission of GHGs. As per modalities and procedures agreed for A/R projects, the A/R project proponents can choose either tCERs or lCERs (UNEP 2005). Temporary credits (tCER) are issued and expire at the end of the commitment period following the one during which it was issued. Once expired, a tCER can be reissued several times during the project as long as the forest exists. On the contrary, a long-term credit (lCER) expires at the end of the crediting period of the overall afforestation or reforestation project (Olschewski and Benıtez 2005), and cannot be reissued (Neef and Henders 2007).

These different types of accounting procedures have impacts not only on the amount of CER awarded to reforestation projects but also on the risk and the value of the certificates (Olschewski and Benıtez 2010). Theoretically, potential buyers of carbon credits would to be indifferent between buying a permanent credit today and buying a non-permanent credit (lCER or tCER) today and replacing it by a permanent one when the initial credit expires (Olschewski and Benıtez 2010). Equation 1 reflects this difference in preference, where *t* is the lifetime of CER units, P_t is the price per tCO_2 for a credit valid for t-years, P_∞ is the price for a permanent credit, and r is the discount rate of potential buyers of the carbon credits (Olschewski et al. 2005):

$$P_\infty = P_t + \frac{P_\infty}{(1+r)^t} \tag{1}$$

From a carbon buyers' perspective, it would make sense to buy a non-permanent CER if its price is lower than that of a permanent CER minus the discounted price of a permanent CER in year *t*, when the expiring temporary CER has to be replaced (Olschewski and Benıtez 2010). Therefore, the maximum price that an enterprise would be willing to pay for temporary carbon credits, P_t, results from Equation (1) as (after Olschewski and Benıtez 2010):

$$P_t = P_\infty \left[1 - \frac{1}{(1+r)^t}\right] \tag{2}$$

Given that lCER and tCER credits only differ with respect to their expiring time, the price derived from Equation 2 is valid for both types of credits. CDM official accounting rules allow for temporary credits with an expiring time of 5 years and A/R projects with a maximum duration of 30 years (Olschewski and Benıtez 2005; Olschewski and Benıtez 2010).

13.5. THE CLEAN DEVELOPMENT MECHANISM IN GHANA

Ghana ratified the United Nations Framework Convention on Climate Change (UNFCCC) in 1995 and the Kyoto Protocol in May 2003. Because it is a developing country, it has no binding targets under the KP but is eligible to participate in projects under the CDM. Ghana has taken steps to take advantage of the CDM by setting up a Designated National Authority (DNA) on CDM within the Environmental Protection Agency (EPA).

13.5.1. Structure and Functions of the Designated National Authority (DNA)

A legislative process has been in the works since 2005 to make the DNA operational, but was never passed by Ghana's Parliament. At present, the DNA in Ghana is working on the basis of ministerial declarations (CDM Ghana 2008). CDM project evaluation is being carried out by ad hoc expert groups who make recommendations to the CDM Governing Council, which is entitled to take the final approval decision (CDM Ghana 2008). The DNA is to be funded from fees charged for projects and also from the Ministry of Environment and donations from public institutions.

The DNA has a governing council that consists of the following:

1. Executive Director (Environmental Protection Agency) Chairman
2. Chief Director (Ministry of Environment, Science, Technology and Innovation) Member
3. Chief Director (Ministry of Energy) Member
4. Chief Director (Ministry of Lands and Forestry) Member
5. Director (External Resource Mobilization Division, Ministry of Finance and Economic Planning) Member
6. Chief Director (Ministry of Trade and Industry) Member
7. National Climate Change Coordinator (Environmental Protection Agency) Member-Secretary

According to CDM Ghana (2008), the functions of the DNA include:

- Receives projects for evaluation and approval as per the guidelines and general criteria laid down in the relevant rules and modalities pertaining to CDM in addition to the guidelines issued by the Clean Development Mechanism Executive Board and Conference of Parties serving as Meeting of Parties to the United Nations Framework Convention on Climate Change. The evaluation process of CDM projects includes an assessment of the probability of eventual successful implementation of CDM projects and evaluation of extent to which projects meet the sustainable development objectives, as it would seek to prioritise projects in accordance with national priorities.
- Recommends certain additional requirements to ensure that the project proposals meet the national sustainable development priorities and comply with the legal framework so as to ensure that the projects are compatible with the local priorities and stakeholders have been duly consulted.
- Ensures that in the event of project proposals competing for the same source of investment, projects with higher sustainable development benefits and that are likely to succeed are accorded higher priority.
- Carries out the financial review of project proposals to ensure that the project proposals do not involve diversion of official development assistance in accordance with modalities and procedures of Clean Development Mechanism and also ensure that the market environment of the CDM project is not conducive to under-valuation of Certified Emission Reduction (CERs) particularly for externally aided projects.

- The DNA carries out activities to ensure that the project developers have reliable information relating to all aspects of Clean Development Mechanism which include creating databases on organisations designated for carrying out activities like validation of CDM project proposals and monitoring and verification of project activities, and to collect, compile and publish technical and statistical data relating to CDM initiatives in Ghana.
- The Member-Secretary of the Governing Council of the DNA is responsible for the day-to-day activities of the Authority including constituting committees, subgroups or ad hoc committees to coordinate and conduct detailed examination of the CDM project proposals.

13.5.2. The CDM Project Approval Process

According to the draft legislation on CDM, the CDM project approval process would involve the following steps (after CDM Ghana 2008):

Application to the DNA: An application for Initial screening must be made to the DNA. This application includes a letter signed by the project proponent and/or the project sponsors and a project Identification Note in the format provided by the DNA.

Initial Screening: This is a voluntary step that allows project developers the opportunity to receive an initial evaluation of their project from the DNA to identify any potential conflicts with the project approval criteria and other government policies. The DNA will conduct an initial evaluation of the sustainable development impacts of the project against the sustainable development criteria. Results of the initial screening will be provided within 15 working days of submission of the application form and Project Identification Note. The comments will be submitted to the project developer in the form of an informal notice of letter-of-no-objection.

Submit Validated Project Design Document (PDD) to DNA: The project proponent's request for final project approval must have: a) a Validated Project Design Document in most recent format published by the Executive Board of the CDM; b) a completed application form; and 3) An Environmental Impact Assessment (EIA) of the project if this is required.

Public Comment: The DNA will post the submitted Project Design Document (PDD) on its website for public consultation, for a period of 21 working days. Project Design Documents will also be made available to any interested parties upon request. The recommendations prepared by the DNA will be sent to the members of the Governing Council of the DNA for consideration at its next meeting. The DNA may also decide to submit the project to the Council via email circulation for comments. Any comments on the PDD received from the period of public consultation will also be circulated to the Council for consideration. The Council of the DNA will evaluate the project and submit its comments to the DNA.

Letter of Approval: The Governing Council of the DNA will make a final decision based on the feedback received from the public. The decision to approve or disapprove a project will take no more than 60 days.

Appeal: Project participants will have the right to appeal the final decision taken by the DNA. In a first step, they may appeal the decision to the Minister of Environment and Science. The Minister will verify the decision taken by the DNA and determine whether it has been produced in accordance with the approval procedures (formal and substantial determination). The Minister will notify the project participants of his/her decision within 60 days. The project participant has the right to appeal the determination of the Minister before the Administrative Courts of Ghana.

13.5.3. Potential Challenges to CDM Implementation

The main challenge to CDM projects in Ghana, especially those related to land use (forestry) is the small size of the projects. As mentioned previously, many of the plantations established in Ghana are small- to medium-sized plantations to serve individual or community needs. It would be difficult for many of these plantations to enter into the CDM offset trading scheme. First, there would likely be a conflict in priorities of local needs for wood versus the environmental need to sequester and store carbon in trees for long periods of time. Secondly, the market for carbon is not frictionless, i.e., there are transaction costs to be borne by project proponents, which these small-scale plantation owners may be unable to bear. Furthermore, there is a need for expertise to manage the plantations according to the procedures of the CDM, which is lacking in Ghana in general and at the community levels in particular. Hiring such expertise may be beyond the financial ability of communities. The obstacles to raising commercial capital to develop projects within the CDM will be identical to those already discussed for forest plantations in general. The difficulty with implementing the CDM is even more so given the potential risks of natural disturbances such as fire, pests and insect destruction and the fact that carbon assets are still being conceived by most financial institutions as "abstract assets." It is likely that only industrial plantations or community plantations with NGO or donor-funded support will be able to participate in the CDM.

There are also risks related to the regulatory environment. At present, there is no legislation that backs the DNA, and large-scale investors may be concerned about the legality and long-term implications of Ministerial declarations to operate a national CDM programme. There are legal issues as well. Land and tree tenure systems do not provide incentives for long-term forestry projects, especially when there is a financial benefit for the standing trees. The ownership of carbon contained in trees cannot be divorced from the ownership of the trees.

Hence, the land and tree tenure obstacles discussed in Chapter 3 are relevant here as well, which does not make Ghana attractive for international investors who seek carbon credits and profits on their investments. In addition, the ability to earn credits is based on the proponent's ability to demonstrate additionality. There is a general lack of data on plantation growth and dynamics for most of the tree species in Ghana to establish credible baselines that would demonstrate additionality and prove that leakage issues can be adequately addressed.

13.6. CARBON SEQUESTRATION AND ACCOUNTING IN PLANTATIONS

13.6.1. Estimating the Amount of Carbon Dioxide Sequestrated by Forest Plantations

We begin this section by describing the process of estimating the amount of CO2 sequestered by a forest plantation. Protocols for quantifying C storage in living trees rely on the use of known allometric relationships and biomass expansion factors, based on measures of stem size, to estimate biomass and C contained in the stem wood, bark, branches, and coarse roots of living trees (Jenkins et al. 2003, Lambert et al. 2005). Under current CDM rules, accounting for changes in C on the forest floor and mineral soil is not mandatory, but could be included in future climate change negotiations. The following steps are used to estimate total above- and below-ground CO2 in a plantation.

1. Estimate the Total (Green) Biomass Contained in One Ha of the Plantation

First, determine the total aboveground biomass for the plantation (in tonnes/ha) either through direct measurement or the use of biomass equations for the species. Most biomass equations give a relationship of the DBH and /or height to aboveground biomass. In general, the root systems of many species weigh about 20% of the aboveground biomass. This is a conservative estimate based on studies by (Cairns et al. 1997, Li et al. 2003) for conifers. In teak, the ratio of belowground to aboveground biomass was found to be about 16% in Panama (Kraenzel 2000). This ratio is smaller than the more general ratio Cairns et al. (1997) produced from a review of tropical forest biomass studies. They found that the average ratio for primary and secondary tropical forests was 24%, with a standard deviation of 14%. If the actual root system to above ground biomass ratio is available for the species, this should be used. Hence to get the total aboveground and belowground estimate of the plantation biomass, multiply the aboveground biomass by the appropriate factor (e.g., a factor of 1.2 for conifers or 1.16 for teak plantations).

If biomass equations are not available, stem wood volume can be estimated using yield equations or yield tables in m^3/ha. This stem wood is then converted to stem wood biomass by multiplying the stem volume by the mean wood specific gravity for the tree species. Stem wood biomass is then converted to total aboveground biomass (i.e., including biomass for branches, leaves, twigs, etc.) if the appropriate relationships exist between stem biomass and total tree biomass. If such equations do not exist, then this technique of going through the yield table to estimate total biomass cannot be applied.

2. Determine the Dry Weight of the Biomass

The next step is to convert the green biomass determined in Step 1 into the dry biomass. For each species, there may be equations that relate green weight to dry weight. This factor should be used if known, or assumed, if unknown.

3. Determine the Weight of Carbon

Total living tree biomass is converted to total carbon by multiplying the dry weight by 0.50, which is the average proportion of carbon in dry plant biomass. That is, about 50% of the dry mass of wood is made up of carbon.

4. Determine the Weight of Carbon Dioxide (CO2) Sequestered

Total CO2 sequestration (i.e., CO_2e contained in living biomass C) is now estimated by multiplying total C by 3.667 (i.e., 1 kg biomass C equals uptake of 3.667 kg CO2). The reason is that we know that CO2 is composed of one molecule of carbon and 2 molecules of oxygen. The atomic weight of carbon is 12.00, and that for oxygen is 16.00. Therefore, the weight of CO2 as 44.00 and a ratio of CO2 to C as 3.667. To estimate the amount (weight) of CO2 sequestered per ha/year, we divide the amount of CO2 sequestered per ha by the age of the trees.

In cases where it is required and where the information exists, carbon sequestered in the soil carbon pool (includes various forms of soil organic carbon (humus) and inorganic soil carbon and charcoal, but excludes soil biomass, such as roots and living organisms) and litter would be included in the accounting framework. Current practice in economic analyses of impacts of carbon sequestration on optimal timber management have often assumed that carbon sequestered in mineral soil, branches, roots and litter are simply recycled into the next rotation of crop trees in order to simplify the data requirements and analyses (e.g., see Van Kooten et al. 1995).

13.6.2. Analysing Subsidies and Penalties for Carbon Management in Plantations

Benefits resulting from the carbon sequestered by plantations fall into the general category of non-timber benefits. Incorporating these benefits follow similar analytical methods as discussed in Chapter 12. In order to evaluate the impact of carbon accounting and credits on plantation management decisions, I incorporate carbon sequestration benefits into economic analyses of plantation management and optimal rotation decisions in a fashion identical to Section 12.6, but introduce more specificity to the analyses. Previously, we saw that the correct optimal rotation age for a forest stand is determined by the Faustmann rotation model. It would also be recalled that when non-timber benefits are included in the analysis, the optimal rotation may change from the Faustmann rotation, depending on the behaviour of the non-timber benefits as stand age increases. Carbon uptake in forests in particular, increases proportionally with stand growth and hence carbon uptake is a function of the biomass and the amount of carbon per m^3 of biomass (van Kooten et al. 1995). Using this as a basis, van Kooten et al. (1995) derived an optimal rotation framework for analysing subsidies and taxes on carbon sequestration. Due to the ability of this framework to tease out the various incentives (and disincentives) for forest management, and its impact on the optimal rotation age, it has been adopted in this section.

Assume a forest stand of volume at time t given as V(t), with α being the amount of carbon/m^3 of volume. The carbon sequestered at any time t, is given by $\alpha V'(t)$, where $V'(t)$ is the increase in volume per unit time (i.e., first derivative of the volume function). If the price of carbon is P_c /m^3 of carbon taken from the atmosphere and r is the discount rate, then the present value of the carbon uptake benefits is given as:

$$PV_c = \int_0^t P_c \alpha V'(t) e^{-rt} dt \tag{3}$$

In Equation (3), it is assumed that the carbon sequestered has a net benefit to society. P_c could be the price paid for each m^3 sequestered in the CDM, a domestic offset trading system or a pure subsidy from governments to encourage carbon sequestration. On the other hand, governments could decide to impose a penalty for emitting carbon from a forest through harvesting or forest fires. In this case, a penalty equal to $P_c \alpha(1-\beta)$ where, β is the fraction of the harvested timber that goes into long-term storage[3] can be imposed. The present value of this penalty is given as:

$$PV_p = P_c \alpha (1-\beta) V(t) e^{-rt} \tag{4}$$

When the timber is harvested at time t, the present value of the wood sold in markets with a price of P_F /m^3 is given as:

$$PV_F = P_F V(t) e^{-rt} \tag{5}$$

Combining Equations 3-5 gives the present value of the timber and carbon benefits for all future rotations using the Faustmann framework as:

$$SEV = \frac{PV_c + PV_F - PV_p}{1-e^{-rt}} = \frac{P_c\left[V(t)e^{-rt} + r\int_0^t V(t)e^{-rt}dt\right] + [P_F - P_c\alpha(1-\beta)]V(t)e^{-rt}}{1-e^{-rt}} \tag{6}$$

From Equation 6, van Kooten et al. (1995) derive the optimal rotation age that considers timber values, carbon benefits and penalties for releasing carbon into the atmosphere by taking the first derivative of Equation 6 with respect to t, and setting the result to zero as:

$$\left[[P_F + P_c\alpha\beta]\frac{V'(t)}{V(t)} + rP_c\alpha\right] = \frac{r}{1-e^{-rt}}\left[(P_F + P_c\alpha\beta) + \frac{rP_c\alpha}{V(t)}\int_0^t V(t)e^{-rt}dt\right] \tag{7}$$

If the price of carbon, P_c, is set to zero, the optimal condition in Equation (7) equals the Faustmann rotation model, while setting β to zero gives the general Hartman (1976) model discussed earlier (Equation 21 in Section 12.6).

[3] Carbon dioxide is only emitted if the wood is burnt or decays. If the wood is used for construction or long-term storage, the carbon is not emitted. Therefore, industry only pays a penalty for the portion of carbon that is emitted.

While the above theoretical framework is useful for analysing the implications of subsidies and taxes, no country has as yet implemented any such policy, and hence so far, this approach has not been applied in practice.

13.7. Case Study: Optimising Joint Production of Carbon and Timber in Teak Plantations

This case study uses information about teak plantations in Ghana to evaluate the implications of accounting for carbon benefits in terms of the potential amount of credits that can be obtained and its implications for plantation management. The following information and assumptions were used. The growth and biomass functions for teak developed by Nunifu (1997) were used to estimate timber volume and above- and belowground biomass of the teak plantations. These equations are given as:

Timber volume:

$$\ln V = 8.10 - 11.13t^{-0.5} \tag{8}$$

Aboveground fresh Biomass:

$$\ln B_{AG} = 7.69 - 10.683t^{-0.5} \tag{9}$$

Aboveground oven-dry biomass:

$$B_{AG} = 1.284 - 0.969D + 0.314D^2 \tag{10}$$

where t is the plantation age in years, and D is the mean DBH of the trees.

To determine the belowground biomass, the aboveground biomass estimated from Equation 9 is multiplied by a factor of 0.16, based on studies by Kraenzel (2000) for teak plantations in Panama. Using this information, the aboveground and belowground carbon sequestered by Site Class I teak plantation in the savannah zone was computed. This information is presented in Table 1. If we assume that the area to which this plantation was planted would have remained savannah grassland without trees, we can project that the amount of carbon that would have been sequestered by the grassland vegetation would be negligible. Hence, we would be able to award all the carbon sequestered in the plantation as credits under the CDM (baseline = 0) for this afforestation project.

From Table 1, the temporary CER credits awarded to the teak plantations between year 5 and 30 years (the maximum number of years allowed under current CDM rules) are summarised in Figure 1. The rules imply that the last period for awarding credits is year 25 of the plantation life. The credits awarded in year 25, together with any credits that were renewed, will all expire in year 30. No credits are given for the growth in carbon in year 30 and beyond.

Table 1. Estimated timber volume, biomass and carbon sequestered in teak plantations in the savannah vegetation zone of Ghana

Age (yrs)	Timber Volume (m^3)	Aboveground Biomass (t/ha)	Belowground* Biomass (t/ha)	Total Biomass (t/ha)	Carbon (tCO_2/ha)
1	0.05	0.87	0.14	1.01	1.86
2	1.26	0.64	0.10	0.74	1.36
3	5.33	2.41	0.39	2.79	5.12
4	12.62	6.26	1.00	7.26	13.31
5	22.70	11.75	1.88	13.64	25.00
6	35.03	18.45	2.95	21.40	39.23
8	64.39	34.07	5.45	39.53	72.47
9	80.64	42.54	6.81	49.35	90.49
10	97.55	51.24	8.20	59.44	108.99
12	132.56	68.93	11.03	79.96	146.61
14	168.24	86.59	13.85	100.44	184.16
15	186.09	95.30	15.25	110.55	202.70
18	239.04	120.76	19.32	140.09	256.85
20	273.49	137.05	21.93	158.98	291.49
25	355.67	175.21	28.03	203.24	372.64
30	431.79	209.80	33.57	243.37	446.22
35	502.04	241.21	38.59	279.80	513.01
40	566.91	269.82	43.17	312.99	573.86
45	626.93	296.00	47.36	343.36	629.55
50	682.65	320.08	51.21	371.29	680.77
55	734.53	342.32	54.77	397.09	728.07
60	782.99	362.95	58.07	421.03	771.95
65	828.40	382.16	61.15	443.31	812.81
70	871.06	400.11	64.02	464.13	850.98
75	911.25	416.94	66.71	483.65	886.77
80	949.21	432.76	69.24	502.00	920.42
85	985.14	447.67	71.63	519.30	952.14
90	1019.22	461.77	73.88	535.65	982.12
95	1051.62	475.12	76.02	551.14	1010.52
100	1082.47	487.79	78.05	565.84	1037.47

* Below ground biomass is the biomass of the root system only, calculated as 16% of the above ground biomass.

As illustrated in Figure 1, during the first five years of the plantation life, a total of 25 tCO_2/ha is accumulated and issued as tCER to the owner at the end of year 5 of the plantation. The credits issued in year 5 expire in year 10 of the plantation life, but can be re-issued for another five years, together with the additional carbon accumulation between year 5 and 10 of 84 tCO_2/ha. This process continues as in Figure 1 until the maximum 30 years is reached.

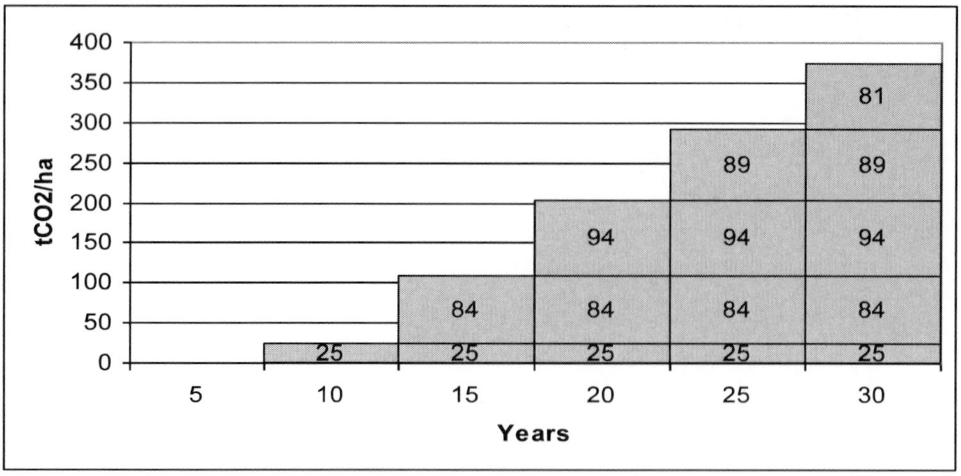

Figure 1. Accumulated carbon credits awarded to a teak plantation over 30 years.

Table 2. Present value of income from carbon credits with changes in tCER prices

Permanent price (Ghc/tCO_2)	Equivalent temporary tCER price (Gh¢/tCO_2)	Present value of income (Gh¢/ha)
10	2	1,338.95
20	4	2,677.90
25	5	3,347.38
30	7	4,686.33
35	8	5,355.80

The flow of income to the landowner can be calculated by estimating the price of temporary carbon credits from those of permanent credits using Equation 2. From the literature, estimates and forecasts of permanent carbon credits vary widely, from \$3/tCO_2 (The World Bank 2003) to about \$35/tCO_2 floated at the European Union Emissions Trading Scheme (PointCarbon 2008). In this analysis, various starting prices of a permanent carbon credit were used, and the corresponding tCER prices calculated using Equation 2, with a discount rate of 5% and t =5 years (Table 2).

One of the main implications of carbon credits from A/R activities is that the carbon for which credits have been awarded must be maintained till the end of the crediting period (permanence). This means that the trees cannot be harvested before the end of the crediting period. In the example in Table 2, if the landowner decides to obtain credits up to the maximum 30 years, then the trees cannot be harvested before the end of the 30 years, otherwise, he would be liable to replace the released carbon for which he has received payment.

We can analyse whether the landowner should apply for credits up to the maximum period of 30 years or for a shorter period by comparing the carbon and timber benefits at each 5-year interval. We will allow the flexibility for the landowner to decide at the end of each 5-year-crediting period whether he should renew the credits already received or terminate the project altogether (harvest the trees). The following assumptions are used in Tables 3 and 4.

A discount rate of 5%, the price of wood of Gh¢100/m^3, and a present value of costs of planting and management of Gh¢4000/ha are assumed. A higher regeneration and management cost is assumed here because managing the trees to generate carbon credits entails more transaction costs (application fees, monitoring, measurements, etc.) than just managing the plantation for timber only. The costs incurred are assumed to be joint costs of producing timber and carbon.

Table 3 presents the results of the present values of carbon benefits for various years and prices of tCER. The calculations were based on the allocation of carbon credits over the crediting periods in Figure 1. Each value in the table represents the amount of money the landowner would receive for the particular year and tCER price. For example, for the tCER price of Gh¢2/tCO_2 and year 10, the present value of the amount is Gh¢336.05. The totals show the present value of the accumulated revenue over the 30 years if the landowner decides to hold the trees for that length. If we now subtract the costs of production of Gh¢4000/ha, we notice that the carbon prices up to Gh¢5/tCO_2 do not make a profit. Carbon prices have to be greater than Gh¢5/tCO_2 for this project to make economic sense. Interestingly, the project would post a loss for all prices in the range used if the full 30-year crediting period is not used. This is seen from the fact that at each 5-year crediting period, the present value of the benefits is smaller than the present value of the costs, for the full range of the tCER prices.

Table 3. Present values of carbon credits for various years and prices of tCER

Year	Temporary credit carbon price Gh¢/tCO_2				
	2	4	5	7	8
5	176.04	352.08	440.10	616.14	704.16
10	336.05	672.09	840.11	1176.16	1344.18
15	363.47	726.95	908.68	1272.16	1453.89
20	339.87	679.75	849.69	1189.56	1359.50
25	299.56	599.11	748.89	1048.45	1198.23
Total	1,338.95	2,677.90	3,347.38	4,686.33	5,355.80
Net value*	-2,661.05	-1,322.10	-652.62	686.33	1,355.80

* Net value is the total benefit minus the costs of production.

Table 4. Timber and carbon values for various years and a tCER price of Ghc5/tCO_2

Year	Present value of Carbon (Gh¢/ha)	Present value of Timber (Gh¢/ha)	Net timber and carbon values (Gh¢/ha)*
5	440.10	-19,131.45	-18691.4
10	840.11	-211.39	628.72
15	908.68	5,288.54	6197.22
20	849.69	6,424.48	7274.17
25	748.89	5,872.61	6621.5

* Net values are calculated by subtracting the present value of costs (Gh¢4000) from the total benefits.

In Table 4, the tCER price of Gh¢5/tCO_2 is used to illustrate the impact of combined management for carbon and timber on the optimal time to terminate the project under the CDM. From Table 4, the NPV of the combined timber and carbon values is maximised in year 20. This is the optimal time for the plantation to be harvested. This suggests that although the plantation could have received additional carbon credits and timber revenue beyond this point, the returns are lower than if the owner decides to terminate the project, harvest and sell the wood and invest the money in a bank for a 5% interest on the proceeds. Even if the tCER of Gh¢8/tCO_2 is used; it results in the same optimal rotation age of 20 years.

The above analysis does not consider the risk of fires to these plantations. Bush fires that burn teak plantations is an annual occurrence in the savannah zones. A landowner could either spend additional resources to ensure that the plantations are protected from these fires, or buy additional insurance to ensure he can be reimbursed if a fire were to occur during the time credits have been received for the carbon sequestered by the plantation. These analyses suggest that carbon management under the current assumptions and CDM rules can be profitable, however, it must be emphasised that just because there is a potential for carbon credits does not mean all plantations would automatically become profitable. At present, there is no approved CDM forestry project in Ghana, and hence the full range of expected costs of managing for carbon is not available. The necessary ex-ante analyses need to be undertaken in each case using the most realistic expected costs and benefits to make well-informed decisions regarding the economic profitability of joint management of the plantations for carbon and timber.

13.8. Recent International Climate Change Negotiations

The outcome of climate change negotiations and agreements have implications on how forests are defined, managed and financed. For example, under the Kyoto Protocol, the definition of forest makes no distinction between planted crops of monoculture perennial woody plants and complex biodiverse natural forests. One area of climate agreements that will have major impacts on forest plantations in Ghana is related to climate funding. Forest plantation development is capital intensive, and Ghana may need to rely on funding from external sources to be successful. Funds made available through these international agreements will be a major source.

13.8.1. Recent Negotiations and Outcomes

The 15th session of the Conference of the Parties (COP-15) was held in December 2009 in Copenhagen, Denmark. The resulting Accord, known as the Copenhagen Accord is not legally binding and does not commit countries to agree to a binding successor to the KP, whose present round ends in 2012. The Accord endorses the continuation of the KP and agrees that the Parties to the Kyoto Protocol would strengthen their existing targets. The Accord recognises "the crucial role of reducing emission from deforestation and forest degradation and the need to enhance removals of greenhouse gas emission by forests," and the need to establish a mechanism (including REDD+) to enable the mobilisation of financial resources from developed countries to help achieve this. In regards to emissions trading, the meeting decided to pursue opportunities to use markets to enhance the cost-effectiveness of, and to promote, mitigation actions.

COP-16 was held in Cancun, Mexico in December 2010 and the major decisions re-affirmed some of the decisions of COP-15. The decisions out of Cancun, known as the Cancun Agreement, provides for action to reduce emissions from both developed and developing countries, embeds pledges from the Copenhagen Accord in a CoP decision, sets out a framework for avoided deforestation and confirms the continuation of emissions trading and project-based mechanisms under the Kyoto Protocol. It establishes a Green Climate Fund to help disburse money needed to help developing countries combat climate change, a Cancun Adaptation Framework and a Technology Network to support the deployment of low-carbon economy.

Following COP16, COP17 was held in Durban, South Africa in 2011. At the Durban conference, developed countries including the European Union (EU), Norway, Australia and New Zealand agreed to a second commitment period of the Kyoto Protocol, to begin from January 2012 to 2020. The second commitment period is weaker than the first (2008-2012) as it will be a voluntary basis not tied to any global carbon reduction targets. In addition, both developed and developing countries committed to completing a new international agreement to reduce greenhouse gas emissions that will come into effect in 2020; termed the Durban Platform for Cooperative Action. The Green Climate Fund that was set up at COP 16 in Cancun, Mexico in 2010 to assist developing countries transition to a low-carbon economy is now operational.

COP18 was held in Doha (Qatar) in December 2012. There were three main outcomes of the conference. The first is the continuation of the Kyoto Protocol based on the agreement in Doha. The second major outcome was an agreement to consider the creation of an international mechanism for loss and damage from extreme weather and slow onset climate impacts in developing countries. Finally, though the need for a plan for long-term finance was reiterated, no firm commitments on scaling up finance towards the agreed US$ 100 billion a year were forthcoming. Climate finance pledges amounting to approximately US$ 10 billion were made by some European countries.

A meeting of the UNFCCC was held in Warsaw, Poland from 11 to 23 November 2013. This was the 19th yearly session of the Conference of the Parties (COP 19) to the 1992 UNFCCC and the 9th session of the Meeting of the Parties (CMP 9) to the 1997 Kyoto Protocol. This conference built on previous meetings of the delegates to continue the negotiations towards a global climate agreement in 2015. At the Warsaw Conference, governments took further essential decisions to stay on track towards securing a universal climate change agreement in 2015. The objective of the 2015 agreement will be to bind nations together into an effective global effort to reduce emissions rapidly enough to chart humanity's longer-term path out of the danger zone of climate change, while building adaptation capacity; and secondly to stimulate faster and broader action now.

The latest meeting of the UNFCCC were the 20th session of the Conference of the Parties (COP 20) and the 10th session of the Conference of the Parties (CMP 10) serving as the Meeting of the Parties to the Kyoto Protocol took place from 1 to 14 December 2014 in Lima, Peru. Though the Lima meeting made important advances, the decisions related to forestry were even more important. Countries meeting in Lima made progress on providing support to avoid deforestation. For example, Colombia, Guyana, Indonesia, Malaysia and Mexico formally submitted information and data on the status of their greenhouse gas emission reductions in the forest sector to the UNFCCC secretariat following a similar submission by Brazil earlier in the year. These baselines are likely to increase the possibility of obtaining international funding under initiatives like Reduced Emissions from Deforestation and Forest Degradation (REDD+). In support of this, the COP President announced that an 'information hub' will be launched on the UNFCCC web site, spotlighting actions by countries carrying out REDD+ activities. The aim was to bring greater transparency on both the actions being undertaken, including safeguards for communities and the payments being made.

13.9. CONCLUSION

International climate change policies continue to recognise the important role of forests in mitigating climate change impacts. This has led to the creation of markets for carbon sequestered in forest plantations. The Kyoto Protocol has developed the Clean Development Mechanism, which provides a global market to trade in carbon credits. Ghana has already taken steps to benefit from this opportunity and has set up a designated national authority within the Ministry of Environment, Science, Technology and Innovation. There are other international initiatives such as REDD+ (reducing emissions from deforestation and forest degradation), which Ghana is seriously pursuing. There are opportunities for Ghana to include forest plantation development into the REDD+ strategy to secure long-term funding to support their development.

Part V. Conclusions

Chapter 14

TOWARDS SUSTAINABLE FOREST MANAGEMENT IN GHANA

14.1. SUSTAINABLE FOREST MANAGEMENT

14.1.1. What Is Sustainable Forest Management (SFM)?

Forest resources, and, in fact, all resources, need to be managed in such a way that they serve the needs of present and future generations. The idea to sustain resources is not new and has been practised in traditional communities for millennia, where people lived in harmony with their environment. With increasing populations and scarcity of resources, natural resources sustainability is becoming more difficult.

The genesis of the concept of SFM can be traced to the *Forest Principles* and Chapter 11 of Agenda 21, adopted at the United Nations Conference on Environment and Development (UNCED) in 1992. Principle 2b specifically states that:

> *"Forest resources and forest lands should be sustainably managed to meet the social, economic, ecological, cultural and spiritual needs of present and future generations."*

Since the UNCED in 1992, the concept of SFM has evolved, with further national and international efforts targeted at developing the practical methods of achieving and monitoring SFM, such as national and international Criteria and Indicators (C&I).

The importance of this Chapter is to emphasise that SFM needs to be practised in both the natural forests and plantations. The principles of SFM are the same whether applied to a natural forest or forest plantations. Practising SFM on natural forests and plantations provide mutually re-enforcing benefits.

14.1.2. Criteria and Indicators of SFM

To implement SFM strategies, Ghana has adopted the Criteria and Indicators (C&I) developed by the International Tropical Timber Organisation (ITTO), although these are not currently being implemented. The ITTO's C&I provide the tools for assessing trends in forest conditions and forest management and provide an interpretation of SFM and a common framework for monitoring, assessing, and reporting on progress. A criterion is an aspect of forest management that is considered important and by which SFM may be assessed. A criterion is accompanied by a set of related indicators and describes a state or situation that should be met to comply with sustainable forest management (ITTO 2005b). An indicator is defined as a quantitative, qualitative or descriptive attribute that, when measured or monitored periodically, indicates the direction of change in a criterion (ITTO 2005b). The indicators identify information needed to monitor change, both in the forest itself (outcome indicators) and as part of the environmental and forest management systems used (input and process indicators). If the values of any indicator are placed in a time sequence, they provide information on the direction of change, either towards or away from SFM. However, the indicators cannot, by themselves, establish whether management is or is not sustainable (ITTO 2005b).

The ITTO has worked with tropical country partners to develop the C&I. The ITTO document specifies the following seven criteria as essential elements of SFM:

Criterion 1: Enabling conditions for sustainable forest management

- *Policy, legal and governance framework*
- *Economic framework*
- *Institutional framework*
- *Planning framework*

Criterion 2: Extent and condition of forests

Criterion 3: Forest ecosystem health

Criterion 4: Forest production

- *Resource assessment*
- *Planning and control procedures*
- *Silvicultural and harvesting guidelines*

Criterion 5: Biological diversity

- *Ecosystem diversity*
- *Species diversity*
- *Genetic diversity*
- *Procedures for biodiversity conservation in production forests*

Criterion 6: Soil and water protection

- *Extent of protection*
- *Protective functions in production forests*

Criterion 7: Economic, social and cultural aspects

- *Socioeconomic aspects*
- *Cultural aspects*
- *Community and indigenous peoples' rights and participation*

There are indicators to assess each of these criteria. The ITTO criteria correspond closely with a global set of 'common thematic areas' of SFM that was agreed at ITTO/FAO-sponsored international conferences on C&I in 2002 and 2004. The common thematic areas are:

- extent of forest resources;
- biological diversity;
- forest health and vitality;
- production functions of forest resources;
- protective functions of forest resources;
- socioeconomic functions; and
- legal, policy, and institutional framework.

14.2 A Results-based Framework Approach to SFM

SFM is currently the widely accepted paradigm for managing forests around the world. SFM has replaced the concept of sustained yield, which was focused on maintaining the continuous supply of only timber products. The principles and approaches to SFM provide a sound basis for managing forests; hence adopting SFM practices would ensure that forests are able to continue to supply all products to meet the social, environmental, economic, cultural and spiritual needs of present and future users.

Achieving SFM is going to be a task of considerable difficulty for Ghana, given that over many decades, Ghana could not even achieve sustained yield of timber products, which was the main goal of the 1948 forest policy, and was obviously easier to attain than SFM. We believe that Ghana's approach to SFM has to begin with the basics: increasing or at least maintaining the current forest area. This can be achieved with two broad policy tools: reduce deforestation and forest degradation, and develop forest plantations. Until these are done, any efforts at achieving SFM at the national level would be a mirage. This is because managing forests for other values in addition to timber presumes that there is a forest to begin with. Therefore, ensuring the existence of forests has to be the starting point for Ghana.

Despite the advantages of SFM and the efforts of several governments, environmental groups, and development agencies to promote SFM, past analyses have shown that almost no logging of natural forests in the tropics can be considered sustainable (Rice et al. 2001). Rice et al. (2001) further argue that returns to investments in SFM are usually lower than those earned from conventional logging or other land uses, hence SFM has not been widely adopted. From an economic perspective, therefore, logging companies have no incentive to adopt SFM practices.

The FC currently uses a strategic planning process that is based on Ghana's Vision – 2020 and the Forest Sector Development Master Plan (1996-2020). Vision – 2020 hopes, among other things, to make Ghana a middle-income country by the year 2020, an open and liberal market economy and accelerate economic growth and narrow the gap between the standards of living of rural and urban people. Within these long-term national objectives, specific actions that need to be undertaken in the forestry sector in the short and medium term periods include (Ghana Forestry Commission 1998):

- co-ordination or close integration of operations between the government ministries/ departments/ agencies for agriculture and forestry, particularly in extension work;
- significant improvement and effective implementation of control measures;

- resource management within forest reserves; and
- introduce off-reserve harvesting controls.

There are some benefits to the current planning process, which includes the fact that it is based on long-term national goals. Secondly, the planning process is implemented in phases to ensure it is manageable. The process is also based on the 1994 Forest and Wildlife Policy, which has the overall aim of "*conservation and sustainable development of the nation's forest and wildlife resources for the maintenance of environmental quality and perpetual flow of benefits to all segments of society.*" However, the four actions listed above are simply focused on control measures and coordination; it is not clear the activities are linked to the vision. Secondly, there seem to be no indicators that measure progress towards the achievement of the national vision, and no plans for evaluating the results based on the implementation plan.

To facilitate Ghana's move towards SFM, it is recommended that the FC move from focusing on inputs to focusing on results by adopting a results-based framework (RBF) approach to planning and implementing forest sector priorities. In this section, we propose a vision for the forestry sector and use it as an example to illustrate how to develop a results-based framework to planning within Ghana's forestry sector. An RBF approach can be used to provide a consistent planning tool for achieving SFM. The RBF process begins by articulating a vision for the forestry sector. This vision becomes the over-arching goal of forest sector decisions and activities. The vision represents the type of results expected from SFM or the final outcomes of SFM practices. SFM should be seen as a *means* to an *end*, the *end* being the vision. The ultimate goal of Ghana's forestry practices could be *to promote sustainable forest management that supports a competitive forest sector, environmental responsibility and community security and sustainability.*

14.3. Strategic Outcomes and Activities

The RBF focuses on three strategic outcomes (SO) of forest sector activities: a) promote a competitive forest sector, b) demonstrate leadership in environmental responsibility through forest sustainability, and c) ensure community security and sustainability by providing forest and environmental benefits to communities (Figure 1). To achieve each SO under the RBF, a strategic outcome (SO) Plan, which articulates a multi-year vision and sets out strategic priorities, must be developed. The SO Plan consists of a stable long-term vision section that sets out the long-term vision (10 years and beyond), including long-term risks and challenges, and an annual priorities section that sets out the strategy for completing the work in the SO (for the next 1 to 3 years). To build a foundation for the long-term vision, each SO Plan sets out short and medium term goals. These goals are referred to as immediate (2-3 years) and intermediate (4-10 years) outcomes in Figure 1.

Each SO Plan is described in the format of a Logic Model, which displays how the activities of a policy or programme are linked to the results or final outcomes. The Logic Model identifies the causal linkages between the activities and the achievement of its outcomes. Immediate and intermediate outcomes measure success in the short and medium terms and are helpful in determining whether the programme or policy is on track to meet its final goals.

The activities are the inputs and are under the direct control of management. These activities are identified in Figure 1 and are described in detail in Sections 5.1 to 5.4. We propose a three-pronged approach to the kinds of activities that would help meet the strategic outcomes: measures to reduce deforestation, initiatives to promote and remove barriers to forest plantation development, and activities to develop and implement a national strategy for SFM practices on existing and future forests. The most effective strategy is to implement all three categories of activities simultaneously.

14.4.1. Measures to Reduce Deforestation

The rate at which Ghana is losing its forests is not going unnoticed by forestry and government officials. Hardly a month passes by without a government official making pronouncements in the media on the continuous decline of Ghana's forest area. A case in point is the August 2009 pronouncement by the Minister of Lands and Natural Resources, that "Ghana's total forest cover, which stood at 8.2 million ha at the turn of the 20th Century, has decreased to about 1.6 million ha[1]," further warning that "if the current depletion rate of 65,000 ha per annum is not halted, and reversed, Ghana's total forest cover would have been depleted within the next 23 years" (Business News 2009).

However, it seems the government is paying lip service to combating deforestation. Hence, these speeches and pronouncements are not backed by serious efforts to implement what is required to reduce deforestation. It is cheaper to avoid deforestation now than to develop plantations after the forests are lost, not to mention the difficulties of replicating the natural forest structure and function with plantations. The following measures, if implemented, would significantly reduce the current rate of deforestation.

Political Will to Combat Deforestation

The Government of Ghana has the largest role in fighting deforestation in Ghana, and the starting point would be the political willingness to do so. Historically, it is not clear that governments have demonstrated leadership and a willingness to reduce deforestation. As discussed above, there have been major policy restructuring over the decades all aimed at managing the forest resource, but without a practical focus on reducing deforestation, SFM cannot be achieved. Political will must be translated into concrete actions in terms of effective policies, provision of the necessary resources, and focusing deforestation issues in the context of national sustainable development goals. There is a lot of political focus on short-term economic gains from forest harvesting at the expense of sustainable management practices. Ghana needs to move away from harvesting trees to managing the forests.

[1] The Minister was referring to the closed forest area, which stands at 1.62 m ha, as opposed to the total forest area, which is 5.52 m ha as at 2009.

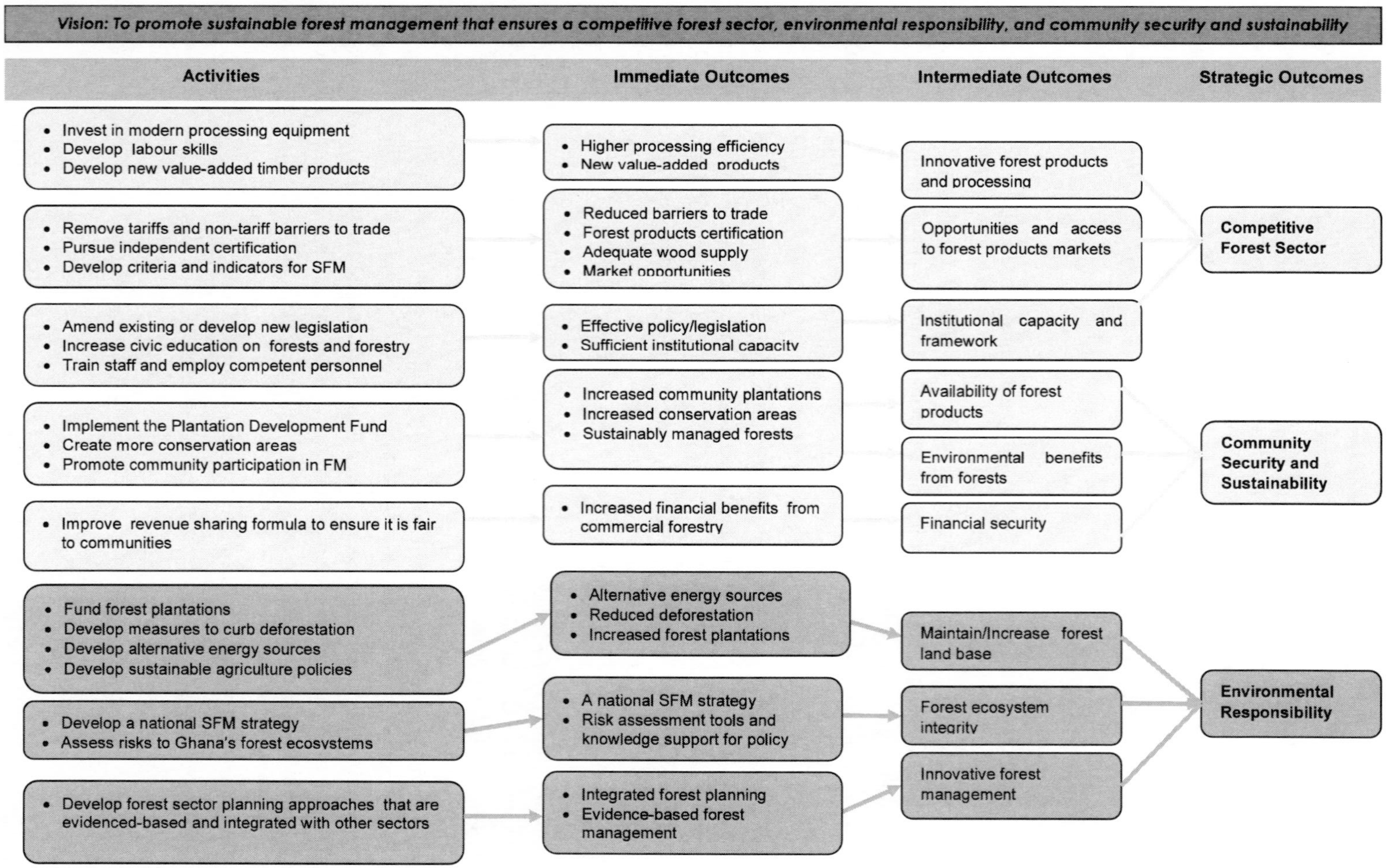

Vision: To promote sustainable forest management that ensures a competitive forest sector, environmental responsibility, and community security and sustainability
Activities
Immediate Outcomes
Intermediate Outcomes
Strategic Outcomes
• Invest in modern processing equipment
• Develop labour skills
• Develop new value-added timber products
• Remove tariffs and non-tariff barriers to trade
• Pursue independent certification
• Develop criteria and indicators for SFM
• Amend existing or develop new legislation
• Increase civic education on forests and forestry
• Train staff and employ competent personnel
• Implement the Plantation Development Fund
• Create more conservation areas
• Promote community participation in FM
• Improve revenue sharing formula to ensure it is fair to communities
• Fund forest plantations
• Develop measures to curb deforestation
• Develop alternative energy sources
• Develop sustainable agriculture policies
• Develop a national SFM strategy
• Assess risks to Ghana's forest ecosystems
• Develop forest sector planning approaches that are evidenced-based and integrated with other sectors
• Higher processing efficiency
• New value-added products
• Reduced barriers to trade
• Forest products certification
• Adequate wood supply
• Market opportunities
• Effective policy/legislation
• Sufficient institutional capacity
• Increased community plantations
• Increased conservation areas
• Sustainably managed forests
• Increased financial benefits from commercial forestry
• Alternative energy sources
• Reduced deforestation
• Increased forest plantations
• A national SFM strategy
• Risk assessment tools and knowledge support for policy
• Integrated forest planning
• Evidence-based forest management
Innovative forest products and processing
Opportunities and access to forest products markets
Institutional capacity and framework
Availability of forest products
Environmental benefits from forests
Financial security
Maintain/Increase forest land base
Forest ecosystem integrity
Innovative forest management
Competitive Forest Sector
Community Security and Sustainability
Environmental Responsibility

Figure 1. A Results-Based Framework to forest sector planning in Ghana.

Sustainable Energy Policy

Forests and woodlands serve as the primary source of fuelwood energy for at least 75% of the total population of Ghana, and about 90% of the rural population (Benhin and Barbier 2001). Fuelwood use, therefore, is one of the major causes of deforestation (Allen and Barnes 1985; Charkraborty 1994). Deforestation has traditionally been considered an issue for the forestry sector alone. More recent studies (e.g., Kant and Redantz 1997) recognise the complex interactions and inter-relationships associated with deforestation. It is clear that successful reductions in deforestation would depend on a national energy policy that significantly reduces the dependence of Ghanaians on fuelwood energy derived from the unsustainable forest and woodland practices. In 2013, wood fuel was estimated at 40 million m^3 of roundwood production in Ghana (FAO 2014). Given an estimated growth rate of Ghana's forests of only 4.6 million m^3 per year, Ghana's level of fuelwood consumption is clearly not sustainable.

Improving energy efficiency is imperative, so is increasing the production and use of clean alternative sources of energy, such as solar, the wind, and hydropower. The government has taken some positive steps towards increasing hydro power production with the construction of the Bui hydroelectric dam, which completed in 2013. However, the Akosombo Dam has suffered from low water levels in the Volta River in the recent past. Forest cover removal and resulting siltation in, and evaporation from, the tributaries of the Volta River upstream (Northern Ghana and Burkina Faso) could have contributed to these low water levels. It is important for government and the FC to recognise the role of forests in protecting critical watersheds in Ghana; otherwise, the Bui dam would suffer from similar low water levels in future. This further highlights another dimension of the inter-relationships between the energy and forestry sectors. Solar power usage is also supported by the government by the removal of import duties on solar energy product imports into the country. Wind power is still in its infancy and has so far been limited to identifying some potential wind farm sites in the Northern and Volta regions. Another practical alternative is to increase fuelwood supplies through developing forest plantations for such purposes, especially at the community levels in the savannah vegetation zone.

Sustainable Agricultural Policies

The agricultural sector as a whole provides employment for over 60% of the Ghanaian population, many of whom are engaged in food crop production. Growing scarcity and the need for more fertile cropland in Ghana have caused expansion of food cropland into forest areas. The need to increase agricultural production to feed the growing population should be a priority for the government. However, care must be taken to ensure that such increases in production are not at the expense of the forests. Intensive agricultural production, rather than extensive production, should be pursued. Agroforestry practices and the modified taungya system could minimise the impacts of increased food production on forests. Providing irrigation facilities especially in Northern Ghana would promote multiple cropping per year and increase food production on the same land base. It is fair to note that when it comes to land use conflicts, agriculture and forestry are the major competitors. This is one area where rational and integrated policy planning and focus would be required, such that proposed development projects are viewed in the context of overall national development goals, rather than narrow sectoral interests.

Curb Illegal Logging

Illegal logging is a menace that has plagued the forest industry in Ghana for many decades. This practice has been fuelled by the shortage of wood products for the domestic market, resulting from government's over-emphasis on promoting export trade in timber and wood products. For example, it is estimated that in 1999, over two-thirds of the wood harvested in the country was done illegally and mainly on off-reserve areas and dominated by the activities of illegal small-scale chainsaw operators (Gayfer et al. 2002; Sarfo-Mensah 2005). To reduce illegal logging, there is a need to increase the supply of wood products to the local market. In the 1980s and 1990s, government's policy response was to impose levies on exported products as a means of increasing the domestic supply. The policy failed because even with the levies, the exported products still earned more revenues than the domestic market for the timber firms. Another attempt in 2005 was when the government directed sawmills to sell at least 20% of their timber production on the domestic market before permits for timber exports are approved. This policy is unlikely to be successful in reducing illegal logging since analyses show that the 20% is less than what the domestic market requires (Sarfo-Mensah 2005).

As the AAC is continuously reduced to reflect the scarcity of the forest resource, it is obvious that the wood supply will experience an acute shortfall. This could lead to increased illegal activities that would deplete the forest. The biggest challenge for the government is how to meet the increasing demand for wood products from a diminishing resource base. In this case, we would argue that we should challenge our conventional thinking that demand for wood will continue to grow. If we think that way, our only response would be to continue to increase wood supply to meet this demand. In short, we should think about how to use less wood even as population grows, i.e., efficient use of the wood we have as a way to mitigate the impacts of impending wood shortages.

With increasing population and demand for wood products in Ghana, the effectiveness of increasing wood supply to the domestic market as a tool for curbing illegal logging is limited. This means that other tools would have to be employed, such as enforcing existing laws, importing logs, efficient use, alternatives to wood, reducing corruption in the forestry sector and enforcing building and other construction codes. If building codes were developed and enforced in Ghana, it would ensure that only wood products that meet the standards (e.g., kiln-dried sawnwood) can be used in construction. This would significantly reduce the demand for illegal sawnwood, which is usually of poor quality and would not meet building standards. If there is no demand for illegal wood in Ghana, there would be no economic incentive for its production. Existing legislation in Ghana is sufficient to ensure that wood processors do not purchase wood from illegal and questionable sources. However, as with many of Ghana's legislations, they are not effectively enforced.

Another important reason for illegal logging is that it may be the only source of employment for those engaged in it. The government should ensure that alternative livelihoods are provided for chainsaw operators, such as absorbing them into reforestation projects, or engaging them in legal ventures such as salvage felling or forest plantation harvesting. Finally, communities must be educated on the dangers of using processed wood from illegal sources as these pose a risk of failure in use. Communities should be advised to alert the authorities to illegal harvesting activities and given financial incentives for their cooperation to reduce illegal logging. A better way to curb illegal logging is to put the responsibility of protecting the forest resource squarely in the hands of those with legal rights to harvest it.

Take advantage of International Climate Change Frameworks

Ghana ratified the United Nations Framework Convention on Climate Change (UNFCCC) in 1995 and the Kyoto Protocol in May 2003. To help countries achieve their targets under the Kyoto Protocol, flexible mechanisms have been incorporated, such as the Clean Development Mechanism (CDM), which allows developed countries to undertake carbon sequestration projects in developing countries, which can help developed countries reach their emission targets. Ghana has taken steps to take advantage of the CDM by setting up a designated national authority on CDM within the Environmental Protection Agency. Deforestation and forest degradation are the second leading cause of global warming, responsible for about 20% of global greenhouse gas emissions, which makes the loss and depletion of forests a major issue for climate change. As a result, the World Bank introduced the Reducing Emissions from Deforestation and forest Degradation (REDD) initiative to compensate developing countries for reducing their emissions from deforestation. As at 2008, Ghana was one of the few countries in Africa that expressed interest in participating in the REDD initiative and submitted a readiness project idea participation note. Ghana should continue to pursue and participate in these initiatives and any future schemes that provide financial resources to reduce deforestation.

Ban Surface Mining within All Forests

Surface mining is a major cause of deforestation because large tracts of land are cleared by heavy machinery to carry out the operations. These operations leave large open areas that are no longer suitable for agricultural production due to the cyanide and other chemicals used in the production process. Especially in the Western Region of Ghana, there have been several mass demonstrations, conflicts and court actions taken by residents to protect what is left of the tropical forests from usually multinational companies seeking huge profits. The biggest tragedy is the government of Ghana shirking its responsibilities to protect the interests and rights of the residents to a safe environment, sacrificing these at the altar of foreign exchange and tax revenues from mining. In 2003, a coalition of public interest, human rights, labour and environmental groups launched a campaign against mining in forest reserves. The coalition expressed outrage at the decision of the government of Ghana to open up some of the forest reserves for surface mining and asked the government to withdraw mining licences that had already been issued to some of the mining companies (Anane 2003). The devastating impacts of surface mining are enormous: forest loss, polluted water sources, biodiversity loss, evacuation of inhabitants, degraded soils, etc. The government should show some courage by enforcing legislation that completely bans surface mining in forested areas across Ghana. Mining companies often claim that they undertake environmental assessments and reclaim and reforest any lands, but the evidence from abandoned mining sites shows otherwise.

Promote Community Participation

Forest resource use conflicts arise where there is a divergence of opinion between local communities and the state regarding access rights, control and benefits that accrue from resource use at the various societal levels. Governments have often relied on strict law enforcement to achieve conservation objectives. It is now widely recognised that this approach is largely ineffective. Community participation allows local communities to have a direct say over the utilisation and benefits from forest resources, which ensures that they value them in a sustainable manner. Involving local communities would reduce deforestation and contribute to SFM. Forestry and forests should be considered integral parts of community development efforts. Currently, community participation in forestry is being facilitated through community forest committees (CFCs) and a Collaborative Forest Management Unit of the FC. In 2003, there were some 100 CFCs across Ghana (ITTO 2005a). However, more CFCs need to be formed to cover the majority of the forest dependent communities. In addition to these CFCs and collaborative forest management initiatives, local communities should be encouraged to develop local regulations that are consistent with their traditional practices to protect vulnerable ecosystems and manage common pool forest resources. When communities are empowered to manage and regulate the use of resources, it reduces the pressure on, and the misuse of, the resources because they recognise that their livelihood depends on them.

Enforce Existing Forestry Legislations

Ghana has several laws, regulations, policies, and procedures that govern the management and use of forest resources in the country. For example, the Planning Branch of the Forestry Department (now the FSD) developed a 'Handbook of Harvesting Rules for Sustainable Management of Tropical High Forest in Ghana' in 1992 to serve as a guide to all forest exploiters to enable them bring their practices in keeping with the SFM plans. However, most of these procedures are not effectively enforced. The most commonly cited reason is the lack of resources (personnel and money). However, the real reason is likely related to connivance among law enforcement agencies, FSD staff and those who harvest timber (illegal loggers or timber utilisation contract/concession holders). Enforcing laws maintains property rights, which is critical to the efficient management of all natural resources. The inability to enforce laws sends the wrong signal that Ghana is a lawless country, and this is a disincentive to foreign and local investors.

Eradicate Corruption from the Forestry Sector

Corruption is a major problem facing many developing countries, and Ghana is no exception. Corruption is a major threat to the sustainable management of forests and seems to be widespread across various levels of the forestry sector in Ghana. It is believed that Forestry officers connive with loggers to manipulate the allowable cuts and /or look the other way as unsustainable harvesting practices are carried out. For example, in 2005, the FC attempted to collect stumpage fees arrears that timber companies had accumulated for over eight years. The Ghana Timber Association (GTA) sued the FC in court to block the collection and threatened to expose long-term collusion of FC officials by allowing companies to cut undersized logs provided they declared the minimum girth on the log measurement and conveyance certificates (Forest Watch Ghana 2006). The out of court settlement involved the FC accepting deferment of arrears for another year interest-free, waiving of minimum girth standards, and relaxing of AAC rules (Forest Watch Ghana 2006).

The FC needs to stamp out the bad nuts in the system to ensure SFM practices. The usual strategy is to transfer corrupt officials to Northern Ghana where there is no commercial harvesting of natural forest timber. This is a poor solution as these people find their way back into the high forest zone as their careers progress, and end up at the helm of affairs in the forestry sector. Beyond the forestry sector, the government needs the political will to combat corruption in all spheres of Ghanaian society. The judicial system does not treat environmental crimes with the seriousness they deserve.

14.4.2. Develop and Implement a National Strategy for SFM

The next logical step once Ghana has maintained or increased the forest area is to ensure that those forests are managed in a sustainable way to provide the goods and services demanded by society. There is an urgent need for Ghana to develop a national strategy to achieve sustainable forest management. Most forestry programmes are designed in Accra, with the help of the international donor community who often provide financial support. A better approach is to determine the key priorities and overall direction for the sustainable management of Ghana's forests through extensive cross-country consultations and public dialogue with the forest communities and all stakeholders.

Rational and Integrated Policy Planning and Focus

One of the key ways of ensuring that forests play a major role in the national development agenda is to introduce a rational and integrated policy planning and focus. At the national level, planning for forestry should not be conducted in isolation but should recognise the impacts of policies in other sectors on forests. For example, policies in the agriculture, infrastructure and energy sectors directly or indirectly affect forests. The Ghana Forest and Wildlife Policy (2012) situates forestry development within the national development agenda, and this is a very good start.

Develop and Implement Criteria and Indicators (C&I)

As discussed above, criteria and indicators are the signposts that show whether and how forest management practices are close to SFM. It is commendable that Ghana is involved and has adopted the C&I from the ITTO. This is a sign of commitment on the part of the government and the forestry sector to SFM principles. But the commitment should not be limited to the principle, but must go further, to implement these C&I with all seriousness. A critical review of the criteria should be undertaken to ensure that they suit our local conditions. Also, it is important to note that efforts and resources should not be diverted to international initiatives at the expense of local initiatives to reduce deforestation and increase the forest area.

Enforce Sustainable Harvesting Standards

Forest harvesting by timber companies does not necessarily have to lead to deforestation. However, most harvesting practices in Ghana are still based on outdated equipment and techniques that result in serious degradation of the residual forest. Harvesting should be conducted in ways that maintain the ecological integrity of the residual forests and promote sustainability. The FC has manuals on procedures for harvesting, and the authorities should enforce these standards and other reforestation requirements on the part of contract holders. There is simply no reason this should not be done.

Pursue Independent Forest Certification

Forest certification is a system for identifying forests that are managed to maintain ecological, economic and social components of the ecosystem. Certification is a market-based mechanism to reward SFM, and in a way resembles a command-and-control approach as those who do not certify their forests could be penalized by the loss of market share for their products. Although many importing countries have not yet made forest certification mandatory, it is clear that countries that wish to have unfettered market access around the world would be compelled to show that their wood is coming from sustainably managed forests. The Forest Stewardship Council (FSC) is the most recognised NGO that issues third-party certification around the world. Ghana has been developing the processes needed to certify its forests, and this is a step in the right direction. The Ghana Forest Management Certification System Project was initiated in 1997 with the assistance of the European Union and the Netherlands to develop draft standards for certification. Field tests on the resulting chain-of-custody and log tracking systems have been carried out since 2002, as has the development of standards for sustainably managed forests (ITTO 2005a). The forest industry in Ghana is dominated by small to medium scale enterprises, leading to over fragmentation of the timber industry. This is a major constraint to pursuing certification, which is often expensive to small timber companies. The concessions of Samartex Timber and Plywood Ltd, of about 110,000 ha in the natural forest production area, were certified in 2003 by SGS for compliance with Ghanaian logging and chain-of-custody standards. However, several challenges remain, especially as it relates to the rights and access to benefits of fringe communities to forest resources. The government should continue to ensure that timber firms pursue an independent certification, using economic and legislative tools.

Implement Natural Resource and Environmental Accounting

Standard computations used in the System of National Accounts (SNA) to calculate the gross domestic products (GDP) of countries usually accounts for depreciation of capital (such as machinery) used in manufacturing. However, no similar consideration is given to the depreciation of natural capital, such as depletion of forests. Once the costs of natural resource degradation are included in the SNA, it may become clear that the GDP growth is less than when depletion is not explicitly taken into account. Any economic growth that is carried out by depleting the natural resource base unsustainably should be avoided. An additional advantage of natural resource accounting is that policymakers can explicitly observe the costs of forest depletion and its implications for the long-term sustainability of forests and the economy.

14.4.5. Implementing the Strategic Outcome Plans

Effective implementation of the SO Plans would require that each year, the FC establish priorities by working with other government agencies, forest sector stakeholders, and international and development partners. It should then align available resources with the chosen priorities. It may be necessary to realign existing programmes to support those priorities, including using programme renewal to redesign and/or develop new programmes.

Finally, performance indicators should be developed at the planning stage to measure results as programmes are implemented.

14.5. CONCLUSION

Sustainable forest management is the current paradigm for managing forests around the world. Despite several policy changes and restructuring in Ghana since 1979, the area of the permanent forest estate is still declining at an alarming rate. For Ghana to achieve SFM, it needs first to be able to increase or at least maintain its current forest area. Without concrete and immediate action, Ghana's forests could disappear in a matter of 50 years from now. Reducing deforestation and developing a national strategy for sustainable forest management are the first two steps. The next Chapter focuses on the third requirement for achieving SFM: establishing forest plantations.

Chapter 15

Necessary Conditions for Successful Plantation Forestry Development

The forestry sector in Ghana relies heavily on the exploitation of the natural forest resource to meet the needs of its population. The sector remains one of the major sources of foreign income and supplies the domestic market with a wide range of timber and non-timber forest products. Presently, plantation forestry plays only a minor role, with the majority of plantations being teak, which were established over many decades across the country. Despite several decades of attempts at plantation forestry and different versions of national plantation programmes, none of these has led to a substantial increase in the land area planted to forest plantations in Ghana. The reasons for this failure may lie in the constraints identified in Chapter 3. Plantation establishment and management are more labour and capital intensive than managing natural forests, and hence require lots of initial and continuous investments of capital, technical expertise, long-term commitment, economic management skills and forward integration to value adding secondary processing to be successful (AFORNET 2008).

In this final chapter, we discuss the conditions that must prevail to achieve a successful plantation programme in Ghana. Some of the requirements are currently being pursued to different degrees in Ghana, while others are not. These recommendations are categorised into four major areas: economic, policy, technical and institutional and are required to promote, encourage and support plantation forestry development in Ghana.

15.1. Economic and Fiscal Environment

Plantation forestry is essentially an economic venture, whether they are grown in private industrial plantations, community or individual woodlots, or for environmental protection. Therefore, the economic benefits derived from the investments play a large part in determining its success. Furthermore, the economic environment and the fiscal and monetary policies in the country work to affect the outcomes of investment decisions. The economic requirements for successful plantation development centre on two main areas: financial ability to establish and manage plantations, and the ability to make a profit from the plantation enterprise. In this case, the major areas for improvement include taxation, financial assistance through incentives or loans, and market availability. The following are recommended:

- The Government should, through its fiscal and monetary policies, promote plantation development. First, economic stability makes investments in Ghana attractive and hence could encourage private sector investments in plantations. Secondly, plantation forestry can be promoted by ensuring that taxation of plantations and plantation forest products are favourable to investment. A favourable environment can be created by providing tax and import duty exemptions for investments in forest plantations. Currently, most of the tax-related incentives are targeted at industrial plantations and foreign investors. However, the majority of plantation forests in Ghana would remain small- to medium-sized community and individual plantations. Therefore, tax exemptions should be expanded to include community, individual and NGO-supported plantations. For example, investors in plantation forests greater than a certain minimum size, say five ha, should be allowed by the tax laws to deduct the costs of planting and management from their annual tax obligations. Finally, tax policy on the forest industry should be targeted at achieving increased domestic processing in the country.
- There is a need for the continued flow of private investment into plantation forestry. It is recognised that the government alone cannot fund plantation forestry. Private investments can only be achieved if the government continues to ensure a climate that promotes private investments. Private investors do not necessarily have to be foreigners. Efforts should be made to make plantation forestry attractive to local businesses, such as banks, timber companies and other financial institutions.
- The government should provide low-interest rates loans or loans with zero interest, or grants to investors in plantation forestry.
- The National Forest Plantation Development Fund (NFPDF) should be streamlined to provide financial support to small-scale farmers and communities that want to establish forest plantations. Although, in theory, there is no limitation on who can apply, the bureaucratic bottlenecks, processes and costs involved in applying for funds from the NFPDF are beyond the abilities of most small-scale farmers. These processes should be streamlined to ensure that small-scale farmers can truly benefit from these incentives.
- The government should continue to provide seedlings and other materials to help in the establishment and management of plantations.
- The government should promote lesser-used plantation species locally and internationally to increase acceptability and marketability. Plantation products that have no chance of being marketed do little to support the economic sustainability of the enterprise. International marketability of plantation forest products would increase the value of the plantation products, and hence their economic profitability.

15.2. Policy and Legislative Changes

Policies affect how investment decisions are made by individuals and firms. Therefore, poorly designed and coordinated policies can be detrimental to any plantation development programme. It is important that the government re-examines the current forest policy and legislative framework to ensure that the policies are not only relevant but also effective and efficient. The following recommendations are made:

- To promote integration of trees into rural and urban environments, the Ministry of Lands and Natural Resources should collaborate with other ministries such as Education, Health, Environment, etc. to direct all public facilities in Ghana to ensure at least a 30% tree canopy cover on their properties.
- There is an urgent need for policy and legislative reform to remove inconsistencies and contradictions in the legislative framework in the forestry sector, and to harmonise these with other sectors of the economy. Forestry laws and regulations should also be consolidated to ensure efficiency. The current practice of superimposing plantation forestry laws/regulations directly on or by amending existing laws that were intended for natural forests creates more confusion. These amendments would remove the ambiguities in the forest legislation as it relates to plantations. For example, the kinds of fees, taxes and export levies applicable to forest plantations and plantation forest products must be clarified.
- Section 21 of the *Forest Plantation Development Fund Act* 2000 (Act 583) requires the Minister responsible for forestry to make regulations for the effective implementation of the Act. After 14 years of passing Act 583, these regulations have not yet been made. These regulations should be made as a matter of urgency, and should include provisions that direct a prescribed percentage of the funds of the Fund to support small-to-medium scale plantation forestry. Even more useful would be regulations that prescribe a streamlined application process that is less bureaucratic for these categories of applicants. In fact, farmers and communities should be able to apply for funds through their District Assembly offices, rather than in Accra.
- Section 7(b) of the *Forest Plantation Development Fund Act* (2000) that empowers the Fund Board to "invest" the Fund money should be amended to remove the wording that has led to the Fund Board investing the funds in money markets rather than in plantation forestry. Furthermore, improved monitoring of the investments made by the Fund Board will ensure that the funds are used appropriately.
- The *Timber Resources Management Act* 1997 (Act 547, section 8(d)) should be amended to remove the requirement for TUC holders to reforest their contract areas after logging. Given that the reforestation requirement has not achieved much success, the regulations should instead impose a plantation forest levy that should go into a specific fund which should be used to hire a private firm to undertake reforestation activities in logged areas.
- Modernisation of the land and tree tenure systems in the country to provide well-defined and enforceable property rights over trees is essential. A new policy, backed by legislation that clearly defines ownership rights over timber and carbon sequestered (and other non-timber benefits) by plantation forestry under the different

types of land ownership and tenurial arrangements in Ghana is urgently needed. It is obvious that people will not invest in plantations if there are ambiguities as to who will get the benefits of the investments. Therefore, changes to the land tenure systems in Ghana are absolutely critical for the success of plantation forestry.

- The government must continue to maintain policies that promote plantation development and to develop the legislative framework to recognise market mechanisms for carbon offset trading within the CDM or a separate domestic offset trading scheme. In fact, it is time for Ghana to develop a domestic carbon offset trading scheme that will recognise and award credits for the carbon sequestered by forests and other greenhouse gas mitigating activities. The availability of carbon markets will stimulate the development of low-carbon energy technologies within the industrial sector.
- Because plantations are long-term investments, it is often the case that initial interest in the projects fades over time. Governments must commit to maintain this interest and encouragement needed to see that plantations become successful.
- The government should introduce favourable trade laws for plantation logs or products that make it easy to market plantation products in domestic and international markets.
- The government should ensure that policies related to the forestry sector minimise, rather than increase corruption. For example, policies that involve lengthy bureaucratic processes are often prone to corruption because people prefer to pay their way to avoid following through the cumbersome processes.
- Policies that support and encourage stakeholder participation in plantation development are essential to ensure that the trees are well tended and protected. Whether communities are the owners of the plantations or live adjacent to plantations, obtaining their buy-in into the project at the beginning of the project and throughout its lifecycle are critical to the success of plantations.
- Ghana is a small country in terms of land area, and hence land use efficiency has to be central to the government's priorities. Policies that reduce urban sprawl and those that increase intensive, rather than extensive agriculture should be pursued. In short, an integrated land-use planning system approach that uses land efficiently should be adopted.

15.3. Technical Requirements

The technical needs of plantation forestry lie mainly in effective research, training and extension. Research is the foundation for developing a successful national plantation programme. Training and extension ensure that the results are effectively disseminated and utilised. These are the areas where the government can make the most direct contribution in terms of funding and support.

- The government and private sector investors should support sustained research into plantation forestry. There should be an efficient mechanism to disseminate the results to those who need it most. Some of the funds accruing to the National Forest

Plantation Development Fund should be used to support research in plantation forestry.

- The research should focus on: seed production and vegetative propagation, species/site matching, species trials, testing multipurpose species for community forestry and industrial agroforestry, economic viability of plantations, local factors affecting plantation adoption, diseases and pest management, growth and yield models, fire management, social research on how to ensure true community participation, etc.
- Training forestry professionals at all levels must be improved, with a focus on plantation forestry and extension training. With the rapid decline of the natural forest resources, it is clear that forestry training that remains focused on natural forest silviculture and management would be inadequate in meeting plantation forestry requirements of the future.
- The government should provide funding to FORIG to augment its seed production and storage capabilities. The basis of any successful plantation programme relies on the availability of good seeds for the targeted species. In addition, FORIG should be funded to establish permanent nurseries that can produce good quality seedlings for plantation development programmes.
- Information dissemination and extension services for plantations are weak and need to be improved. More forestry technical officers need to be deployed to targeted communities, more forestry offices in strategic districts in the country, and the promotion of forestry as an environmentally sustainable land use option would work together to ensure continuous technical support to tree growers and increase plantation forestry acceptance.

15.4. Institutional Reform

There is a strong link between institutions and policy, as the former helps in setting, interpreting and enforcing the rules of society. Ghana's institutional framework can shape forest plantation development through the policy process; as a result, the following recommendations related to institutions are made:

- First and foremost, forest sector governance reform that promotes multi-sector and stakeholder support and participation is needed to encourage plantation development efforts.
- The government of Ghana must commit to ensuring that plantation forestry remains a national policy priority for the next several decades. In this regard, the importance of the Plantation Department of the FSD should be elevated to ensure that it plays a much bigger role in the forestry sector.
- The planning and management functions of the Forestry Commission should be re-examined, with a view to eliminating any inefficiencies in the existing structure. The FC should remain the central agency for planning while implementation of plantation forestry programmes can be out-sourced to another public or private organisation.

- The Forestry Commission has made public the Ghana Forest Plantation Strategy for 2015-2040. The planting targets should be legislated so that all future governments should be obliged to fund adequately and implement the plan and report to Parliament each year on the achievements of the programme. Together with legislated targets, institutional changes are also required to ensure that the FSD receives adequate funding to support plantation activities. Consequently, plantation forestry funding should be allocated in the annual national budget.
- The government of Ghana should directly undertake tree planting to target environmentally sensitive and poor areas in the country. These should include ecologically fragile watersheds, major rivers and streams (such as the tributaries of the Volta River) mountainous regions and other ecologically sensitive areas.
- District Assemblies should be encouraged to use some of their Common Fund allocation to support community and individual growers as a way to improve community energy supplies and incomes. Some of their Common Fund should be set aside for tree planting to protect critical watersheds and waterways, such as streams, rivers and dams.
- The government should go into partnerships with industry or the NGO-community to develop plantations. For example, the government could pay upfront for the establishment costs and later hand over the management of the plantation to the private entity, in return for a percentage equity in the plantation enterprise.
- The government should seek international funding to support community-based and individual-based plantation forestry programmes.
- Industry-community partnerships should also be encouraged, whereby industrial investors can work with communities to establish and maintain forest plantations that meet the needs of the communities and those of the industrial partners.

REFERENCES

Abildtrup, J. (1999). Optimal thinning of forest stands: An option value analysis – preliminary results. *Scandinavian Forest Economics*, *37*, pp. 2-1 – 2-13.

Adegbehin, J. O. (1982). Preliminary results of the effects of spacing on the growth and yield of *Tectona grandis*, Linn F. *Indian Forester*, *108*, pp. 423-430.

Adegbehin, J. O., Abayomi, J. O. & Nwaigbo. L. B. (1988). *Gmelina arborea* in Nigeria. *Commonwealth Forestry Review*, *67*(2), pp. 159–166.

Adjei, S. & Kyereh, B. (1999). Land suitability assessment of some degraded forest reserves and headwaters for the establishment of *Ceiba pentandra* plantations in the dry semi-deciduous forest in Ghana. *Journal of the Ghana Science Association*, *1*(2), pp. 110-124.

Adu-Nsiah, K. (2009). Ghanaians consume $205 million of bushmeat annually. *Report of Ghana News Agency*, August, 25, 2009. www.ghanaweb.com.

AFORNET, (2008). Plantation forestry in Sub-Saharan Africa. Lessons learnt on Sustainable Forest Management in Africa Policy brief No 4. African Academy of Sciences, Nairobi, Kenya.

Agyeman, V. K., Marfo, K. A., Kasanga, K. R., Danso, E., Asare, A. B., Yeboah, O. M. & Agyeman, F. (2003). Revising the taungya plantation system: new revenue-sharing proposals from Ghana. *Unasylva*, *45*, pp. 40-43.

Ahmed, S. (1984). Use of neem materials by Indo-Pakistani farmers: Some observations. *In* Saxena R.C and S. Ahmed (Eds.). Proc. Res. Planning Workshop, *Botanical Pest Control Project. Int. Rice Res. Inst.*, Los Banos, Philippines.

Ahmed, S. A. & Koppel, B. (1985). Plant extracts for pest control: village level processing and use by limited-resource farmers. Paper presented, Amer. Assoc. Advancem. Sci., annual meeting, Los Angeles CA, May, pp. 26-31.

Ahmed, S. A. & Grainge, M. (1985). Use of indigenous plant resources in rural development; potential of the neem tree. *International Journal for Development Technology*, *3*(2), pp. 123-130.

Ahmed, S. A. & M. Grainge. (1986). Potential of the neem tree (*Azadirachta indica*) for pest control and rural development. *Economic Botany*, *40*(2), pp. 201-209.

Aidoo, J. B. (1996). Our Common Estate Tenancy and the Land Reform Debate in Ghana. *The Royal Institution of Chartered Surveyors*, London, England. p. 13.

Akindele, S. O. (1991). Development of site index equations for teak plantation in South-western Nigeria. *J. Trop. For. Sci.*, *4*, pp. 162-1 69.

Akobundu, I. O. (1986). Allelopathic potentials of selected legume species. In: *IITA: Resource and Crop Management Program. Annual Report*, 1986, pp. 15-19. IITA, Ibadan, Nigeria.

Alam, M. K., Siddiqi, N. A. & Das, S. (1985). Fodder trees of Bangladesh. Bangladesh Forest Research Institute, Chittagong, Bangladesh. 167 p.

Alder, D. (1980). *Forest volume estimation and yield prediction*, Vol. 2, Yield prediction. FAO forestry paper, 22. Rome. FAO.

Allan, G. G., Gara, R. I. & Wilkins, R. M. (1973). Phytotoxity of some systemic insecticides to Spanish cedar. *International Pest Control*, *15*(l), pp. 4-7.

Allen, J. C. & Barnes, D. F. (1985). The Causes of Deforestation in Developing Countries. *Annals of the Association of American Geographers*, *75*(2), pp. 163-184.

Anane, M. (2003). Gold Rush in Ghana's Forest Reserves Resisted. Environment News Service. http://www.ens-newswire.com/ens/may2003/2003-05-13-01.asp. Accessed July, 2009.

Anderson, F. J. (1976). Control theory and the optimum timber rotation. *For. Sci.*, *22*, pp. 242 - 246

Anderson, F. J. (1992). Natural resources in Canada. Economic theory and policy. 2nd edition. Nelson Canada. Scarborough. p. 292

Anderson, D. (1986). Declining tree stocks in African countries. *World Development*, *14*(7), pp. 853-863.

Anderson, D. & Fishwick, R. (1984). *Fuelwood Consumption and Deforestation in African Countries*, World Bank Staff Working Paper No., *704*, Washington, D.C.

Anning, A. K. & Yeboah-Gyan, K. (2006). Diversity and distribution of invasive weeds in Ashanti Region, Ghana. *African Journal of Ecology.*, *45*(3), pp. 355-360.

Anonymous, (1952). Effect of hoeing in plantations of *Melia Azadirachta.* Report For. Adm. Nigeria. 1950/51 (22-3).

Anonymous, (1992). Summary of survey data for teak. Ghana forestry Department records, Planning Branch, Kumasi, Ghana (Unpub.).

Anonymous, (2004). Características y usos de 30 especies del bosque latifoliado de Honduras. Fundacion Cuprofor, Proecen, Proinel, Eap-Zamorano.

Anonymous, (2009). Growing teak under farm forestry for posterity http://planning.up.nic.in/innovations/inno3/fw/teak.htm.

Apetorgbor, M. M., Siaw, D. & Gyimah, A. (2003). Decline of *Ceiba pentandra* seedlings, a tropical timber species, in nurseries and plantations. *Ghana Journal of Forestry*, *11*(2), pp. 51-62.

Appiah, M., Blay, D., Damnyag, L., Dwomoh, F. K., Pappinen, A. & Luukkanen, O. (2009). Dependence on forest resources and tropical deforestation in Ghana. *Environment, Development and Sustainability*, *11*(3), pp. 1573-2975.

Asante, M. S. (2005). Deforestation in Ghana: Explaining the chronic failure of forest preservation policies in a developing country. United Press of America, Maryland, USA.

Asare, R. (2004). Agroforestry initiatives in Ghana: a look at research and development . A presentation made at the World Cocoa Foundation conference in Brussels April, pp. 21 – 22, 2004. Danish Centre for Forest, Landscape and Planning - KVL, *Horsholm Kongevej*, *11*, DK-2970.

Aseidu, J. B. K. (2013).Technical Report on Reclamation of Small Scale Surface Mined Lands in Ghana: A Landscape Perspective. *American Journal of Environmental*

Protection, 2013, *Vol. 1, No. 2, 28-33* Available online at http://pubs.sciepub.com/env/1/2/3 © Science and Education Publishing DOI:10.12691/env-1-2-3

Assmann, E. (1955). Die Bedeutung des "erweiterten Eichhorn'schen Gesetzes" für die Konstruktion von Fichten-Ertragstafeln . *Forstwiss. Centralbl.*, *74* , pp. 321 – 330.

Assmann, E. (1959). Höhenbonität und wirkliche Ertragsleistung *Forstwiss. Centralbl. 78*, pp. 1 – 20.

Assmann, E. V. (1970). The principles of forest yield study. Pergamon Press, Oxford U. K. p. 506.

Asuming-Brempong, S. (2003). Roles of Agriculture Project. National Report Ghana. Agricultural and Development Economics Division (ESA) Food and Agriculture Organisation of the United Nations. International Conference held October, pp. 20-22, 2003 Rome, Italy.

Avery, T. E. & Burkhart, H. E. (1994). Forest measurements. 4th ed. McGraw-Hill Co., New York. p. 408

Avery, T. E., & Burkhart, H. E. (2002). Forest measurements, 5th edn. McGraw-Hill, New York.

Cochran, W. G. (1977). Sampling techniques. 3rd Edition. Wiley, New York

Ayamga, R. A. (1997). Evaluation of community participation in the Community Afforestation programme in Northern Ghana. A case study of the Tolon-Kumbungu District. BSc (Tech) Thesis, *Department of Renewable Natural Resources*, University for Development Studies, Tamale, Northern Region.

Ayine, D. (2008). Social responsibility agreements in Ghana's forestry sector. Developing legal tools for Citizen Empowerment Series, *IIED*, London.

Bailey, R. L. & Dell, T. R. (1973). Quantifying diameter distributions with the Weibull function. *Forest Science*, *19*(2), pp. 97-104.

Bailey, J. D. & Harjanto, N. A. (2005). Teak (*Tectona grandis* L.) tree growth, stem quality and health in coppiced plantations in Java, Indonesia. *New Forests*. *30*(1), pp. 55-65

Bailey, M. D. & Sporleder, T. S. (2000). The real options approach to evaluating a risky investment by a new generation cooperative: further processing. *A Paper Presented at the Annual Meetings of NCR-194 Research on Cooperatives*, Las Vegas, NV, Dec. pp. 12-13.

Baldwin, C. & Ruback, R. (1986). Inflation, uncertainty, and investment. *Journal of Finance*, July, pp. 657-669.

Baskerville, G. L. (1972). Use of logarithmic regression in the estimation of plant biomass. *Can. J. For. Res.*, *2*, pp. 49-53.

Battaglia, M., Bruce, J., Brack, C. & Baker, T. (2009). Climate Change and Australia's plantation estate: analysis of vulnerability and preliminary investigation of adaptation options. Forest and Wood Products. Australia (FWPA) Project No. PNC068-0709, CSIRO.

B.C. Ministry of Forests. (1995). Forest practices code (1995). B.C. Ministry of Forests, Government of British Columbia, Victoria, BC, Canada.

Beard, J. S. (1942). Summary of silvicultural experience with cedar, *Cedrela mexicana* Roem. in Trinidad and Tobago. *Caribbean Forester*, *3*(3), pp. 91-102.

Beauchamp, J. J. & Olson, J. S. (1973). Corrections for bias in regression estimates after logtharithmic transformation. *Ecology*, *54*(6), pp. 1402-1407.

Beck, D. E. & Trousdell, K. B. (1973). Site Index. Accuracy of prediction. Research paper SE-108 Asheville, North Carolina, *Southeastern Forest and Range Experiment Station*, U. S. Forest Service.

Becker, H. & Vanclay, F (eds.) (2003). The International Handbook of Social Impact Assessment: *Conceptual and Methodological Advances*. Edward Elgar, Cheltenham, UK.

Beers, T. W. (1962). Components of forest growth. *Journal of Forestry*, *60*, pp. 245-248

Behre, C. E. (1924). Computation of total cubic contents of trees. *J. For.*, *22*(6), pp. 62-63.

Bell, F. W. (1991). Critical silvics of conifer crop species and selected competitive vegetation in Northwestern Ontario. Ont. Min. Nat. Resour., *Northwestern Ont. For. Tech. Dev. Unit Rep. No.*, *19*. p.177.

Benhin, J. K. A. & Barbier, E. B. (2001). The Effects of the Structural Adjustment Programme on Deforestation in Ghana. *Agricultural and Resource Economics Review*, *30*(1), pp. 66-80.

Benhin, J. K. A. & Barbier, E. B. (2004). Structural Adjustment Programme, Deforestation and Biodiversity Loss in Ghana. *Environmental and Resource Economics*, *27*, pp. 337–366.

Benneh, G. (1989). The dynamics of customary land tenure and agrarian systems in Ghana. In: Ghana, P. (Ed.), WCARRD; Ten Years of Follow-Up, The Dynamics of Land Tenure and Agrarian Systems in Africa, *Case Studies from Ghana, Kenya, Madagascar and Togo*. FAO, Rome, Italy, pp. 34– 97.

Berger, J. J. (2006). Ecological Restoration and Non-indigenous Plant Species: A Review. *Ecological Restoration*, *1*(2), pp. 74-82

Bergeuschbacher, R. J. (1990). Natural Forest Management in the Humid Tropics: Ecological, Social, and Economic Considerations. *Ambio*, *19*(5), pp. 253-258.

Betancourt, A. (1972). The growth of *Azadirachta indica* in Cuba. *Baracoa*, *2*(2), pp. 17 - 23.

Bhati, U. N., Klijn, N., Curtotti, R., Dean, M. & Stephens, M. (1991). Impediments to the development of commercial forest plantations in Australia. The Role of Trees in Sustainable Agriculture, National Conference, Albury, New South Wales, 30 September - 3 October, 1991. Australian Bureau of Agricultural and Resource Economics. *GPO Box*, 1563, Canberra, 2601.

Bickford, A. C. (1956). Proposed design for continuous inventory: A system of perpetual forest survey for the North US Forest Service Eastern Techniques meeting, *Forest survey*, Cumberland Falls, K. Y. October, 8-13,1956.

Bjerksund, P. & Ekern, S. (1993). Contingent claims evaluation of mean-reverting cash flows in shipping. In L. Trigeorgis (Ed.). Real options in capital investment (New Contributions) New York, NY, Praeger.

Black, F. & Scholes, M. (1973). The pricing of options and corporate liabilities. *Journal of Political Economy*, *3*, pp. 637-654.

Bliss, C. I. & Reinker, K. A. (1964). A lognormal approach to diameter distributions in even-aged stands. *Forest Science*, *10*, pp. 350-360.

Binkley, C. S. (1987). When is the economic rotation longer than the rotation of maximum sustained yield? *Journal of Environmental Economics and Management.*, 14, pp. 152-158.

Boadu, F. O. (1992). Contingent valuation for household water in rural Ghana. *Journal of Agricultural Economics*, *43*(3), pp. 458-465.

Boardman, A. E., Greenberg, D. H., Vining, A. R. & Weimer, D. L. (2001). Cost-benefit analysis: concepts and practice. 2nd Edition. Upper Saddle River, NJ. Prentice-Hall.

Boateng, E. A. (1966). *A geography of Ghana*. 2nd Ed. Cambridge. Cambridge Univ. Press. p. 212.

Bodie, Z. & Merton, R. C. (2000). Finance. Prentice Hall. Upper Saddle River, New Jersey.

Bolfrey-Arku, G. E. K., Onokpise, O. U., Carson, A. G., Shilling, D. G. & Coultas, C. C. (2006). The Speargrass (Imperata cylindrica (L) Beauv.) menace in Ghana: Incidence, farmer perceptions and control practices in the forest and forest-Savannah transition Agro-ecological Zones of Ghana. *West African Journal of Applied Ecology. 10*(1).

Boni, S. (2006). Ghanaian Farmers' Lukewarm Reforestation: Environmental degradation, the timber option and ambiguous legislation. Colloque international "Les frontières de la question foncière – *At the frontier of land issues*," Montpellier, 2006.

Borota, J. (1991). *Tropical forests: Some African and Asian studies of composition and structure*. New York. Elsevier Publishers.

Bosu, P. P. & Krampah, E. (2005). Triplochiton scleroxylon K.Schum. In: Louppe, D., Oteng-Amoako, A.A. and Brink, M. (Eds.). Prota, 7(1), Timbers/Bois d'œuvre, 1. [CD-Rom]. PROTA, Wageningen, Netherlands.

Boulding, K. (1955). Economic analysis. 3rd Edition. New York. Harper and Brothers.

Braathe, P. (1957). Thinnings in even-aged stands: *A summary of the European literature. Fac. For. Univ.* New Brunswick, Fredericton., p. 92.

Bradley, R. T. (1963). Thinning as an instrument of forest management. *Forestry*, *36*, pp. 181-194.

Brack, C. L. (1997). Forest Inventory in the, 21st Century. Proceedings of the Australian and New Zealand Institute of Foresters Conference "Preparing for the, 21st Century." 21 - 24 April, 1997. Canberra, ACT. pp. 329 - 335.

Branch, K., Hooper, D., Thompson, J. & Creighton, J. (1984). Guide to Social Assessment Boulder, Colorado. Westview Press.

Bradley, G. (1995). Urban Forest Landscapes: Integrating Multidisciplinary Perspectives. Seattle: University of Washington Press.

Brealey, R., Myers, S. & Marcus, A. (2001). *Fundamentals of corporate finance*. Boston, MA. McGraw-Hill Irwin.

Brechling, F. (1975). Investment and employment decisions. Manchester University Press. Manchester, UK.

Brennan, M. J. & Schwartz, E. S. (1985). Evaluating natural resource investments. *Journal of Business*, *58*(2), pp. 135-157.

Briscoe, C. B. & Ybarra-Conorodo, R. (1971). Increasing the growth of established teak. Res. Notes. ITF-13, Rio Piedras, PR. USDA For. Serv. Inst. of tropical forestry.

Broadhead, J. S., Durst, P. B & Brown, C. L. (2009). Climate change: will it change how we manage forests? Pp:57-65. *In*: Van Bodegom, Arend Jan, Herman Savenije and Marieke Wit (eds). (2009). *Forests and Climate Change: adaptation and mitigation.* Tropenbos International, Wageningen, The Netherlands. xvi + p. 160.

Brown, C. (2000). The Global Outlook for Future Wood Supply from Forest Plantations. *Global Forest Products Outlook Study Working Paper Series*. FAO, Rome. p. 164

Brown, T. C., Peterson, G. L. & Tonn, B. E. (1995). The value jury to aid natural resource decisions. *Land Economics*, *71*(2) (May), pp. 250-260.

Browne F. G. (1968). *Pests and Diseases of Forest Plantation Trees*. Clarendon Press, Oxford, UK.

Brownlee, K. A. (1967). Statistical theory and methodology in science and engineering. 2nd ed. John Wiley and Sons, New York. p. 400

Bruce, D. & Schumacher, F. X. (1950). *Forest mensuration*. McGraw-Hill Co., New York. pp. 484

Bruce, D. & Reineke, L. H. (1931). Correction alignments charts in forest research. *U.S. Dept. Agric. Tech. Bull., 210*. p. 87

Bruijnzeel, L. A. (2004). Hydrological functions of tropical forests: not seeing the soil for the trees? Agriculture, *Ecosystems and the Environment., 104*, pp. 185-228

Budelman, A. (1988). The performance of the leaf mulches of *Leucaena leucocephala, Flemingia macrophylla* and*Gliricidia sepium* in weed control. *Agroforestry Systems, 6*(1), pp. 137-145.

Buongiorno, J. & Giless, J. K. (1987). *Forest management and economics*. New York. MacMillan Publishing Co.

Burdge, R. (1994). *A community guide to social impact assessment*. Middleton, Wisconsin. Social Ecology Press.

Bury, K. V. (1975). *Statistical models in applied science*. John Wiley and Sons Inc., pp. 625, New York, U. S. A.

Buschbacher, R. J. (1990). Natural Forest Management in the Humid Tropics: Ecological, Social, and Economic Considerations. *Ambio, 19*(5), pp. 253-258.

Business News. (2009). Malfeasance hits forest plantation fund. Business News of Tuesday, 18 August, 2009. .http://www.ghanaweb.com/GhanaHomePage/NewsArchive/artikel.php?ID=166993.

CAB International. (2004). Forestry Compendium Global Module. Wallingford, UK: CAB International. Accessed online at: http://www.cabi.org/compendia/fc/index.asp.

CAB International. (2005). *Forestry Compendium*. Wallingford, UK: CAB International.

Cabaret N. & Nguessan K. (1988). *Gmelina arborea spacing trial*, 1985. CTFT-CI.

Cairns, M. A., Brown, S., Helmer, E. H. & Baumgardner, G. A. (1997). Root biomass allocation in the world's upland forests. *Oecologia, 111*, pp. 1-11.

Calish, S., Fight, R. & Teeguargen, D. (1978). "How do non-timber values affect Douglas-fir rotation?" *Journal of Forestry, 76*(4), 217 - 221.

Camirand, R. (2002). Guidelines for Forest Plantation establishment and management in Jamaica. Jamaica: Trees for Tomorrow Project Phase II. *Tecsult International, 4700*, Boulevard Wilfrid-Hamel Québec, Québec Canada.

Cannell M. G. R. & Last, F. T. (Eds.). (1976). *Tree physiology and yield improvement*. Academic Press, London.

Carle, J. & Holmgren, P. (2003). *Definitions Related to Planted Forests*. Forest Resources Assessment Programme. *Forest Resources Development Service Forestry Department. FAO Working paper* No. 79. FAO. Rome, Italy.

Carpenter, J. F. (1998). Internally Motivated Development Projects: A potential tool for biodiversity conservation outside protected areas. *Ambio, 27*, 3, pp. 211- 216.

Carr, P. (1988). The valuation of sequential exchange opportunities. *Journal of Finance*, December, pp. 1235-1256.

Castedo-Dorado, F., Crecente-Campo, F., Álvarez-Álvarez, P. & Barrio-Anta, M. (2009). Development of a stand density management diagram for radiate pine stands including assessment of stand stability. *Forestry*, *82*(1), pp. 1-16.

Cavers, S., Navarro, C. & Lowe, A. J. (2004). Targeting genetic resource conservation in widespread species: a case study of *Cedrela odorata L. Forest Ecology and Management*, *197*(1-3), pp. 285-294.

CDM Ghana. (2008). The CDM process. http://www.epa.gov.gh/cdm/DNA/dna1.htm.

Centeno, J. C. (2009). The management of teak plantations. http://www.treemail.nl/teakscan.dal/files/mngteak.htm#stan. Accessed January, 2010.

Chachu, R. (1989). Allowable cut from the forest. *In* Wong, J., ed., *Ghana Forest Inventory Seminar proceedings*, 29–30 Mar, 1989, Accra, Ghana.UK Overseas Development Administration, London, UK.

Chacko, K. C. (1995). Silvicultural problems in management of teak plantations. Proc. 2nd Regional Seminar on Teak 'Teak for the Future' Yangon, Myanmar May, 1995 *FAO* (Bangkok) pp. 91-98.

Chang, S. J. (1984). Determination of the optimal rotation age: a theoretical analysis. *Forest Ecology and Management*, *8*, 137 – 147.

Chaplin, G. E. (1980). Progress with provenance exploration and seed collection of *Cedrela spp. In Proceedings*, *Commonwealth Forestry Conference*, *Port-of-Spain*, *Trinidad*, September, 1980. p. 17

Chapman, D. G. (1961). Statistical problems in population dynamics. *In Proc. Fourth Berkeley Symp. Math Stat. and Prob. Univ. Calif.* Press, Berkeley.

Chapman, G. W. & Allan, T. G. (1978). Establishment techniques for forest plantations. Forest Resources Division. Forestry Department. *FAO Forestry Paper*, *8*. Rome.

Chapman, H. H. & Meyer, W. H. (1949). *Forest mensuration*. McGraw-Hill Co., New York. p. 522

Charkraborty, M. (1994). An Analysis of the Causes of Deforestation in India. In K. Brown, and D. W. Pearce (Eds.), *The Causes of Tropical Deforestation* (pp. 226-238). London: UCL Press.

Chavangi, N. (1992). Household based tree planting activities for fuelwood supply in rural Kenya; the role of the Kenya Woodfuel Development Programme. In *Development form Within: survival in rural Africa.* Taylor, D.R.F. and F. Mackenzie, 1992. Routledge, London and New York. pp. 148-169.

Chen, C. M. & Rose, D. W. (1978). Direct and indirect estimation of height distributions in even-aged stands. *Minnesota Forestry Research Notes*. No. *267*. January, 1978. p. 3.

Chundamannii, M. (1998). Teak plantations in Nilambur - an economic review. KFRI Research Report No. 144. Kerala Forest Research Institute, Peechi, Kerala, India. p. 71.

Cintron B. B. (1990). *Cedrela odorata* L. *Cedro hembra*, Spanish cedar, pp. 250-257. *In:* Burns R. M. H.and Barbara H. (Eds.), Silvics of North America, *2*, Hardwoods. *Agricultural Handbook*, *654.* United States Department of Agriculture, Washington, DC. Vol. 2. pp 250-257.

CITES. (2007). Convention on International Trade in Endangered Species of wild Fauna and Flora. Fourteenth meeting of the Conference of the Parties *The Hague* (Netherlands), 3-15 June, 2007.

Cline-Cole, R. A., Main, H. A. C. & Nichol, J. E. (1990). On Fuelwood Consumption, Population Dynamics and Deforestation in Africa. *World Development*, *18*, 4, pp.513-527.

Clutter, J. L. (1963). Compatible growth and yield models for loblolly pine. *For. Sci.*, *9*, pp. 354-371.

Clutter, J. L. & Bennett, F. A. (1965). Diameter distributions in old-field slash pine plantations. *Georgia For. Res. Council Rep.*, *13*, p. 9

Clutter, J. L., Fortson, J. C., Piennar, L. V., Brister, G. H. & Bailey, R. L. (1983). *Timber management - a quantitative approach.* John Wiley and Sons, Toronto. p. 333

Coates, D. & Haeussler, S. (1986). A preliminary guide to the response of major species for competing vegetation to silvicultural treatments. British Columbia Min. For., *Info. Serv.* Br. p. 88

Cochran, W. G. (1953). Sampling Techniques. John Wiley and Sons Inc., New York. p. 330

Cochran, W. G. (1977). *Sampling Techniques*. New York. John Wiley and Sons.

Cohen, A. C. Jr. (1965). Maximum likelihood estimation in the Weibull distribution based on complete and censored samples. *Technometrics*, *7*, pp. 579-588.

Coilie, T. S. (1938). Forest classification of forest sites with special reference to ground vegetation. *Journal of Forestry.*, *36*, pp. 1062-1066.

Conrad, J. M. & Clark, C. W. (1987). *Natural resource economics: notes and problems.* Cambridge University Press. Cambridge. p. 231

Cossalter, C. & Pye-Smith, C. (2003). Fast-Wood Forestry—Myths and Realities. *Center for International Forestry Research*, Jakarta, Indonesia.

Cotrone, V. (2008). The Role of Trees & Forests in Healthy Watersheds: *Managing Stormwater, Reducing Flooding,and Improving Water Quality: Pen State Extension.* http://www.dcnr.state.pa.us/cs/groups/public/documents/document/dcnr_009116.pdf. p.7.

Cox, J., Ross, R. & Rubinstein, M. (1979). Option pricing: a simplified approach. *Journal of Financial Economics*, *7*(4), pp. 71-90.

Cox, J. & Ross, S. (1976). The valuation of options for alternative stochastic processes. *Journal of Financial Economics*, January, pp. 145-166.

CQFA, (2009). Centrel Queensland Forest Association. Plantation Establishment http://www.cqfa.com.au/resources/agroforestry/plantation-establishment. Accessed January, 2010.

Craib, I. J. (1939). Thinning, pruning and management studies on the main exotic conifers grown in South Africa. *Dept. of Agric. And For. Sci. Pretoria. Bull.* No. *196*. p. 179

Crow, T. R. (1971). Estimation of biomass in an even-aged stand - regression and "mean tree" techniques. pp. 35-50 *in* Young, H. E (ed.) Forest Biomass Studies. XVth IUFRO Congress, Univ. of Florida, March, 15-20 1971. p. 205

Crush, J. S. & Namasasu, O. (1985). Rural rehabilitation in the Basotho Labour Reserve. *Applied Geography*, *5*.

CSPS. (2008). Canada School of Public Service. *Cost Benefit analysis and risk analysis.* Unpublished course notes.

CSPS. (2009). Canada School of Public Service. *Regulatory performance measurement and evaluation.* Unpublished course notes.

Csurhes, S. (2008). Pest plant risk assessment: the neem tree (*Azaradirachta indica*). *The State of Queensland*, Department of Primary Industries and Fisheries. PR08-3685.

Cunia, T. (1964). Weighted least squares method and the construction of volume tables. *For. Sci.*, *10*, pp. 180-191.

Curtis, R. O. (1967). Height-diameter and height-diameter-age equations for second growth Douglas fir. *Forest Science*, *13*(4), pp. 365-375.

Daniel, T. W., Helms, J. A. & Baker, F. S. (1979). Principles of silviculture. 2nd Ed. New York: McGraw-Hill.

Daniel, W. W. (1978). *Applied nonparametric statistics*. London. Houghton Mifflin Co.

Darkwa, E. O., Johnson, B. K., Nyalemegbe, K., Yangyuoru, M., Oti-Boateng, C., Willcocks, T. J. & Terry, P. J. (2001). Weed management on Vertisols for small-scale farmers in Ghana. *International journal of pest management*, *47*(4), pp. 299 – 303.

Datta, S. K. de. (1978). Fertiliser management for efficient use in wetland rice soils. *IRRI, Soils and Rice*, pp. 671-701.

Dauda, C. (2009). Interview transcript. Interview with the Minister of Lands and Natural Resources. Upper Reach. www.upper-reach.com.

Davidson, J. (1985). Assistance to the forestry sector of Bangladesh. Species and sites What to plant and where to plant. *Field Doc. No.* 5, UNDP/FAO/BGD/79/017. p. 50

Davis, L. S., Johnson, K. N., Bettinger, P. S. & Howard, T. E. (2001). *Forest management: to sustain ecological, economic, and social values.* 4th Edition. McGraw Hill. New York. p. 804

Day, R. J. (1985). Crop plans in silviculture. *Canadian Pulp and Paper Association*, Woodlands Section Index, 2975. p. 55

Day, R. J. (1996). A manual of silviculture. School of Forestry, Lakehead University, *Thunder Bay*, ON, Canada. p. 344

Day, R. J. & Nanang, D. M. (1997). Principles of thinning for improved growth, yield, and economic profitability of lodgepole and jack pine. Pages, 1-13 *in* Proceedings of a Commercial Thinning Workshop, Whitecourt, Alberta, 17-18 October, 1996. *FERIC, Vancouver. Spec. Rep.*, SR-122.

De Jong, I. L. (1991). Social Forestry in Mali: is participation by the rural population difficult to realise? *AT-Source*, *17*, 1, pp.17-20.

Deacon, R. T. (1994). Deforestation and the rule of law in a cross-section of countries. *Land Economics*, *70*, pp. 414-430.

Dean, T. J. & Baldwin, V. C. 1996 Crown management and stand density. In *Growing Trees in a Greener World: Industrial Forestry in the, 21st century; 35th LSU Forestry Symposium* . M.C. Carter (ed.) Louisiana State University Agricultural Center, Louisiana *Agricultural Experiment Station* , Baton Rouge, LA. pp. 148 – 159.

Dei, G. S. (1990). Deforestation in a Ghanaian Rural Community. *Anthropologica*, *32*, pp. 3-27.

Dei, G. S. (1992). A Forest Beyond the Trees: Tree Cutting in Rural Ghana. *Human Ecology*, *20*(1), pp. 57-88.

Deleporte, P., Laclau J. P., Nzila, J. D., Kazotti, J. G., Marien, J. N., Bouillet, J. P. & Szwarc, M., D'Annunzio, R. & Ranger, J. (2008). Effects of slash and litter management practices on soil chemical properties and growth of second rotation eucalypts in the Congo. *In*: Nambiar, E.K.S. (ed.) Site management and productivity in tropical plantation forests: workshop proceedings, 22-26 November, 2004 Piracicaba, Brazil, and, 6-9 November, Bogor, Indonesia, 5-22. *Center for International Forestry Research*, Bogor, Indonesia.

Disperati, A. A., Ferreira, C. A., Machado, C., Gonçalves, J. L. M. & Soares, R. V. (1995). Proceedings of 1° Seminario Sobre Cultivo Mínimo do Solo em Florestas. Curitiba, Brazil, p. 162

Dixit, A. & Pindyck, R. S. (1994). *Investment under uncertainty*. Princeton University Press. Princeton, New Jersey.

Djarbeng, V. & Ameyaw, D. S. (2002). *ADRA's agroforestry development programme in Ghana gives farmers new chances*. Project document.

Donkor, B. N. (2003). Evaluation of government interventions in Ghana's forest product trade: a post-intervention impact assessment and perceptions of marketing implications. *Unpublished Ph.D. Thesis*. The School of Renewable Natural Resources, Louisiana State University. p. 178

Donkor, B. N., Vlosky, R. P. & Attah, A. (2006). *Evaluation of government interventions on increasing value-added wood product export from Ghana*. Louisiana *Forest Products Development Center. Working Paper #77*. p. 16

Drechsel, P. & Zech, W. (1994). DRIS evaluation of teak (*Tectona grandis* Linn F.) Mineral nutrition and effects of nutrition and site quality on teak growth in West Africa. *For. Ecol. and Manage.*, *70*, pp. 121 - 133.

Drew , T. J. & Flewelling , J. W. (1979) Stand density management: an alternative approach and its application to Douglas-fir plantations . *For. Sci.*, *25*, pp. 518 – 532.

Drew, T. J. & Flewelling, J. W. (1977). Some recent Japanese theories of yield–density relationships and their application to Monterey pine plantations. *For. Sci.*, *13*, pp. 39–53.

Duke, J. A. (1983). Handbook of energy crops. Unpublished. Available on http://www.hort.purdue.edu/newcrop/duke_energy/Gmelina_arborea.html#Yields%20and%20Economics.

Duku-Kaakyire, A. & Nanang, D. M. (2004). Application of real option theory to forest investment analysis. *Forest Policy and Economics*, *6*, pp. 539– 552.

Dutta, S., Rajaram, R. & Robinson, B. Chapter, 5, Mineland Reclamation, in Sustainable Mining Practices -- A Global Perspective, V. Rajaram and S. Dutta, Editors. 2005, A. A. Balkema Publishers, a member of Taylor & Francis Group: Leiden, The Netherlands. pp. 179-191.

Duvall, C. S. (2009). *Ceiba pentandra* (L.) Gaertn. [Internet] Record from Protabase. Brink, M. & Achigan-Dako, E. G. (Editors). PROTA (Plant Resources of Tropical Africa / Ressources végétales de l'Afrique tropicale), Wageningen, Netherlands. < http://database.prota.org/search.htm>. Accessed, 21 February, 2010.

Dye, T. R. (1998). Understanding public policy. 9th Edition. New York. John Wiley.

Dzanku, F. M. (2004). Contribution of wildlife to the economy of Ghana. ISSER, University of Ghana. Unpublished notes. Available on: http://are.berkeley.edu/~dwrh/IPALP_Web/Meetings/Dakar0506/Presentation-Dzanku-Ghana.pdf.

Echenique-Marique, R. & Plumptre, R. A. (1990). A guide to the use of Mexican and Belizean timbers. *Tropical Forestry Papers*, *20*. Oxford Forestry Institute.

Elton, E. J., Gruber, M. J. (1995). *Modern portfolio theory and investment analysis*. Fifth edition. John Wileyand Sons, Inc.

Enters, T. (2001). Incentives for soil conservation. In: *Response to land degradation*, E.M. Bridges, I.D. Hannam, L.R. Oldeman, F.W.T. Penning de Vries, S.J. Scherr and Samran Sombatpanit (Eds.), 351-360. New Delhi and Calcutta: Oxford and IBH Publishing Co. Pvt. Ltd.

Enters, T., Durst, P. B. & Brown, C. (2003). What Does it Take? The role of incentives in forest plantation development in the Asia-Pacific Region. *UNFF Intersessional Experts Meeting on the Role of Planted Forests in Sustainable Forest Management*, 24-30 March, 2003, New Zealand.

Enters, T., Durst, P. B., Brown, C., Carle, J. & McKenzie, P. (2003). What Does it Take? The Role of Incentives in Forest Plantation Development in the Asia-Pacific Region. *FAO. RAP PUBLICATION*, 2004/28. Rome.

Errington, J. C. History and Future of Mine Reclamation in British Columbia, Proceedings of the, 16th Annual British Columbia Mine Reclamation Symposium, 1992 [cited, 2012 July, 20]; The Technical and Research Committee on Reclamation. Available from: https://circle.ubc.ca/bitstream/handle/2429/14193/1992%20-%20Errington%2c%20History%20and%20Future%20of%20Mining%20Reclamation.pdf?sequence=1

European Commission/VPA. (2009). *FLEGT Voluntary Partnership Agreement Between European Commission Delegation in Ghana and VPA Secretariat in Ghana.*

Evans, J. (1989). Community Forestry in Ethiopia: The Bilate Project. *Rural Development in Practice*, 1, *4*, 7-8, 25.

Evans, J. (1992). Plantation forestry in the tropics. 2nd ed. Oxford Univ. Press. pp. 403

Evans, J. (2001). Sustainability of productivity in successive plantations. In: *Proceedings of the International Conference on Timber Plantation Development*, Manila, the Philippines, 7-9 November, 2000. *Quezon City, the Philippines, Department of Environment and Natural Resources.*

Evans, J. & Turnbull, J. W. (2004). Plantation forestry in the tropics: *The role, silviculture and use of planted forests for industrial, social, environmental and agroforestry purposes.* Third edition. Oxford University Press. p. 488

Faculty of Renewable Natural Resources. (2013). Rport on SADA – ACICL Afforestation Project. Draft Final Report. http://www.citifmonline.com/wp-content/uploads/2014/04/SADA_DraftFinal-Report.compressed.pdf.

Faleyimu, O. I. & Akinyemi, O. (2010). The role of trees in soil and nutrient conservation. *African Journal of General Agriculture.*, *6*(2), pp. 77-82

FAO & UNEP. (1981). *Forest resources of Tropical Africa. Part II: Country briefs.* Tropical forest resource assessment project. GERMS, UN *23*(6). *Tech. Report*, *2*. Rome. FAO.

FAO. (1956). *Tree planting practice in tropical Africa.* Rome, Italy. FAO.

FAO. (1981). *Map of the Fuelwood Situation in the Developing Countries.* Scale, *1*, 25 000 000.

FAO. (1983). *Growth and yield of plantation species in the tropics.* Rome. FAO. Forest Resource Div. Rome.

FAO. (1983). Fuelwood Surveys. Forestry for local community development programme - GCP/INT/365/SWE. Food and agriculture Organization of the United Nations, Rome.

FAO. (1985). Intensive multiple-use forest management in the tropics: Analysis of case studies from India, Africa, Latin Arnerica and the Caribbean. Rome. FAO.

FAO. (1989). Arid zone forestry: A guide for field technicians. Publication #20. Food and Agriculture Organization of the United Nations, Rome, Italy.

FAO. (1992). *Mixed and pure forest plantations in the tropics and subtropics.* FAO Forestry Paper no. 102. Rome. FAO.

FAO. (1993). Vegetative propagation. *Field Manual No.5. UNDP/FAO Project* RAS/91/004. Rome, Italy.

FAO. (1995). Preparing to tropical plant trees. http://www.fao.org/docrep/006/ad229e/ad229e00.htm

FAO. Rome. p. 250

FAO. (1995). Forest resource management - Project findings and recommendations. Terminal Report. Project FO:UTF/GHA/025/GHA. Rome.

FAO. (1997). Update on sustainable forest management and certification: Example from a developing country – Ghana. *Advisory committee on paper and wood products*. Thirty-eighth session. Rome, 23 - 25 April, 1997.

FAO. (1999). *Incentive systems for natural resource management. Environmental Reports Series, 2.* FAO Investment Centre. Rome, Food and Agriculture Organisation of the United Nations.

FAO. (2002a). *Hardwood plantations in Ghana.* Forest Plantations Working Paper, 24. *Forest Resources Development Service, Forest Resources Division.* Rome.

FAO. (2002b). Forest plantation productivity. Report based on the work of W. J. Libby and

C. Palmberg-Lerche. Forest Plantation Thematic Papers, Working Paper, *3. Forest Resources Development Service, Forest Resources Division. FAO*, Rome (*unpublished*).

FAO. (2005). *The State of the World's Forests.* (2005). Rome. Italy.

FAO. (2009). Forestry Statistics. http://www.fao.org/corp/statistics/en/.

FAO. (2011). Global Forest Resource Assessment, 2010. http://www.fao.org/news/story/en/item/40893/icode/. FAO. Italy. Rome.

Farmer, R. H. (1972). *Handbook of hardwoods.* 2nd Ed. London. H. M. Stationery Office.

Faustmann, M, 1849. Calculation of the value which forest land and immature stands possess for forestry (Transl. in M. Gane and W. Linnard in "Martin Faustmann and the evolution of discounted cash flow: two articles from the original German of 1849." Comm. For. Inst, Univ. Oxon, Inst. Pap. No. 42, 1968). *Reprinted in Journal of Forest Economics*, 1996, *1*(1), pp. 7 - 44.

Fernandez-Cornejo, J., Gempesaw II, C. M., Elterich, J. G. & Stefanou, S. E. (1992). Dynamic Measures of Scope and Scale Economies: An Application to German Agriculture American Journal of Agricultural Economics, Vol. 74, No. 2 (May, 1992), pp. 329-34.

Field, B. C. (2001). *Natural resource economics: An introduction.* New York. McGraw-Hill.

Finger, S. E., Church, S. E. & P.v. Guerard. Chapter F: Potential for Successful Ecological Remediation, Restoration, and Monitoring, Integrated Investigations of Environmental Effects of Historical Mining in the Animas River Watershed, San Juan County, Colorado, 2008 [cited, 2012 July, 20]; U. S. Geological Survey Professional Paper, 1651. Available from: http://pubs.usgs.gov/pp/1651/downloads/Vol2_combinedChapters/vol2_chapF.pdf.

Finney, D. J.1941. On the distribution of a variate whose logarithm is normally distributed. *R. J. Statist. Soc. Supplement.* 7, pp. 155-158.

Fisher, I. (1930). *The theory of interest rates.* MacMillan, New York, NY pp. 566

Fisher, N. M. (1984). The impact of climate and soil on cropping systems and the effect of cropping systems and weather on the stability of yield. pp. 55-70 *in* Steinner, K. G. (ed.) *Report on the On-Farm-Experimentation Training Workshop*, Nyankpala Ghana, July, 3-13 1984. pp. 120

Foley, G. (1987). Exaggerating the Sahelian woodfuel problem? *Ambio*, *16*(6), pp. 367-371.

Ford-Robertson, F. C. (Ed.). (1971). Terminology of forest science, technology, practice and products. *Multilingual For. Terminol. Ser. No. 1. Soc. Am. For.*, Washington, DC.

Forest Research Programme. (2006). Chainsaw milling and logging in Ghana: *Background study report.* Available on: http://www.illegal logging.info/uploads/FRP_ Chainsaw_Logging_Ghana.pdf.

Forest Watch Ghana. (2006). Forest governance in Ghana. An NGO perspective. *A report produced for FERN by Forest Watch Ghana*, March, 2006.

Forest Watch Ghana. (2014). (as repoted by the Ghana News Agency) http://edition.radioxyzonline.com/pages/news/05102014-0907/19583.stm

Forestry/Fuelwood Research and Development Project (F/FRED). (1994). Growing multipurpose trees on small farms, module, 9, *Species fact sheets* (2nd ed.). Bangkok, Thailand: Winrock International. p. 127

Fortson, J. C. (1972). Which criterion? *Effect of the choice of criterion on forest management plans. For. Sci.*, *18*, pp. 292 - 297.

Frank, R. M. (1973). The course of growth response in released white spruce-10 year results. US Dep. Agric., For. Serv., Northeastern Forest Experiment Station, Broomall, PA. *Research Paper NE*-268. p. 6.

Freeman, A. Myrick. III. (1993). The measurement of environmental and resource values: *Theory and methods.* Washington D. C. Resources for the Future.

Freese, F. (1962). Elementary Forest Sampling. *Agriculture Handbook*, *232.*, *US Department of Agriculture*.Friends of the Urban

Forest. (2015). Benefits of Urban Greening. http://www.fuf.net/benefits-of-urban-greening/Accessed July, 10, 2015.

Friday, K. S. (1987). Site index curves for teak (*Tectona grandis* Linn F.) in the limestone regions of Puerto Rico. *Commonw. For. Review*, *66*, 239-252.

Fries, J. (1991a). From village forestry towards farming systems – the development of the Swedish Sahel Programme. *IDRCurrents* No.1, pp. 32-35.

Fries, J. (1991b). Management of natural forests in the semiarid areas of Africa. *Ambio*, *20*, 8, pp.395-400.

FSC. (2010). Principles and criteria for forest certification. Forest Stewardship Council. http://www.fsc.org/fsc-rules.html. Accessed March, 2010.

Furnival, G. M. (1961). An index for comparing equations used in constructing volume tables. *For. Sci.*, *7*, pp. 337-341.

Gadow, K. von. (1983). Fitting distributions in *Pinus patula* stands. *South African Forestry Journal*, *126*, pp. 20-29.

Gayfer, J., Sarfo-Mensah, P. & Arthur, E. (2002). Gwira Banso – Joint Forest Management Project: Mid Term Review. *Technical Report Prepared for CARE Ghana.*

Ghana Export Promotion Council. (2008). Ghana: *Non-traditional exports post strong growth*. Reported on www.modernghana.com.

Ghana Forestry Commission. (1998). Manual of procedures: Forest resource management planning in the HFZ. Section A - strategic planning. http://www.fcghana.com/ publications/manuals/hfz/StratPlanMan.htm#_Toc526929766.

Ghana Forestry Commission. (2005). *National forest plantation development programme. 2005-Annual Report.*

Ghana Forestry Commission. (2006a). Report on export of wood products. *Timber Industry Development Division*, *Ghana Forestry Commission*. Unpublished Report.

Ghana Forestry Commission. (2006b). *National forest plantation development programme.* 2006-Annual Report.

Ghana Forestry Commission. (2007a). Report on export of wood products. *Timber Industry Development Division, Ghana Forestry Commission.* Unpublished Report.

Ghana Forestry Commission. (2007b). *National forest plantation development programme.* 2007-Annual Report.

Ghana Forestry Commission. (2008a). *National Forest Plantation Development Programme-Anuual Report*, 2008. http://www.fcghana.com/publications/index.htm.

Ghana Forestry Commission. (2008b). Report on export of wood products. *Timber Industry Development Division, Ghana Forestry Commission.* Unpublished Report, 2008. http://www.fcghana.com/publications/index.htm.

Ghana Forestry Commission. (2009a). Report on export of wood products. Jan-May, 2009. Timber Industry Development Division, Ghana Forestry Commission. http://www.fcghana.com/publications/industry_trade/export_reports.htm/year_2009/may_2009.pdf.

Ghana Forestry Commission. (2009b). New procedures for stumpage collection and disbursement. (Published in conjunction with the in Association with the Office of Administrator of Stool Lands). http://www.fcghana.com/library_info.php?doc=55 and publication: New Procedure for Stumpage Disbursement.

Ghana Forestry Commission. (2010). Report on export of wood products. Jan-Dec, 2009. *Timber Industry Development Division, Ghana Forestry Commission.* http://www.fcghana.com/publications/industry_trade/export_reports.htm/year_2009/jan_2010.pdf.

Ghana Forestry Commission. (2011). Report on export of wood products. Jan-Dec, 2010. Timber Industry Development Division, Ghana Forestry Commission. http://www.fcghana.com/publications/industry_trade/export_reports.htm/year_2010/jan_2011.pdf

Ghana Forestry Commission. (2012). *National Forest Plantation Development Programme-Anuual Report*, 2008. http://www.fcghana.com/publications/index.htm.

Ghana Forestry Commission. (2014). *National Forest Plantation Development Programme-Anuual Report*, 2012. http://www.fcghana.com/publications/index.htm.

Ghana News Agency. (2006). Forest Plantation Development Fund Board calls for support. http://www.modernghana.com/news/99414/1/forest-plantation-devt-fund-board-calls-for-suppor.html.Ghana.

Ghana News Agency (2009). Thousands of jobs to be created through reforestation. News Report from the GNA. November, 30, 2009. http://www.ghanaweb.com/GhanaHomePage/NewsArchive/artikel.php?ID=172684andcomment=5308073#com.

Ghana News Agency, (2012). I am gratified at Ghana's tourism contribution – Ms Sena Dansua. Reported by the Ghana News Agency at http://vibeghana.com/2012/02/02/i-am-gratified-at-ghanas-tourism-contribution-ms-sena-dansua/.

Ghana Statistical Service. (2000). Ghana Living Standards Survey IV. Accra. Available on http://www.worldbank.org/html/prdph/lsms/country/gh/docs/G4Qprice.pdf.

Ghebremichael, A., Nanang, D. M. & Yang, R. (2005). Economic analysis of growth effects of thinning and fertilisation of lodgepole pine in Alberta, Canada. *Northern Journal of Applied Forestry*, *22*(4), 254-261.

Ghosh, R. C. & Singh, S. P. (1981). Trends in rotation. *Indian Forester*, *107*, pp. 336-347.

Gilbertson, T. & Reyes, O. (2009). Carbon trading: How it works and why it fails. Critical currents No. 7. November, 2009. *Dag Hammarskjöld Foundation Occasional Paper Series*. Uppsala.

Gonçalves, J. L. M., Barros, N. F., Nambiar, E. K. S. & Novais, R. F. (1997). Soil and stand management for short rotation plantations. *In*: Nambiar, E.K.S. & Brown, A.G. (Eds.) Management of soil, water and nutrients in tropical plantation forests, 379-417. *Australian Centre for International Agricultural Research (ACIAR*), Monograph, 43, Canberra.

Gonçalves, J. L. & N. F. Barros. (1999). Improvement of site productivity for short-rotation plantations in Brazil. BOSQUE, *20*(1), 89-106.

Gonçalves, J. L. M., Wichert, M. C. P., Gav3, J. L. & M. I. P. Serrano. (2008). Effects of Slash and Litter Management Practices on Soil Chemical Properties and Growth of Second Rotation Eucalypts in the Congo pp., 51-62 In: Nambiar, E.K.S. (Editor). Site Management and Productivity in Tropical Plantation Forests. Proceedings of Workshops in Piracicaba (Brazil) 22-26 November, 2004 and Bogor (Indonesia) 6-9 November, 2006. Bogor, Indonesia: Center for InternationalForestry Research (CIFOR), 2008.

Gove, J.H. & Fairweather, S. E. (1989). Maximum-likelihood estimation of Weibull function parameters using a general interactive optimiser and grouped data. *For. Ecol. Manage.*, *28*, pp. 61-69.

Gove, J. H. (2004). Structural stocking guides: a new look at an old friend. *Can. J. For. Res.*, *34* , pp. 1044 – 1056 .

GPRS. (2002). *Ghana Poverty Reduction Strategy*, pp. 2002-2004.

Grainge, M., S. Ahmed, W. C. Michell, and J. W. Hylin. (1985). Plant species reportedly possessing pest control properties: *An EWC/UH Database. Resource Systems Institute*, East-West Centre, Honolulu, HI.

Greenhill, A. G. (1881). Determination of the greatest height consistent with stability that a vertical pole or post can be made, and of the greatest height to which a tree of given properties can grow. *Proceedings of the Cambridge Philosophical Society IV*. Part II, pp. 65-73. Cambridge, England.

Paul Gregg. 'Soil erosion and conservation - Biological control of erosion', Te Ara - the Encyclopedia of New Zealand, updated, 13-Jul-12 URL: http://www.TeAra.govt.nz /en/soil-erosion-and-conservation/page-7

Gregersen, H. M. (1984). Incentives for forestation: a comparative assessment. In K. F. Wiersum, (Ed.) *Strategies and designs for afforestation, reforestation and tree planting*. Wageningen, the Netherlands, Wageningen Agricultural University.

Grijpma, P. (1976). Resistance of *Meliaceae* against the shootborer *Hypsipyla* with particular reference to *Toona ciliata* M. J. Roem. var. *australis* (F. v. M.) DC. *In:* Tropical trees. Variation breeding and conservation. J. Burley, and B. T. Styles, (Eds.) pp. 69-79. Academic Press, Oxford.

Haener, M. K. & Luckert. M. K. (1998). Forest Certification: Economic Issues and Welfare Implications. Canadian Public Policy, 24, Supplement, 2, pp. S83-S94

Hafley, W. L. & Schreuder, H. T. (1976). Some non-normal bivariate distributions and their potential for forest application. *International Union of Forest Research Organisations. XVI World Congress Proceedings*, Div. VI, pp. 104-114 (Oslo, Norway; June, 20 – July2, 1976).

Hafley, W. L. & Schreuder, H. T. (1977). Statistical distributions for fitting diameter and height data in even-aged stands. *Canadian Journal of Forest Research*, *7*, pp. 481-487.

Hall, D. O. (1983). Financial maturity for even-aged and all-aged stands. *For. Sci.*, *29*(4), pp. 833 - 836.

Hall J. B., Pandey D. & Hirai, S. (1999). Global Overview of teak plantations. Paper presented to the Regional Seminar *Site, Technology and Productivity of Teak Plantations* Chiang Mai, Thailand, 26-29 January, 1999.

Hall, J. B. & Bada, S. O. (1979). The Distribution and Ecology of Obeche (*Triplochiton Scleroxylon*) *Journal of Ecology*, Vol. *67*(2), pp. 543-564.

Hall, J. B. & Swaine, M. D. (1981). *Distribution and ecology of vascular plants in a tropical rainforest. Forest Vegetation in Ghana*. Geobotany, 1. The Hague. Springer.

Haltia, O. & Keipi, K. (1997). *Financing forest investments in Latin America: the issue of incentives.* Washington, DC, USA, Inter-American Development Bank.

Harper, J. L. (1977). *Population biology of plants*. Academic Press. London. p. 892

Hartemink, A. E. (2003). Soil fertility decline in the tropics with case studies on plantations. pp. 360 *ISRIC-CABI, Wallingford.*

Hartman, R. (1976). The harvesting decision when a standing forest has value. *Economic Inquiry*, pp. 14, 52-58.

Hawthorne, W. D. & Abu-Juam, M. (1995). *Forest Protection in Ghana. IUCN, Gland*, Switzerland and Cambridge. p. 203.

Hawthorne, W. D. (1995). Ecological profiles of Ghanaian forest trees. *Trop. For. Pap. 29*, Oxford Forestry Institute, pp. 345.

Healey, S. P. & Gara, R. I. (2003). The effect of teak (*Tectona grandis*) plantation on the establishment of native species in an abandoned pasture in Costa Rica. *For. Ecol. and Manage.*, *176*, pp. 497-507.

Hearne, D. A. (1975). *Trees for Darwin and northern Australia*, Department of Agriculture, Forestry and Timber Bureau, Australian Government Publishing Service, Canberra.

Hedegart, T. (1976). *Breeding systems, variation and genetic improvement of teak (Tectona grandis Linn F.) in tropical trees*. London. Academic Press Inc.

Hepburn, G. (1989). Pesticides and drugs from the neem tree. *Ecologist*, *19*(1), pp. 31-32.

Hiley, W. E. (1930). *The economics of forestry*. The Clarendon Press, Oxford, U. K., p. 256

Hitch, C. J. & McKean, R. N. (1960). *The economics of defence in the nuclear age*. Cambridge, Massachusetts. Harvard University Press.

Holdridge, L. R. (1976). Ecología. de las Meliáceas Latinoamericanas. Studies on the shootborer *Hypsipyla grandella* Zeller. vol. 3. J. L. Whitmore (Ed.), Centro Agronómico Tropical de Investigación y Enseñanza, Miscellaneous Publication, *1*. Turrialba, Costa Rica. p. 7.

Holm L. R., Pluncknett D. L., Pancho J. V. & Herberger J. P. (1977). *Imperata cylindrica* (L.) Beauv. In *The world's worst weeds: Distribution and biology*. pp. 62–71. University Press of Hawaii, Honolulu, USA.

Honu, Y. A. K. & Dang, Q. L. (2000). Response of tree seedlings to the removal of *Chromolaena odorata* Linn. in a degraded forest in Ghana. *For. Ecol. Manage*. *137*, pp. 75-82.

Honu, Y. A. K. & Dang, Q. L. (2002). Spatial distribution and species composition of tree seed and seedlings under the canopy of the shrub, *Chromolaena odorata* Linn. in Ghana. *For. Ecol. Manage.*, *164*, pp. 185-196.

Horne, J. E. M. (1966). *Teak in Nigeria. Nigerian Information Bulletin* (New Series) No.16.

Hossain, K. L.(1999). *Gmelina arborea*: A popular plantation species in the tropics. A quick guide to multipurpose trees from around the world. *Fact Sheet*, pp. 99-05.

Howland, P., Bowen, M. R., Ladipo, D. P. & Oke, J. B. (1977). The study of clonal variation in *Triplochiton scleroxylon* K. Schum, as a basis for selection and improvement. *Proceedings of the Joint IUFRO Workshop*, pp. 6–13.

Hunter, J. L. Jr. (1990). Wildlife, forests, and forestry. *Principles of managing forests for biological diversity*. Englewood Cliffs. New Jersey: Prentice Hall.

Husch, B. (1963). Forest mensuration and statistics. The Ronald Press Co., New York. p. 474.

Husch, B., Miller, C. I. & Beers, T. W. (1982). *Forest mensuration*. 3rd ed. John Wiley and Sons. New York. p. 402.

Husch B, Beers T. W. & Kershaw, J. A. (2003) Forest mensuration, 4th edn. Wiley, New Jersey Loetsch, F. & Haller, K.E. (1973). Forest Inventory. Vol II, München, Germany, BLV Verlagsgesellschaft mbH.

Hyde, W. F. (1980). *Timber supply, land allocation and economic efficiency*. The Johns Hopkins University for Resources for the Future. Baltimore, MD.

Ingersoll. J. E. & Ross, S. A. (1992). Waiting to invest: investment and uncertainty. *Journal of Business*, *65*(January), pp. 1-29.

Iowa State University. (1997). Farmstead Windbreaks: Planning. (1997). Iowa State University of Science and Technology, Ames, Iowa

IPCC, 2007, Climate Change, 2007, Synthesis Report. Contribution of Working Groups I, II and III to the Fourth Assessment Report of the Intergovernmental Panel on Climate Change [Core Writing Team, Pachauri, R.K and Reisinger, A. (Eds.)]. *IPCC*, Geneva, Switzerland, p. 104.

ITTO. (2005a). *Status of tropical forest Management. ITTO Technical Series* Note, *24*, Ghana (pp. 98-104).

ITTO. (2005b). Revised ITTO Criteria and Indicators for the sustainable management of tropical forests including reporting format. *ITTO Policy Development Series* No 15.

ITTO, (2005c). *Tropical timber market report*, 1–15 February, 2005. International Tropical Timber Organisation, Yokohama, Japan.

ITTO. (2008). Developing forest certification: Towards increasing the comparability and acceptance of forest certification systems. ITTO Technical Series No 29.

Jack, S. B. & Long, J. N. (1996). Linkages between silviculture and ecology: an analysis of density management diagrams. *For. Ecol. Manage.*, *86*, pp. 205–220.

Jacobson, M. (1958). Insecticides from plants: A review of the literature, 1941-53. *USDA Handbook, 154*. Washington, DC.

Jacobson, M. (1975). Insecticides from plants: A review of the literature, 1952-72. *USDA Handbook, 461*. Washington, DC.

Jackson, D. H. (1980). *The microeconomics of the timber industry*. Westview Press, Boulder, Co.

Jadhav, B. B. & Gaynar, D. G. (1994). Effect of Tectona grandis (L.) leaf leachates on rice and cowpea. *Allelopathy Journal*, *1*, pp. 66-69.

James, T., Vege, S., Aldrich, P. & Hamrick, J. L. (1998). Mating systems of three tropical dry forest tree species. *Biotropica*, *30*(4), pp. 587-594.

Jenkins, J. C., D. C. Chojnacky, L. S. Heath & R. A. Birdsey. (2003). National-scale biomass estimators for United States tree species. *For. Sci.*, *49*, pp. 12-35.

Jensen, M. (1995). Trees commonly cultivated in Southeast Asia; Illustrated field guide. *RAP Publication:* 1995/38, FAO, Bangkok, Thailand. p. 93.

Jessen, R. J. (1955) Determining the fruit count on a tree by randomized branch sampling. *Biometrics*, *11*, pp. 99-109.

Joet, A., Jouve, P. & Banoin, M. (1998). Le defrichement ameliore au Sahel. Une pratique agroforestiere adoptee par les paysans. (Improved forest clearance: an agroforestry method adopted by the people in the Sahel) *Bois-et-Forets-des-Tropiques*, *225*, pp. 31-44.

Johnson E & Miyanishi, K. (2007). *Plant disturbance ecology: the process and the response.* Academic Press/Elsevier, Burlington, MA.

Johnson, N. L. (1949a). Systems of frequency curves generated by methods of translation. *Biometrika.*, *36*, pp. 149-176.

Johnson, N. L. (1949b). Bivariate distributions based on simple translation systems. *Biometrika.*, *36*, pp. 297-304.

Johnson, N. L. & Kotz, S. (1972). *Distributions in statistics: Continuous multivariate distributions.* John Wiley and Sons Inc., New York, U. S. A. p. 333.

Johnson, N. L. Kotz, S. & Balakrishnan, N. (1994). *Continuous univariate distributions.* Vol. I. 2nd Edition. John Wiley and Sons Inc., New York, U. S. A. p.756

Johnstone, W. D. (1997). The effect of commercial thinning on the growth, and yield of lodgepole pine. Pages, 13-23 *in* Proceedings of a Commercial Thinning Workshop, Whitecourt, Alberta, 17-18 October, 1996. *FERIC, Vancouver. Spec. Rep.* SR-122.

Jøker, D. & Salazar, R. (2000). Seed Leaflet No. 22. *Ceiba pentandra* (L.) Gaertn. Danida Forest Seed Centre. CATIE.

Jones, N., (1974). Records and comments regarding the flowering of *Triplochiton scleroxylon* K. Schum. *Commonwealth For. Rev.*, *53*, pp. 52–56.

Jones, R. (1969). Review and comparison of site evaluation methods. U. S. Forest Service Research paper RM-51.

JoyFM. (2014) .SADA: GHc32m afforestation project go waste? http://www.myjoyonline.com/news/2014/april-15th/sada-ghc32m-afforestation-project-go-waste.php

Judd, M. P. (2004). Introduction and Management of Neem (*Azadirachta indica*) in Smallholder's Farm Fields in the Baddibu Districts of The Gambia, West Africa. *MSc in Forestry Thesis.* Michigan Technological University.

Kadambi, K. (1972). *Silviculture and management of teak.* Bul. 24. Nacogdoches, TX, Stephen F. Austin State University School of Forestry.

Kahurananga, J., Alemayehu, Y., Tadesse, S. & Bekele, T. (1993). Informal surveys to assess social forestry at Dibandiba and Aleta Wendo, Ethiopia. *Agroforestry Systems*, *24*, 57-80.

Kant, S. & Redantz, A. (1997). An econometric model of tropical deforestation. *Journal of Forest Economics*, *3*(1), pp. 51-86.

Kassier, H. W. & Bredenkamp, B. V. (1994). Modelling diameter and height distributions through dispersion statistics in even-aged pine plantations. *South African Forestry Journal*, *171*, pp. 21-27.

Kelty, M. J. (1992). Comparative productivity of monocultures and mixedspecies stands. In: Kelty, M.J., Larson, B.C., Oliver, C.D. (Eds.), The Ecology and Silviculture of Mixed-Species Forests. Kluwer Academic Publishers, Dordrecht, Boston, pp. 125–141.

Kelty, M. J. (2006). The role of species mixtures in plantation forestry. *Forest Ecology and Management*, *233*(2006) pp. 195–204

Kemma, A. (1993). Case studies on real options. *Financial Management, Autumn*, pp. 259-270.

Kengen, S. (1997). Forest Valuation for decision making: *Lessons of experience and proposals for improvement.* Rome. Food and Agriculture Organisation.

Kensinger, J. (1987). Adding the value of active management into the capital budgeting equation. *Midland Corporate Finance Journal*, *5*(Spring), pp. 31-42.

Keogh, R. M. (1982). Teak (*Tectona grandis.* Linn F.) provisional site classification chart for the Caribbean, Central America, Venezuela and Colombia. *For. Ecol. and Manag.*, *4*, pp. 143-153.

Kerr, G. (1999). The use of silvicultural systems to enhance the biological diversity of plantation forests in Britain. *Forestry*, *72*(3), pp. 191-205.

Kester, W. C. (1984). Today's options for tomorrow's growth. *Harvard Business Review.* March-April, pp. 153-160.

Kester, W. C. (1993). *Turning growth options into real assets.* In: Aggarwal, A. (Ed.), Capital budgeting under uncertainty Englewood Cliffs, NJ, Prentice-Hall: pp. 187-207.

Ketkar, C. M. (1976).Utilisation of neem (*Azadirachta indica*) and its by-products. Directorate of Non-Edible Oils and Soap Industry, Khadi and Village Industries Commission, Pune, India.

Ketkar, C. M. (1984). Crop experiments to increase the efficiency of urea fertiliser nitrogen by use of neem products. pp. 507-518 *in* Schmutterer, H. and K. R. S Ascher (eds.) *Proc. 2nd. Int. Neem Conf. Rauischholzhausen*, Federal Republic of Germany, May, 25-28 1983. p. 587.

Khandiya, S. D. & Goel, V. L. (1986). Patterns of variability in some fuelwood trees grown on sodic soils. *Indian Forester*, *112*(2), pp. 118-123.

Kimmins, J. P. (2003). *Forest Ecology*, 3rd edition. Prentice-Hall, Upper Saddle River, NJ.

King, D. M. & Mazzotta, M. J. (2000). Ecosystem valuation. U. S. Department of Agriculture, Natural Resources Conservation Service and National Oceanographic and Atmospheric Administration. Website: http://www.ecosystemvaluation.org/default.htm. Accessed December, 2008.

Kira, T., Ogawa, H. & Shinozaki, K. (1953). *Interspecific competition among higher plants.* 1. J. inst. Polytech. Osaka City University. D. *4*, pp. 1-16.

Knoebel, B. & Burkhart, H. (1991). A bivariate distribution approach to modelling forest diameter distributions at two points in time. *Biometrics*, *47*, pp. 241-298.

Korankye-Gyamera, Y. (1997). Preliminary studies into methods of propagating dawadawa (parkia biglobosa). *BSc (Tech) Thesis, Department of Renewable Natural Resources*, University for Development Studies, Tamale, Northern Region.

Kouch, T., Preston, T. R. & Hieak, H. (2006). Effect of supplementation with Kapok (*Ceiba pentandra*) tree foliage and Ivermectin injection on growth rate and parasite eggs in faeces of grazing goats in farmer households. Livestock Research for Rural Development. *Volume, 18, Article #87.* Retrieved February, *23*, 2010, from http://www.lrrd.org/lrrd18/6/kouc18087.htm.

Kozak, A. (1988). A variable-exponent taper equation. *Can. J. For. Res.*, *18*, pp. 1363-1368.

Kraenzel, M. B. (2000). Carbon Storage of Panamanian Harvest-Age Teak *(Tectona grandis)* Plantations. *Masters Thesis. Department of Biology*, McGill University, *Montreal*, Canada.

Krebs, C. (1989). Ecological Methodology. A collection of methods commonly used in ecology including a section on sampling

Krishnapillay, B. (2000). Silviculture and management of teak plantations. *Unasylva*, *51*, available at http://www.fao.org/docrep/x4565e/x4565e04.htm#P0_0. Accessed July, 2009.

Kulatilaka, N. (1993). The value of flexibility: the case of a dual fuel industrial steam boiler. *Financial Management*, *22* (Autumn), pp. 271-280.

Kulatilaka, N. & Trigeorgis, L. (1994). The general flexibility to switch: real options revisited. *International Journal of Finance*, *6*(Spring), pp. 778-798.

Kunkle, S. H. (1978). Forestry support for agriculture through watershed management, windbreaks and other conservation measures. Proceedings of the, 8th World Forestry Congress, *Jakarta*. Vol, *3*., Pp. 113-46

Kumar, B. M., Long, J. N. & Kumar, P. (1995). A density management diagram for teak plantations of Kerala in peninsular India. *Forest Ecology and Management.*, *74*, 125-131.

Kuuzegh, R. S. (2009). Case Study: Ghana's climate change vulnerability assessment. Special Session of AMCEN on Climate Change, Nairobi 25th – 29th May, 2009.

Kuypers, H., Mollema, A. & Topper, E. (2005). Agromisa Foundation, Wageningen, 97 pp.

Lamb, A. F. A. (1968). *Fast growing timber trees of the lowland tropics*. No. 2 *Cedrela odorata*. Commonwealth Forestry Institute, Dept. of Forestry, University of Oxford. p. 46.

Lambert, M. C., Ung, C. H. & Raulier, F. (2005). Canadian national tree aboveground biomass equations. *Can. J. For. Res.*, *35*, pp. 1996-2018.

Lamers J. P. A., Michels, K. & Vandenbeldt, R. J. (1994). Trees and windbreaks in the sahel: establishment, growth, nutritive, and calorific values. *Agroforestry Systems*, *26*, 171-184.

Lamprecht, H. (1989). Silviculture in the tropics: *tropical forest ecosystems and their tree species..*

Landsberg, J. J. (1997). The biophysical environment. *In*: Nambiar, E.K.S. and Brown, A.G. (eds.) Management of soil, water and nutrients in tropical plantation forests, 65-96. *Australian Centre for International Agricultural Research (ACIAR)*, Monograph, *43*, Canberra.

Langyintuo, A. S., Ntoukam, G., Murdock, L., Lowenberg-DeBoer, J. & Miller, D. J., (2004). Consumer preferences for cowpea in Cameroon and Ghana. *Agricultural Economics*, *30*(3), pp. 203–213.

Laurie, M. V. (1974). Tree planting practices in African savannas. *Forestry Development Paper* No. 19. FAO. Rome.

Laurie, M. V. & Ram, B. S. (1940). Yield and stand tables for teak (*Tectona grandis* L. F) plantations in India and Burma. *Indian Forest Records*. No.1. Vol. IV-A. p. 115.

Lawson, G. W. (1968). Ghana. pp 81-86 In Inga, H. And H. Olov (Eds.). *Conservation of vegetation in Africa south of the Sahara*. Proc. of a symposium of the, 16th planetary meeting of the "Association pour I'etude Taxanomique de la Flore d'Afrique Tropicale" in Uppsala, Sweden, Sept, 12-16. 1966. p. 320.

Lawson, G. W., Jenik, J. & Armstrong-Mensah, K. O. (1968). A study of a vegetation catena in the Guinea Savannah at Mole Game Reserve (Ghana). *J. Ecol.*, *56*, pp. 505-522.

Leach, G. & Mearns, R. (1988). *Beyond the Woodfuel Crisis: People, Land and Trees in Africa*. Earthscan Publications. London.

Li, Z., Kurz, W. A., Apps, M. J. & Beukema, S. J. (2003). Belowground biomass dynamics in the Carbon Budget Model of the Canadian Forest Sector: recent improvements and implications for the estimation of NPP and NEP. *Can. J. For. Res.*, *33*, pp. 126-136.

Lim, T. T. & Huang, X. (2006). Evaluation of kapok (*Ceiba pentandra* (L.) Gaertn.) as a natural hollow hydrophobic-oleophilic fibrous sorbent for oil spill cleanup. *Chemosphere. 66*(5), pp. 955-963.

Little, E. L. Jr. (1983). *Common fuelwood crops: a handbook for their identification.* McClain Printing Co., Parsons, WV.

Locatelli, B. & Vignola, R. (2009). Managing watershed services of tropical forests and plantations: Can meta-analyses help?. *Forest Ecology and Management*, *258*, pp. 1864-1870.

Loetsch, F. & Haller, K. E. (1973). Forest Inventory. Vol II, München, Germany, BLV Verlagsgesellschaft mbH.

Logu, A. E.. Brown, S. & Chapman J. (1988). Analytical review of production rates and stemwood biomass of tropical forest plantations. *For. Ecol. and Manag.*, *23*, pp. 179-200.

Long, J. N. (1985). A practical approach to density management. *For. Chron.*, *23*, pp. 23 – 26 .

Long, J. N. & Shaw, J. D. 2005 A density management diagram for even-aged ponderosa pine stands . *West. J. Appl. For.*, *20*, pp. 205 – 215.

Lowe. R. G. (1976). Teak (*Tectona* grandis. Linn F.) Thinning experiment in Nigeria. *Commonw. For. Rev.*, *55*, pp. 189-202.

Lowery, R. F., Lamberto, C. C., Endo, M. & Kane. M. (1993). Vegetation management in tropical forest plantations, *Can. J. For. Res.*, *23*, pp. 2006-2014.

Lowry, R. F. & Gjerstad, D. H. (1991). Chemical and mechanical site preparation. In M. L. Duryea and P. M. Dougherty (Eds.), *Forest Regeneration manual*, pp. 251-261. The Netherlands: Klumer Academic Publishers.

Lovasi, G. S., Quinn, J. W., Neckerman, K. M., Perzanowski, M. S. & Randle, A. (2008). Children living in areas with more street trees have lower asthma prevalence. Journal of *Epidemiology and Community Health* doi:10.1136/jech.2007.071894

Ludwig, J, Tongway, D. & Noble, J. (2002), Landscape ecology, function and management: principle from Australia's rangelands: CSIRO, Melbourne.

Luehrman, T. A. (1998). Investment opportunities as real options: getting started on the numbers. *Harvard Business Review*, July-August.

Lugo, A. E. & Brown, S. (1982). Conversion of Topical Forests: A Critique. *Interciencia*, *7*(2), pp. 89-93.

Luo, Q., Bellotti, B., Bryan, B. & Williams, M. (2003). Risk analysis of possible environmental change and future crop production in South Australia: proceedings of the Australian Society of Agronomy, 11th Australian Agronomy Conference, Geelong, Victoria.

Lutz, E., Pagiola, S. & Reiche, C. (1994). Cost-benefit analysis of soil conservation: the farmers' viewpoint. *The World Bank Research Observer*, *9*, pp. 273-295.

Mackay, J. H. (1952). Notes on establishment of neem plantations in Bornu Province of Nigeria. *Farm and For.*, *11*, pp. 9-14.

Mackenzie, J. A., (1959). Phenology of *Triplochiton scleroxylon.* Technical Note, 1. Department of Forest Research of Nigeria, pp. 5.

Madsen, K. H. & Streibig, J. C. (2003). Benefits and risks of the use of herbicide-resistant crops. In: R. Labrada (Ed.), *Weed management for developing countries*. pp. 245-269. FAO. Rome.

Malende, Y. H. & Temu, A. B. (1990). Site Index curves and volume growth of teak (*Tectona grandis*) at Mtibwa, Tanzania. *For. Ecol. and Manage.*, *31*, pp. 91-99.

Malimbwi, R. E. (1978). *Cedrela* species international provenance trial (CFI at Kwamsambia, Tanzania). *In* Progress and problems of genetic improvement of tropical forest trees. p. 910. Commonwealth Forestry Institute, Oxford.

Manley, B. & Maclaren, P. (2010). Potential impact of carbon trading on forest management in New Zealand. *Forest Policy and Economics. doi*: 10.1016/j.forpol.2010.01.001.

Manshard, W. (1992). Problems of deforestation in tropical Africa: Fuelwood extraction, agroforestry and sustainable development. *Agricultural Change, Environment and Economy.* Mansell, New York. pp. 203-222.

MarMoller, C. (1954). The influence of thinning on volume increment. 1. Results of investigations. Pages, 5-32 *in* Thinning Problems and Practices in Denmark. SUNY Coll. For. at Syracuse, *World For. Sev. Bull.* No. 1, Tech. Pub. No.76.

Marshall, A. W. (1955) A note on randomized branch sampling. The RAND Corporation. Note P-725.

Mason, G. (2004). Overview of cost-effectiveness analysis and cost-benefit analysis and their application to labour market and social development policies. *Conference on cost effectiveness in evaluation* held on June, 17, 2004. Ottawa, Ontario. Human Resources and Social Development Canada.

Mason, S. P. & Merton, R. C. (1985). The role of contingent claims analysis in corporate finance. In: *Recent advances in corporate finance*. Altman, E., Subrahmanyam, M. (Eds.), Homewood, IL, Richard D. Inin : 7-54.

Matthews, J. D. (1989). *Silvicultural systems.* Oxford University Press, Oxford.

Mathur, H. N., Babu, R., Joshie, P. & Singh, B. (1976). Effect of clearfelling and reforestation on runoff and peak rates in small watersheds. *Indian Forester*, *102*, pp. 219-26

Maydell, H-J. von. (1990). *Trees and shrubs of the sahel and their characteristics and uses.* Verlag Josef Margraf Scientific Books. p. 600.

McCarter , J. B. & Long , J. N. (1986). A lodgepole pine density management diagram . *West. J. Appl. For.*, *1*, 6 – 11.

McConnell, K. E., Daberkow, J. N. & Hardie, I. W. (1983). Planning timber production with evolving prices and costs. *Land Economics*, *59*(3), pp. 292-299.

McDonald, R. & Siegel, D. (1985). Investment and the valuation of firms when there is an option to shut down. *International Economic Review*, June pp. 331-349.

McFadden, D. (1981). Econometric models of Probabilistic Choice. In C. F. Manski and D. McFadden (Eds.). *Structural Analysis of Discrete Data with Econometric Applications.* Cambridge, Mass. MIT Press.

McPherson, E. Gregory; Simpson, James R.; Peper, Paula. J.; Crowell, Aaron M.N.; Xiao, Qingfu. (2010). Northern California coast community tree guide: benefits, costs, and strategic planting. Gen. Tech. Rep. PSW-GTR-228. Albany, CA: U.S. Department of Agriculture, Forest Service, Pacific Southwest Research Station. p. 118.

Mengel, K. & Kirkby, E. A. (1978). *Principles of Plant Nutrition.* International Potash Institute.

Mensah, J. A. (2009). Tourism raked in $US1.3 in 2008 – Minister. Ghana News Agency report of April, 29, 2009. Ghana Business News. Accessed April, 2009.

Miller, A. D. (1969). *Provisional Yield tables for teak in Trinidad.* Trinidad and Tobago Government Printery.

Miller, J. J., Perry, Jr. J. P. & Borlaug, N. E. (1957). Control of sunscald and subsequent Buprestid damage in Spanish cedar plantations in Yucatan. *Journal of Forestry*, *55*, pp. 185-188.

Miller, C. G. Financial Assurance for Mine Closure and Reclamation. (2005) [cited, 2012 July, 31]; International Council on Mining and Metals. Available from: http://www.icmm.com/page/1158/financial-assurance-for-mine-closure-and-reclamation.

Ministry of Environment and Science. (2002). *National Biodiversity Strategy for Ghana.* Ministry of Environment and Science, Accra.

Ministry of Forests and Range. (2009). Growth and yield modelling: about G and Y prediction models. Research Branch. Government of British Columbia, Canada http://www.for.gov.bc.ca/hre/gymodels/GY-Model/about.htm.

Ministry of Forests. (1998). The management and prospects of teak in Vietnam - Ministry of Forestry. Forest Science Sub-Institute of Southern Vietnam. Teak for the future. *Proceedings of the second regional seminar on teak.* Kashio, M. and K. White (Eds.) Rap Publication -1998/05 p. 249.

Ministry of Lands and Forestry. (2004). Criteria and indicators for sustainable management of natural tropical forests. Reporting Questionnaire for Indicators at the National Level – *Report for Ghana.* Submitted to ITTO, March, 2004. Ghana Forestry Commission, Ministry of Lands and Forestry, Accra, Ghana. Unpublished.

Mitchell, K. J. (1969). Simulation of the growth of even-aged stands of white spruce. Yale University Press, Yale Univ. School of Forestry Bull. No 75, New Haven, CT.

Mitchell, K. J. (1975). Dynamics and simulated yield of Douglas-fir. For. Sci. Monogr. 17.

Mitra, T and Wan, Jr. H. Y. (1985). On the Faustmann solution to the forest management problem. *Journal of Economic Theory*. 40 (2), pp. 229 - 249.

Mitzutani, J. (1999). Selected allelochemicals. *Crit. Rev. Plant Sci.*, *18*, pp. 653-671.

MOFA/AFU. (1986). *The National Agroforestry Policy*. Ministry of Food and Agriculture Accra, Ghana

Morck, R., Schwartz, E., Stangeland, D. (1989). The valuation of forestry resources under stochastic prices and inventories. *Journal of Financial and Quantitative Analysis*, 24, pp. 473-487.

Munslow, B., Katerere, Y., Ferf, A. & O'Keefe, P. (1988). *The Fuelwood Trap: A Study of the SADCC Region.* Earthscan Publications. London.

Murugan, K. & Kumar, N. S. (1996). Host plant biochemical diversity, feeding, growth and reproduction of teak defoliator *Hyblaea puera* (Cramer). *Indian Journal of Forestry*, *19*, pp. 253-257.

Myers, S. C. (1977). Determinants of corporate borrowing. *Journal of Financial Economics*, November, pp. 147-176.

Nagaveni, H. C., Ananthapadmanbha, H. S. & Rai, S. N. (1987). Note on extension of viability of *Azadirachta indica*. Myforest, *23*(4), pp. 245-250.

Nair, P. K. R. (1993). An introduction to agroforestry. Kluwer Academic Publishers. The Netherlands. pp. 489.

Nair, K. S. S. (2007). Tropical forest insect pests: *ecology, impact and management.* Cambridge University Press.

Nambiar, E. K. S. (1990). Interplay between nutrients, water, root growth and productivity in young plantations, *For. Ecol. Manage. 30*, pp. 213-232.

Nambiar, E. K. S. & Kallio, M. H. (2008). Increasing and Sustaining Productivity in Subtropical and Tropical Plantation forests: Making a Difference through Research Partnership

In: Nambiar, E. K. S. (ed.) Site management and productivity in tropical plantation forests: workshop proceedings, 22-26 November, 2004 Piracicaba, Brazil, and, 6-9 November, Bogor, Indonesia, pp. 205-227. *Center for International Forestry Research*, Bogor, Indonesia.

Nanang, D. M. (1996). The silviculture, growth and yield of neem (*Azadirachta indica* A. Juss) plantations in Northern Ghana. Unpublished M.Sc.F Thesis, Lakehead University, Ontario, Canada.

Nanang, D. M. (1998). Suitability of the normal, lognormal and Weibull distributions for fitting diameter distributions of neem (*Azadirachta indica* A. Juss.) plantations in Northern Ghana. *Forest Ecology and Management, 103*(1), pp. 1–7.

Nanang, D. M. (2010). Analysis of export demand for Ghana's timber products: A multivariate co-integration approach. *Journal of Forest Economics, 16*(1), pp. 47-61.

Nanang, D. M. & Owusu, E. H. (2010). Estimating the economic value of recreation at the Kakum National Park, Ghana. In D. M. Nanang and T. K. Nunifu (Eds.). *Natural resources in Ghana: Management, policy and economics.* New York. Nova Science Publishers.

Nanang, D. M. & Nunifu, T. K. (1999). Selecting a functional form for anamorphic site index curve estimation. *Forest Ecology and Management, 118*, pp. 211-221.

Nanang, D. M. & Asante, W. (2000). Effect of neem (*Azadirachta indica* A. Juss) on food crop production and quality in Northern Ghana. *Final Project Report. National Agricultural Research Programme*, Accra, Ghana

Nanang, D. M. & Yiridoe, E. K., 2010. Analyses of the Causes of Deforestation in Ghana: An Econometric Approach. In: Nanang, D. M. & Nunifu, T. K. (Eds.). *Natural resources in ghana: management, policy and economics.* Nova Science Publishers, New York.

Nanang, D. M., Day, R. J. & Amaligo, J. N. (1997). Growth and yield of neem (*Azadirachta indica* A. Juss.) plantations in Northern Ghana. *Commonwealth Forestry Review. 76*(2), pp. 103 -106.

National Academy of Sciences. (1980). Firewood crops. *Shrubs and tree species for energy production.* National Academy of Sciences. Washington D.C. pp. 237.

Nautiyal, J. C. (1988). Forest Economics: *Principles and Application*, Canadian Scholars' Press Inc.

Navrud, S. & Vondolia, G. K. (2005). Using contingent valuation to price ecotourism sites in developing countries, *Tourism, 53*(2), pp. 115-125.

Neef, T. & Henders, S., 2007. *Guidebook to Markets and Commercialisation of Forestry CDM projects.* Tropical Agricultural Research and Higher Education Centre (CATIE).

Nelson, T. C. (1964). Diameter distribution and growth of loblolly pine. *Forest Science, 10*(1), pp. 105-114.

Newman, D. H. (1988). The optimal forest rotation: a discussion and annotated bibliography. Gen. Technical Report SE - 48. Asheville, NC: U.S Dept. of Agric., *Forest Service.* Southeastern Forest Experiment Station. p. 47

Newman, D. H. & Yin, R. (1995). A note on the tree-cutting problem in a stochastic environment. *Journal of Forest Economics, 1*, 181-190.

Newton, P. F. (1997). Stand density management diagrams: review of their development and utility in stand-level management planning. *For. Ecol. Manage. 98*, pp. 251 – 265.

Newton, P. F. & Weetman , G. F. (1994). Stand density management diagram for managed black spruce stands. *For. Chron., 70*, pp. 65 – 74 .

Newton, P. F., Lei, Y. & Zhang, S. Y. (2005). Stand-level diameter distribution yield model for black spruce plantations . *For. Ecol. Manage., 209*, pp. 181 – 192.

New Zeanland Farm Forestry Association. (2014). http://www.nzffa.org.nz/farm-forestry-model/resource-centre/tree-grower-articles/tree-grower-february-2006/understanding-the-way-trees-reduce-soil-erosion/

Nketiah, T., Newton, A. C. & Leakey, R. R. B. (1998). Vegetative propagation of *Triplochiton scleroxylon* K. Schum in Ghana. *Forest Ecology and Management*, 105 _1998. pp. 99–105.

Norby, N. J, DeLucia, E. H, Gielen, B., Calfapietra, C., Giardina, C. P., John, S., King, J. S., Ledford, J., McCarthy, H. R., Moore, D. J. P., Ceulemans, R., De Angelis, P., Finzi, A. C., Karnosky, D. F., Kubiske, M. E., Lukac, M., Pregitzer, K. S., Scarascia-Mugnozzan, G. E., Schlesinger, W. H. & Oren, R. (2005), 'Forest response to elevated CO_2 is conserved across a broad range of productivity', Proceedings of the National Academy of Sciences, vol. *102*, pp. 18052–18056.

Nowak, R. S., Ellsworth, D. S & Smith, S. D. (2004), 'Functional responses of plants to elevated atmospheric CO_2, do photosynthetic and productivity data from FACE experiments support early predictions?', *New Phytologist*, vol. *162*, pp. 253-280.

NRC. (1992). National Research Council. *Neem: a tree for solving global problems*, report of an ad-hoc panel of the Board on Science and Technology for International Development National Academy Press, Washington, DC.

Ntiamoa-Baidu, Y. (1998). *Sustainable harvesting, production and use of bushmeat.* Wildlife Department, Ministry of Lands and Forestry, Accra, Ghana.

Nunifu K. T. & Murchison, H. G. (1999). Provisional yield models of Teak (*Tectona grandis* Linn F.) plantations in Northern Ghana. *For. Ecol. and Manage., 120*, pp. 171-178.

Nunifu, K. T. (1997). The growth and yield of teak (*Tectona grandis* Linn F.) *plantations in Northern Ghana*. M.Sc.F Thesis, Faculty of Forestry, Lakehead University, pp. 101.

Nunifu, T. K. (2010). Growth and management of teak (tectona grandis linn f.) Plantations in ghana. In: nanang, d. M. & nunifu, t. K. (eds.). Natural resources in ghana: management, policy and economics. Nova science publishers, new york.

Nwoboshi L. C. (1994). Development of Gmelina arborea under the Subri Conversion Technique: First three years. *Ghana Journal of Forestry, 1*, pp. 12–8.

Nyland, R. D. (1996). Silviculture: *Concepts and Applications.* McGraw-Hill, New York, p. 631.

Nyland, R. D. (2007). Silviculture: *Concepts and Applications.* Second edition. McGraw-Hill, New York.

O'Hara, K. L. (2004). Forest Stand Structure and Development: Implications for Forest Management. *USDA Forest Service Gen. Tech. Rep.*, PSW-GTR-193.

O'Keefe, P. & Raskin, P. (1985). Crisis and opportunity: fuelwood in Africa. *Ambio*, *14*, 4-5, pp. 220-224.

Oboho, E. E. G & Ali, J. Y. (1985). Preliminary investigation of the effect of seed weight in early growth characteristics of some savannah species. pp. 144-159 *in* Okojie, J.A and Okoro, O.O (eds.) *Proc. 15th Annual Conf. of the Forestry Association of Nigeria*, Yola Nigeria, Nov. pp. 25-29 1985.

Odoom, F. K., (1999). Securing land for forest plantations in Ghana. *International Forestry Review*, *1*(3), pp. 182– 188.

Odoom, M. & Vlosky, R. P. (2007). A Strategic Overview of the Forest Sector in Ghana. Louisiana Forest Products Development Center Working Paper #81. School of Renewable Natural Resources, Louisiana State University Agricultural Center. Baton Rouge. USA.

OECD. (2006). *Good practice guidance on applying strategic environmental assessment (SEA) in development co-operation. Final Draft.* Organisation for Economic Co-operation and Development. Paris. p. 116.

Oliver, C. D. & Larson, B. C. (1996). *Forest stand dynamics*, 2nd edition. John Wiley and Sons, New York.

Olschewski, R. & Benıtez, P. (2010). Optimising joint production of timber and carbon sequestration of afforestation projects. *Journal of Forest Economics*, *16*, pp. 1–10.

Olschewski, R. & Benitez, P. (2005). Secondary forests as temporary carbon sinks? The economic impact of accounting methods on reforestation projects in the tropics. *Ecological Economics*, *55*(3), pp. 380–394.

Olschewski, R., Benıtez, P., de Koning, G. H. J. & Schlichter, T. (2005). How attractive are forest carbon sinks? Economic insights into supply and demand of certified emission reductions. *Journal of Forest Economics*, *11*, pp. 77–94.

Omoyiola, B. O. (1973). Initial observation on *Cedrela odorata* provenance trial in Nigeria. *In Tropical provenance and progeny research and international cooperation*. pp. 250-254. Commonwealth Forestry Institute, Oxford.

Ormerod, W. G. (1973). A simple bole model. *For. Chron.*, *49*, pp. 136-138.

Orwa, C., Mutua, A., Kindt, R. , Jamnadass, R. & Anthony, S. (2009). Agroforestree Database:a tree reference and selection guide version, 4.0 (http://www.world agroforestry.org/sites/ treedbs/treedatabases.asp).

Oregon State University. Undated. Ecophysiology of Tree Growth - Oregon State University, http://www.cof.orst.edu/cof/teach/for442/cnotes/sec15/eco.htm (accessed July, 06, 2015).

Ott, S. H. & Thompson, H. E. (1996). Uncertainty outlay in time to build problems. *Management and Decision Economics*, *17*(1), pp. 1-16.

Owen, D. B. & Wiesen, J. M. (1959). A method of computing bivariate normal probabilities with application to handling errors in testing and measuring. *The Bell System Technical Journal*, *38*, pp. 553-572.

Owubah, K., LeMaster, D. C., Bowker, J. M. & Lee, J. G., (2001). Forest tenure systems and sustainable forest management: the case of Ghana. *Forest Ecology and Management*, *149*, pp. 253–264.

Owusu, J. H. (1998). Current Convenience, disparate deforestation: Ghana's adjustment programmes and the forestry sector. *Professional Geographer*, *50*(4), pp. 418-436.

Paddock, J. L., Siegel, D. R. & Smith, J. L. (1987). Valuing offshore oil properties with option pricing models. *Midland Corporate Finance Journal*, *5*(spring), pp. 22-30.

Pandey, D. & Brown, C. (2000). Teak: a global overview. *Unasylva*, *51*, available at http://www.fao.org/docrep/x4565e/x4565e03.htm#P0_0. Accessed July, 2009.

Peace FM (2014). SADA: government terminates guinea fowl, tree-planting contracts. http://elections.peacefmonline.com/pages/politics/201401/188046.php (accessed July, 04, 2015).

Pearce, D., Atkinson, G. & Mourato, S. (2006). *Cost benefit analysis and the environment: recent developments*. Organisation for Economic Co-operation and Development. Paris. OECD Publishing.

Pearse, P. H. (1992). *Introduction to forestry economics*. University of British Columbia Press. Vancouver.

Pearson, K. (1931). *Tables for statisticians and biometricians*. Part II. Cambridge University Press. Cambridge, England. p. 262.

Pelissier, F. & Souto, X. C. (1999). Allelopathy in northern temperate and boreal semi-natural woodland. *Crit. Rev. Plant Sci.*, *18*, pp. 637-652.

Perera, W. R. H. (1962). The development of forest plantations in Ceylon since the seventeenth century. *Ceylon Forester*, *5*, pp. 142-147.

Perhutani, P. (1992). Teak in Indonesia. In Wood H (ed.) Teak in Asia FORSPA publication, 4. *Proc. regional seminar Guangshou*, China March, 1991. FAO (Bangkok).

Pétry. F.(1991). *Energy for sustainable rural development projects* - Vol.1, A reader IFAD, Africa Group. Training materials for agricultural planning, 23/1. Food and Agriculture Organization of the United Nations Rome.

Pfund, J. L. & Robinson, P. (Editors). (2005). A Workshop on Non-Timber Forest Products: Introduction. Pp. 4-7, In: Pfund, J. L. & P. Robinson (Editors). *Non-Timber Forest Products between poverty alleviation and market forces*. p. 53

Phillips, H. (2004). *Thinning to improve stand quality*. Silviculture / Management No. 10. Coford Connects.

Pienaar, L. V. & Turnbull, K. J. (1973). The Chapman-Richards generalisation of von Bertalanffy's growth model for basal area growth and yield in even-aged stands. *For. Sci. 19*, 2-22.

Pindyck, R. (1988). Irreversible investment, capacity choice, and the value of the firm. *American Economic Review*, *78*(December), pp. 969-985.

Plantinga, A. J. (1998). Optimal harvesting strategies with stationary and non-stationary prices: An option value analysis. *Forest Science*, *44*, pp. 192-202.

PointCarbon. (2008).EUAHistoricPrices.Availableat: /http://www.pointcarbon.com.

Poleno, Z. (1981). Development of mixed forest stands. Prace VULHM, 59, pp. 179– 202 (in Czech with English summary).

Poore, M. E. D. & Fries, C. (1985). The ecological effects of eucalyptus. *FAO Forestry Paper*, *59*. FAO, Rome, Italy.

Prah, E. A. (1994). *Sustainable management of tropical high forests of Ghana*. London. Commonwealth Secretariat, IDRC, Vol. *3*.

Price, C. (1989). *The theory and application of forest economics*. Basil Blackwell Ltd. London. p. 402.

Price, T. & Wetzstein, M. (1999). Irreversible investment decision in perennial crops with yield and price uncertainty. *Journal of Agricultural and Resource Economics*, *24*(1), pp. 173-185.

Prokop'ev, M. N. (1976). Mixed plantings of pine and spruce. Lesnoe Khozyaistvo, *5*, pp. 37–41 (in Russian with English summary).

Purcell, L. (2012). Drought could have lasting effect on trees, specialist says. Purdue University.="http://www.youtube.com/embed/5U9gLnzIhGI?rel=0"

Radosevich, S. R. & Osteryoung, K. (1987). Principles governing plant-environment interactions . In: J. D. Walstad and P. J. Kuch (Eds.), *Forest vegetation management for conifer production*. Ch. 5. pp 105-156. New York. Wiley.

Radwanski, S. & Wickens, G. E. (1981). Vegetative fallows and potential value of the neem tree (*Azadirachta indica*) in the tropics. *Econ. Bot., 35*, pp. 398-414.

Reed, W. J. & Haight, R. (1996). Predicting the present value distribution of a plantation investment. *Forest Science*, *42*, 378-388.

Reineke, L. (1933). Perfecting a stand density index for evenaged forests. *J. Agric. Res.*, *46*, pp. 627–638.

Reineke, L. H. (1926). The determination of tree volume by planimeter. *J. For.*, *24*, pp. 184-189.

Rendle, B. J. (1969). *World timbers*. Volume, 2, North and South America. University of Toronto Press.

Reynolds, M. R., Burk, T. E. & Haung, W. C. (1988).Goodness-of-fit tests and model selection procedures for diameter distribution models. *Forest Science*, *349*(2), pp. 373-399.

Rhodey-Bowman, T. (2007). *Mechanical High Pruning*. Agriculture Notes. State of Victoria, Department of Primary Industries. AG 1014.

Rice, R. E., Sugal, C. A., Ratay, S. M. & Fonseca, G. A. (2001). Sustainable forest management: A review of conventional wisdom. *Advances in Applied Biodiversity Science*, No. *3*, pp. 1-29. Washington, DC: CABS/Conservation International

Richards, F. S. (1959). Flexible growth function for empirical use. *J. Exp. Botany*, *10*, pp. 290-300.

Richards, M. (1995). The role of demand side incentives in fine grained protection: a case study of Ghana's tropical high forest. *Forest Ecology and Management*, *78*, pp. 225-241.

Richards, P. W. (1996). The Tropical Rain Forest, 2nd Edition. Cambridge University Press, pp. 525.

Ross, D. W. & Walstad, J. D. (1986). *Vegetative competition, site preparation and pine performance: a literature review with reference to south-central Oregon*. Oregon State Univ., Corvallis. Paper, p. 21.

Rudolf, P. O. (1974). *Larix* Mill. larch. Pages, 478-485. In C. S., Schopmeyer, (Ed.) Seeds of woody plants in the United States. 450. U. S. Department of Agriculture, *Forest Service*. Washington, DC. Agric. Handb. pp. 478-485.

SADA (2010). Savannah Accerelated Development Authority. Strategy and workplan, 2010 – 2030. Main document. pp.125.

Salifu, K. F. (1997). Physico-Chemical Properties of Soil in The High Forest Zone of Ghana Associated with Logged Forest and with Areas Converted to Teak *(Tectona grandis* Linn. F). MscF Thesis, Lakehead University. Thunder Bay, ON. Canada. pp. 105.

Sampong, E. (2004). A Review of the Application of Environmental Impact Assessment in Ghana. *A Report prepared for the United Nations Economic Commission for Africa* in December, 2004.

Sarfo-Mensah, P. (2005). Exportation of timber in Ghana: the menace of illegal logging operations. The Fondazione Eni Enrico Mattei Note di Lavoro Series Index: http://www.feem.it/Feem/Pub/Publications/WPapers/default.htm. NOTA DI LAVORO, 29.2005.

Savill, P., Evans, J., Auclair, D. & Falk, J. (1997). *Plantation silviculture in Europe*. Oxford University Press, Oxford.

Saxena, A. K., Nautiyal, J. C. & Foot, D. K. (1997). Analysing Deforestation and Exploring Policies for its Amelioration: A Case Study of India. *Journal of Forest Economics*, *3*(3), pp. 253-289.

Sawyer, J. (1993). *Plantations in the tropics – environmental concerns. The World Conservation Union (IUCN)*, Gland, Switzerland and Cambridge, UK in collaboration with UNEP and WWF.

Schiffel, A. (1899). Form and Inhalt der Fichete (Form and volume of spruce). Mitt.ausd. forstl. Versuchsan. *Osterreiche*, *24*.(cited in Husch et al., 1982).

Schlaegel, B. E. (1981). Testing, reporting and using biomass estimation models. pp. 95-112 *in* Greshan, C. A (ed.) Proc. 1981 Southern *Forest Biomass Workshop*. The Belle W. Baruch Forest Science Institute of Clemson University, Clemson USA, June, 11-12, 1981. pp. 127.

Schmutterer, H. (1982). Ten years of neem research in the Federal Republic of Germany. pp 21-32. *In* Schmutterer, .H.,K. R. S. Asher, and H. Rembold (eds.) Natural pesticides from the neem tree. *Proc.First Int. Neem Conf. Rottach-Egern.*, Eschborn Germany

Schofield, J. A. (1987). *Cost-benefit analysis in urban and regional planning*, Boston, MA. Urwin and Allen.

Schreuder, H. T. & Hafley, W. L. (1977). A useful bivariate distribution for describing stand structure of tree heights and diameters. *Biometrics*, *33*, pp. 471-478.

Schreuder H. T., Gregoire T. G., Wood G. B. (1993) Sampling methods for multiresource forest inventory.Wiley, New York.

Schumacher, F. X. (1939). A new growth curve and its application to timber yield studies. *J. of For.*, *37*, pp. 819-820.

Schwarz, C. J. (2011). Sampling, regression, experimental design and analysis for environmental scientists, biologists, and resource managers. Department of Statistics and Actuarial Science, Simon Fraser University.

Seth, S. K. & Yadav, J. S. P. (1959). Teak soils. *Indian Forester*, *85*, pp. 2 – 16.

Shifley, S. & Lentz. E. (1985). Quick estimation of the three-parameter Weibull to describe tree size distribution. *For. Ecol. and Manage.*, *13*, pp. 195-203.

Shinozaki, K. & Kira, T. (1956). Interspecific competition among higher plants. VII. Logistic theory of the C-D effect. *J. Inst. Polytech.*, Osaka Cy Univ., *7*, pp. 35-72.

Shiver B. D., Borders, B. E. (1996) Sampling techniques for forest resource inventory. Wiley, New York

Shugart, H., Sedjo, R. & Sohngen, B. (2003). Forests and global climate change: potential impacts on US forest resources. *Prepared for the Pew Centre on Global Climate Change.*, USA. p. 64.

Sinha, K. C., Riar, S. S., Tiwary, R. S., Dhawan, A. K., Bardhan, J., Thomas, P., Kain, A. K. & Jain, R. K. (1984). Neem oil as a vaginal contraceptive. *Indian J. Med. Res.*, *79*, pp. 131-136.

Skoupy, J. (1991). People's Participation in Planting Trees. Insert Focus on reforestation by Skoupy, J. T. Dida, T. Cecchini, G. Nasser Al Homaid, M.H. Khan and M. Sadiq (1991). *In Desertification Control Bulletin*, *19*, pp. 33-60.

Skovsgaard, J. P. & Vanclay, J. K. (2007). Forest site productivity: a review of the evolution of dendrometric concepts for even-aged stands. *Forestry*, *81*(1), doi:10.1093/forestry/cpm041.

Smith D E, Larson, B. C., Kelty, M. L. & Ashton, P. M .S. (1997). *Practice of silviculture: applied forest ecology*. John Wiley and Sons, New York. p. 537.

Smith, D. M. (1986). *The practice of silviculture*. 8th Edition. New York. Wiley

Smith, E. K. (1999). Developments and setbacks in forest conservation: The new political economy of forest resource use in southern Ghana. Natural Resources Management Programme, *Ministry of Lands and Forestry Technical paper*, Accra, Ghana.

Smith, J. (1939). Germination of neem seed. *Indian Forester*, *65*(3), pp. 457-459.

Smith, J. & McCardle, K. (1998). Valuing oil properties: integrating option pricing and decision analysis approaches. *Operations Research*, March-April: pp. 198-217.

Smith, P. (1998). The use of subsidies for soil and water conservation: a case study from Western India. Network Paper No. 87. *Agricultural and Research Extension Network.* London: Overseas Development Institute.

Smith, V. K. (1989). Taking stock of progress with travel cost recreation demand models: Theory and implementation. *Marine Resource Economics*, *6*, pp. 279-310.

Smith, V. G. (1984). Asymptotic site-index curves, fact or artifact? *For. Chron.* 60, pp. 150-156.

Sokal, R. R. & Rohlf, F. J. (1981). *Biometry*. 2nd Edition. W. H. Freeman and Co., San Francisco. U. S. A. p. 857.

Solomon, D. S. & Zhang, L. (2002). Maximum size–density relationships for mixed softwoods in the northeastern USA. *Forest Ecology and management. 155*, pp. 163-170

Spaargaren, O. C. & Deckers, J. (1998). The world reference base for soil resources - an introduction with special reference to soils of tropical forest ecosystems. *In*: Schulte, A and Ruhiyat D. (eds.) *Soils of tropical forest ecosystems - characteristics, ecology and management*, 21-28. Springer, Berlin.

Spurr, S. H. (1952). *Forest inventory*. The Ronald Press Co., New York. p. 476.

Stanton, K. (2003). Parliament needs law monitoring mechanism. *Ghana News Agency report* of March, *23*, 2003.

State Government of Victoria (2009). Growing plantation fuelwood. Note number AG1106. http://www.dpi.vic.gov.au/forestry/private-land-forestry/forest-products-and-processing/ag1106-growing-plantation-firewood. Department of Environment and Primary Industries, Victoria, Australia.

Stone, S. W., Kyle, S. C. & Conrad, J. M.1993. Application of the Faustmann principle to a short-rotation tree species: an analytical tool for economists, with reference to Kenya and leuceana. *Agroforestry Systems*, *21*, pp. 79-90.

Strang, W. (1984). On the optimal harvesting decision. *Economic Inquiry.*, *14*, pp. 466-492.

Streets, R. J. (1962). *Exotic forest trees in the British Commonwealth*. Claredon Press, Oxford. p. 765.

Styles, B. T. (1972). The flower biology of the *Meliaceae* and its bearing on tree breeding. *Silvae Genetica*, *21*, pp. 175-183.

Sutton, R. F. (1985). Vegetative management in Canadian forestry. *Can. For. Serv., Great Lakes For. Res. Cr., Info. Rep.* \no. O-X-369. p. 35.

Suresh, K. K. & Rai, R. S. V. (1987). Studies on the allelopathic effects of some agroforestry tree crops. *International Tree Crops Journal, 4*, pp. 109-115.

Swaine, M. D. & Hall, J. B., 1988. The mosaic theory of forest regeneration and the determination of forest composition in Ghana. *J. Trop. Ecol. 4*, pp. 253–269.

Taylor, C. J. (1952). *The vegetation zones of the Gold Coast.* Government Printer, Accra, Ghana. p. 57.

Taylor, C. J. (1960) Synecology and silviculture in Ghana. Accra, Ghana, Legon University; Edinburgh, U. K., Thomas Nelson and Sons Ltd., p. 418.

Teeguarden, D. E. (1982). Multiple services. *In*: Duerr, W. A., Teeguarden, D. E., Christiansen, N. B and Guttenberg, S. *Forest resource management*: decision-making principles and cases. Corvallis, *Oregon.*, pp. 276-290.

Tepper, H. B. & Bamford, G. T. (1959). Hardwoods on poorly drained sites do not respond to low thinning. Forest Research Note NE-92. Upper Darby, PA: U.S. Department of Agriculture, *Forest Service*, Northeastern Forest Experiment Station. p. 3.

Tetteh, E. N. (2010). Evaluation of land reclamation practices at Anglogold Ashanti, Iduapriem Mines Ltd.,Tarkwa. Msc Thesis, Faculty of Renewable Natural Resourcs, Kwame Nkrumah University of Science and Technology, Kumasi, Ghana.

Thérivel, R. & Partidario, M. (1996). *The practice of strategic environmental assessment.* London. EarthScan Publications.

Thompson, D. (2009). *Frequently Asked Questions (FAQs) on the use of herbicides in Canadian forestry*. Unpublished Notes.

Thomson, T. A. (1992). *Optimal forest rotation when prices follow a diffusion process.* Land Economics, *68*, pp. 329-342.

Thorson, B. J. (1999). Afforestation as a real option: some policy implications. *Forest Science, 45*(2), pp. 171-178.

Tiarks, A., Nambiar, E. K. S. & Cossalter, C. (1998). Site management and productivity in tropical forest plantations. *CIFOR Occasional Paper* No. 16.

Tilander, Y. (1993). Effects of mulching with *Azadirachta indica* and *Albezia lebbeck* leaves on the yield of sorghum under semi-arid conditions in Burkina Faso. *Agroforestry Systems, 24*(3), pp. 277-293.

Timbilla, J. A. & Braimah, H. (2000). Successful biological control of *Chromolaena odorata* in Ghana: the potential for regional programme in Africa. *In.* Zachariades C, Muniappan R and Strathie L.W. (Eds.), *Proceedings of the fifth international workshop on biological control and management of Chromolaena odorata*, Durban, South Africa, 23-25 October, 2000, pp. 66-70.

Tomforde, M. (1995). *Compensation and incentive mechanisms for the sustainable development of natural resources in the tropics*: their socio-cultural dimension and economic acceptance. Eschborn: Gesellschaft für Technische Zusammenarbeit (GTZ).

Tosi, J. A., Jr. (1960). Zonas de vida natural en el Perú. Memoria explicativa. sobre el mapa ecológico del Perú. Instituto Interamericano de las Ciencias Agriicolas de la E.E.A. Boletín Técnico, *5.*, Zona Andina, Lima, Penú. p. 271.

Tourinho, O. (1979). *The option value of reserves of natural resources.* Working Paper No. 94, University of California at Berkeley.

Townson, I. M. (1995). Patterns of Non-Timber Forest Products Enterprise activity in the Forest Zone of Southern Ghana. *Report to the ODA Forestry Research Programme*. Oxford Forestry Institute, Oxford.

Treadway, A. B. (1971). On the multivariate flexible accelerator. *Econometrica*, *39*, pp. 845–855.

Treadway, A. B. (1974). The globally optimal flexible accelerator. *Journal of Economic Theory*, *7*, pp. 17–39.

Treasury Board of Canada. (2009). Regulatory Impact Analysis Template. http://www.regulation.gc.ca/documents/nriast-nmrir/nriast-nmrir-eng.asp.

Trigeorgis, L, 1993. Real options and interactions with financial flexibility. *Financial Management*, *22*(3), pp. 202-224.

Trigeorgis, L. (1998). A conceptual options framework for capital budgeting. *Advances in Futures and Options Research*, *3*, pp. 145-167.

Trigeorgis, L. (1999). *Real options: managerial flexibility and strategy in resource allocation*. The MIT Press, Cambridge, Massachusetts.

Trigeorgis, L. & Mason, S. P. (1987). Valuing managerial flexibility. *Midland Corporate Finance Journal*, *5* (Spring), pp. 14-21.

Tripathi, S., Tripathi, A. & Kori, D. C. (1999). Allelopathic evaluation of *Tectona grandis* leaf, root and soil aqueous extracts on soybean. *Indian Journal of Forestry*, *22*, pp. 366-374.

Troup, R. S. (1921). *The silviculture of Indian trees*. Volume, 1. Claredon Press, Oxford. Pp. 678.

Turnbull, K. J. (1963). *Population dynamics in mixed forest stands*. Ph.D. dissertation, Univ. of Washington. pp. 186.

Ulibarri, C. A. & Wellman, K. F. (1997). *Natural resource valuation: A primer on concepts and Techniques. United States Department of Energy*, p. 86.

UNEP. (2005). Baseline Methodologies For Clean Development Mechanism Projects: A guide book. *The UNEP project CD4CDM*. UNEP Risø Center, Denmark.

UNFCCC. (2002). *Report of the conference of the parties on its Seventh Session*, held at Marrakesh from, 29 October to 10 November, 2001. FCCC/CP/2001/13/Add.2.

United Nations. (1992). Report of the United Nations Conference on Environment and Development - *Agenda, 21, Chapter, 11, Combating Deforestation*.

Unnikrihnan, K. P. & R. Singh. (1984). Construction of volume tables - a general approach. *Indian Forester*, *110*(6), 561-576.

University of Montana. Undated. Double Sampling (Chapter, 14), http://www.math.montana.edu/~parker/PattersonStats/Double.pdf (accessed July, 07, 2015).

University of Arizona. (1998). Arizona Master gardener manual: an essential reference manual for gardening in the desert southwest. Co-operative Extension, College of Agriculture

USDA. (2010). Wood Technical Fact Sheet: *Triplochiton scleroxylon. USDA Forest Service* Forest Products Laboratory.

Valentine, H. T. & Hilton, S. J. (1977) Sampling oak foliage by the randomized-branch method. *Canadian Journal of Forest Research*, *7*, pp. 295-298.

Valentine, H. T., Tritton, L. M. & Furnival, G. M. (1984) Subsampling trees for biomass, volume, or mineral content. Forest Science, *30*(3) pp. 673-681.

Van Cleve, K. & Zasada, J. C. (1976). Response of 70-year-old white spruce to thinning and fertilisation in interior Alaska. *Can. J. For. Res.*, *6*, pp. 145-152.

Van Kooten, G. C., Binkley, C. S. & Delcourt, G. (1995). Effect of carbon taxes and subsidies on optimal forest rotation age and supply of carbon services. *American Journal of Agricultural Economics*, *77*(2), pp. 365-374.

Vandermeer, J. (1989). The Ecology of Intercropping. Cambridge University Press, Cambridge, p. 237.

Vega, L. (1974). Influencia de la silvicultura sobre el comporta miento de *Cedrela* en Surinam. Instituto Forestal Latinoamericano de Investigación y Capacitación, *Boletín*, 46-48. Mérida, Venezuela. pp. 57-86.

Verinumbe, I. (1991). Agroforestry development in northeastern Nigeria. *For. Ecol. Manage.* *45*, pp. 309-317.

Wadsworth, F. H. (1960). Datos de crecimiento de plantaciones forestales en México, Indias Occidentales y Centro y Sur América. Segundo Informe Anual de Is Sección de Forestación, Comité Regional sobre Investigación Forestal, Comisión Forestal Latinoamericana, Organiziación de las Naciones Unidas para la Agricultura y Alimentación. *Caribbean Forester*, 21 (supplement). p. 273.

Wagner, M. R., Cobbinah, J. R. & Bosu, P. P. (2008). *Forest Entomology in West Tropical Africa: Forests Insects of Ghana.* Springer. Netherlands.

Walters, D. K., Gregoire, T. G. & Burkart, H. E. (1989). Consistent estimation of site index curves fitted to temporary plot data. *Biometrics*, *45*(1), pp. 24-33.

Ware, K. D. & T. Cunia. (1962). Continuous forestry inventory with partial replacement for sarnples. Forest science Manograph, 3.

Warthen, J. D. Jr. (1979). *Azadirachta indica*: A source of insect feeding inhibitors and growth regulators. *Agric. Rev. and manuals ARM-NE-4, USDA*, Washington DC.

Watterson, K. G. (1971). Growth of teak under different edaphic conditions in Lancetilla valley, Honduras. Turr., *21*, pp. 222-225.

Way D. A. & Oren R. (2010). Differential responses to changes in growth temperature between trees from different functional groups and biomes: a review and synthesis of data; *Tree Physiol.*, (30), 669-688.

Weaver, P. L. (1993). Teak (Tectona grandis Linn F.). Res. Notes SO-ITF-SM-64. Rio Piedras, PR. USDA Forest Serv. Inst. of Tropical forestry. pp. 18.

Weibull, W. (1951). A statistical distribution function of wide applicability. *Journal of Applied Mechanics*. *18*, pp. 293-297.

Wengert, E. (2001). Eliminating wood problems - An industry review: 10 ways of eliminating wood problems. Wood Processing Department of Forestry, University of Wisconsin, Madison (www.woodweb.com).

West, P. W. (2003). *Tree and Forest measurement*. Springer. p. 167.

Westoby, M. (1984). The self-thinning rule. *Adv. Ecol. Res.*, *14*, pp. 167–225.

White, F. (1983). *The Vegetation of Africa.* 3 maps, 1 chart. AETFAT Vegetation Map Committee, UNESCO, AETFAT, UNSO.

Whiteman, A. (2003). Money doesn't grow on trees: a perspective on prospects for making forestry pay. *UNASYLVA*, *54*(212), pp. 3 – 10.

Whitmore, J. L. (1976). Myths regarding *Hypsipyla* and its host plants. *In:* Studies on the shootborer *Hypsipyla grandella* Zeller Lep. Pyralidae. vol. 3. p. 54-55. Centro

Agronómico Tropical de Investigación y Enseñanza, Miscellaneous Publication, 1. Turrialba, Costa Rica.

Whittington, D. L., Donald, T., Wright, A. M., Choe, K., Hughes, J. A. & Swarna, V. (1993). Household demand for improved sanitation services in Kumasi, Ghana: A contingent valuation study. *Water Resources Research. 29*(6), pp. 1539-1560.

Wickens, G. E., Sief-El-Din, A. G., Sita, G. & Nahal, I. (1995). Role of Acacia Species in the rural economy of dry Africa and the Near East. *FAO Conservation Guide*, No. 27, Rome, Italy, p. 56.

Williams, J. (2001). Financial and other incentives for plantation forestry. In: *Proceedings of the International Conference on Timber Plantation Development*, Manila, the Philippines, 7-9 November, 2000, (pp. 87-101). Quezon City, the Philippines, Department of Environment and Natural Resources.

Williams, T. M. (1989). Site preparation on forested wetlands of the south-eastern coastal plains. In: D. D. Hook and R. Lea (Eds.). The Forested wetlands of the Southern United states, pp.67-71. Proc. Symp., July, 12-14, 1988. Orlando, *FL. US Forest Ser. Gen. Tech. Rep.* SE-50.

Wilson, E. R. & Leslie, A. D. (2008). The development of even-aged plantation forests: an exercise in forest stand dynamics. *Journal of Biological Education*, *42*(4), pp. 170-176.

Wilson, F. G. (1946). Numerical expression of stocking in terms of height . *J. For.*, *44* , pp. 758 – 761

Wingfield, M. J. & Robison, D. J. (2004). Diseases and insect pests of *Gmelina arborea*: real threats and real opportunities. *New Forests.*, *28*(2-3), pp. 227-243

Winpenny, J. T. (1992). The economic valuation of tropical forests: Its scope and limits. *In* F. R. Miller and K. L. Adam (Eds.). *Wise management of tropical forests.* (pp. 125-138). Proceedings of the Oxford Conference on Tropical Forests, Oxford, 30 March-1 April, 1992, Oxford: Oxford Forestry Institute, University of Oxford.

Wong, C. Y. & Jones, N. (1986). Improving tree form through vegetative propagation of *Gmelina arborea. Commonwealth Forestry Review*, *65*(4), pp. 321–324.

World Agroforestry Centre (ICRAF). (1993). International Centre for Research in Agroforestry: *Annual Report*, 1993. Nairobi, Kenya. P. 208.

World Agroforestry Centre. (2007). http://www.worldagroforestrycentre.org/sea/Products/.

World Bank. (1988). Staff appraisal report. *Ghana Forest Resource Management Project.* Report No.7295-GH. Nov., *17*, 1988.

World Bank. (1998). *Integrated Coastal Zone Management Strategy for Ghana.* Findings. Africa Region. Number, 113. Washington D. C. The World Bank.

World Bank. (2003). *Ghana - Land Administration Project.* Washington, DC: World Bank. http://documents.worldbank.org/curated/en/2003/07/2434956/ghana-land-administration-project

World Bank. (2003). *Basics of the BioCarbon Fund for Project Proponents.* Available at: /www.unfccc.intS.

World Bank. (2004). *Implementation completion report on a Loan/Credit/Grant to the Republic of Ghana for a Natural Resource Management Project*, Phase I. January, 2003. Report No: 27231

World Bank. (2007). *World Development Indicators.* The World Bank. Washington, D. C.

www.un.org/esa/sustdev/documents/agenda21/english/agenda21chapter11.htm (accessed May, 2009).

Yaping, D. (1999). The Use of Benefit Transfer in the Evaluation of Water Quality Improvement: *An Application in China. Economy and Environment Programme for Southeast Asia.* International Development Research Centre, Ottawa, Canada. http://www.idrc.ca/eepsea/ev-8426-201-1-DO_TOPIC.html.

Yaro, J. A., Dogbe, T. D., Bizikova, L., Bailey, P., Ahiable, G., Yahaya, T. & Salam, K. A. (2010). *Development and Climate Change: The Social Dimensions of Adaptation to Climate Change in Ghana*. Discussion Paper Number, 15, December, 2010. Washington, DC; World Bank.

Yin, R. & Newman, D. H. (1995). Optimal timber rotations with evolving prices and costs revisited. *For. Sci. 41*(3), pp. 477 - 490.

Yoda, K. Kira, T., Ogawa, H. & Hozumi, K. (1963). Interspecific competition among higher plants XI. Self-thinning in overcrowded pure stands under cultivated and natural stands. *J. Biol.* Osaka City University, *14*, pp. 107-129.

Zankis, S. H. (1979). A simulation study of some simple estimation for the three parameter Weibull distribution. *J. Stat. Comp. Simul.*, *9*, pp. 260-116.

Zarnoch, S. J. & Dell, T. R. (1985). An evaluation of percentile and maximum likelihood estimators of Weibull parameters. *For. Sci.*, *31*, pp. 260-268.

Zeide, B. (1987). Analysis of the, 3/2 power law of self-thinning. *For. Sci. 33*, pp. 517–537.

Zhang, D. & Owiredu, E. A. (2007). Land tenure, market, and the establishment of forest plantations in Ghana. *Forest Policy and Economics*, *9*, pp. 602-610.

Zhang, D. & Pearse, P. H., 1996. Differences in silvicultural investment under various types of forest tenure in British Columbia. *Forest Science*, *42*(4), pp. 442– 449.

Zobel, B. J. van Wyk, G. & Stahl, P. (1987). *Growing exotic forests*. New York. Wiley.

Zobel, B. J. & Talbert, J. (1984). *Applied forest tree improvement*. New York. Wiley

Zöhrer, F. (1969). Ausgleich von Haufigkeitsverteilungen mit Hilfe der Beta-Funktion. *Forstarchiv*, *40*(3), pp. 37-42.

AUTHORS' CONTACT INFORMATION

Dr. David Mateiyenu Nanang,
Director General,
Natural Resources Canada
68 Fields Square, Sault Ste. Marie,
Ontario, Canada P6B 6H2
Email: dmnanang@yahoo.ca

Dr. Thompson K. Nunifu,
Environmental Statistician,
Alberta Environment and Sustainable Development
3539-15 Avenue NW
Edmonton, Alberta,
Canada T6L 4B1
Email: tnunifu@yahoo.ca

INDEX

#

A

B

C

D

E

F

G

H

I

J

K

L

M

N

O

P

Q

R

S

T

U

V

W

Y

Z